NAND FLASH MEMORY
TECHNOLOGIES

NAND FLASH MEMORY TECHNOLOGIES

SEIICHI ARITOME

IEEE Press Series on Microelectronic Systems

IEEE PRESS

WILEY

Published by John Wiley & Sons, Inc., Hoboken, New Jersey. All rights reserved
Published simultaneously in Canada

Limit of Liability/Disclaimer of Warranty: While the publisher and author have used their best efforts in
preparing this book, they make no representations or warranties with respect to the accuracy or
completeness of the contents of this book and specifically disclaim any implied warranties of
merchantability or fitness for a particular purpose. No warranty may be created or extended by sales
representatives or written sales materials. The advice and strategies contained herein may not be suitable
for your situation. You should consult with a professional where appropriate. Neither the publisher nor
author shall be liable for any loss of profit or any other commercial damages, including but not limited to
special, incidental, consequential, or other damages.

For general information on our other products and services or for technical support, please contact our
Customer Care Department within the United States at (800) 762-2974, outside the United States at (317)
572-3993 or fax (317) 572-4002.

Wiley also publishes its books in a variety of electronic formats. Some content that appears in print may
not be available in electronic formats. For more information about Wiley products, visit our web site at
www.wiley.com.

Library of Congress Cataloging-in-Publication Data is available.

ISBN: 978-1-119-13260-8

Printed in the United States of America

10 9 8 7 6 5 4 3 2 1

CONTENTS

FOREWORD

I had an idea regarding NAND flash memory when I was in Washington, D.C. in 1986.

My stay in Washington, D.C. at that time was long. All of Japan's DRAM manufacturers were sued by the United States' International Trade Commission (ITC) with the demand that they be barred from exporting to the United States because they were infringing upon Texas Instruments' DRAM patents. I was dispatched to Washington, D.C. as the engineer handling the ITC lawsuit. Court session at the ITC was tough to hold continuously from 6 am to midnight. However, the court wasn't held every day. In those days, I had plenty of free time. I thought deeply about future semiconductor memory, so that NOR flash memory that I invented a few years earlier was too weak to drive out magnetic disks, and we needed to further reduce the bit cost and further shrink the per-bit space occupied. The answer was "NAND flash memory." I immediately wrote up the patent in 1986, and I filed the application on April 24, 1987 in Japan. The US patent for NAND Flash memory was registered as USP 5,245,566.

After returning to Japan, I started the development of NAND flash memory at the VLSI Research Center, Toshiba Corporation, in 1987. I involved several engineers to make the development team for NAND flash. The basic data of reading and writing were obtained in a short time. We immediately submitted a paper to the 1987 IEDM. For further acceleration of NAND flash development, I assigned several members—including Dr. Aritome, who is the author of this book—to a flash team from a DRAM team.

After that, I aggressively drove the development. I proposed the design of a prototype device of 4-Mbit NAND flash memory. However, unfortunately, there was not enough budget at the research center to proceed with this project. It was at a crisis point to suspend the development of NAND flash. I had to somehow find someone to fund the NAND project so that it could continue. I first visited the division developing computers in Toshiba, but their response was that they wouldn't fund a dream-like project for replacing magnetic memory on a semiconductor. Meanwhile, I explained to Tajiri, the Director of the Consumer Electronics Laboratory, that if we managed to produce 4-Mbit NAND flash memory, cameras would no longer need film. I was indeed explaining that the digital cameras we know today would become possible. As a result, the Consumer Electronics Laboratory bore the development costs. We successfully developed 4-Mbit NAND flash memory in 1988 and announced 4-Mbit NAND flash memory at the ISSCC in February 1989. Thereafter, Consumer Electronics Laboratory Director Tajiri used 4-Mbit NAND flash memory to launch the world's first digital camera to replace conventional film with NAND flash memory. At the time, the price of the world's first flash-memory-based camera was high, more than two million yen ($20K), and thus it didn't sell well.

In 1992, a production of NAND flash was started. The first device was 0.7-μm rule 16-Mbit memory. The production volume was really small; however, it was an important milestone. For large volume production, we had to wait 4 or 5 years to create the market for flash memory cards mainly for digital cameras. After producing memory cards, the market for the NAND flash was amazingly huge. It was the disruptive innovations. Music players with cassette tape were replaced by flash-memory-based MP3 portable music players. USB memory arrived, and thus the floppy disk disappeared. Smartphone and tablet PC were designed based on the existence of NAND flash. Nowadays, the NAND flash memory became the standard nonvolatile memories that are used everywhere by everybody. However, the dream to replace magnetic memory (HDD, etc.) is along the way. I am expecting that the SSD will replace the HDD in the future.

I left Toshiba Corporation and was transferred to Tohoku University as a professor in 1994. I proposed the SGT (surrounding gate transistor) NAND flash memory, and I also started fundamental development of SGT for three-dimensional (3D) NAND flash. *Forbes* magazine published a structural drawing of the SGT NAND flash memory with my photograph in the cover of the June 24, 2002 edition. The cell structure of SGT NAND flash memory is currently used in 3D NAND flash memories that are in mass production. All NAND suppliers are intensively developing the next advanced 3D NAND flash based on SGT structure.

As the market for NAND flash memory expanded, engineers who are engaged in the development of NAND flash and its related products rapidly increased. This book is good for understanding the history, basic structure and process, scaling issues, 3D NAND flash, and so on. Dr. Aritome is one of the original members of the NAND flash development team and has over 27 years of experience as an engineer of development and production of NAND flash memory. I hope this book will contribute to the coming NAND flash technologies and products, including SSD.

Finally, I am grateful to the original team members of NAND flash development. NAND flash memory could not be realized without their contributions. I am very happy and lucky that I could collaborate with them so that we could devote ourselves to developing NAND flash memories.

Fujio Masuoka

CTO of Semicon Consulting Ltd.
Professor Emeritus at Tohoku University

BIOGRAPHICAL NOTE

Dr. Fujio Masuoka is Chief Technology Officer of Semicon Consulting Ltd., which forms part of the "New Scope Group," a largely respected international group of companies actively pursuing and supporting advanced Research & Development of Breakthrough Technologies and translating these into commercial reality. He is also Professor Emeritus of the Research Institute of Electrical Communication at Tohoku University in Japan. He is the inventor of flash memory. He has spent most of his career working on the research and development of numerous kinds of semiconductor memory including flash memory, programmable read-only memory, and random access memory. He also possesses considerable knowledge in image sensing devices (such as charge-coupled devices) and high-speed semiconductor logic. He filled the original patents for both NOR and NAND flash memories, published the first paper on flash memory at the 1984 IEDM, and published the first paper on NAND flash memory at the 1987 IEDM.

CAREER PROFILE

1966   Graduated from Faculty of Engineering, Tohoku University
1971   Completed the doctoral course, Tohoku University
1971   Joined Toshiba Corporation
1994   Appointed Professor at Tohoku University
2005   Accepted as the Chief Technology Officer of Unisantis Electronics (Japan) Ltd.
2007   Appointed Honorary Professor at Tohoku University

AWARDS AND RECOGNITION

1977   Awarded the Watanabe Prize during the year of its inception
1980   Awarded the invention award, National Invention Awards
1985   Awarded the encouraging award for invention, Kanto district
1986   Awarded the encouraging award for invention, Kanto district
1988   Awarded the encouraging award, twice in the year, for invention, Kanto district
1991   Awarded the encouraging award for invention, Kanto district
1995   IEEE Fellow

1997   Awarded the IEEE Morris N. Liebmann Memorial Award
2000   Awarded the Ichimura-Sangyo Prize (major award)
2002   Awarded the 2002 SSDM Award
2005   Awarded Innovation Award by the *Economist*
2007   Awarded Medal with Purple Ribbon from Emperor Akihito of Japan
2009   Flash memory recognized in the IEEE as one of "25 Microchips That Shook The World"
2010   Computer History Museum
2011   Consumer Electronics Hall of Fame
2012   The winner of the Progress Medal, the highest honor of the Photographic Society of America (PSA)
2013   Awarded the Flash Memory Summit Lifetime Achievement Award
2013   Bunkakorosha (Person of Cultural Merits of Japan)

PREFACE

NAND flash memory became a standard semiconductor nonvolatile memory. Everyone in the world has widely used NAND flash memory in many applications, such as digital cameras, USB drives, MP3 music players, smartphones, and tablet PCs. The cloud data server starts to use SSD (Solid State Drive), which is based on NAND flash memory. Recently, three-dimensional (3D) NAND flash memory was developed and started mass production for reducing bit cost. By using 3D NAND flash memory, an advanced SSD has been intensively developed for high performance and low power consumption to avoid damaging the ecological environment.

As the production volume of NAND flash memory has increased, the number of engineers who are engaged in development and production of NAND flash memory has also increased. And a lot of people who are working for storage device are joining the industry of NAND flash memory. This book on NAND flash memory technologies was written to provide detailed views of NAND flash technologies for such individuals who are not only engineers of developing NAND flash memory, but are also NAND flash users, product engineers, application engineers, marketings, managers, technical managers, engineers for developing and producing SSD, engineers of other NAND flash-related storage devices such as data servers, and so on.

This book is also suitable for new engineers and graduate students to quickly study and to be familiar with NAND flash memory technologies. I expect this book to encourage newcomers to contribute to future NAND flash memory technologies and products.

The contents of this book include the starting history, memory cell technologies, basic structure and physics, principles of operations, history and trend of memory cell scaling, advanced operations for multilevel cells (2, 3, 4 bits/cell), scaling challenges, reliability, 3D NAND flash memory cell, scaling challenge of 3D NAND flash memory cell, and future prospects of NAND flash memory.

After describing the background, the starting history of NAND flash is introduced in Chapter 1. The basic device structures and operations are described in Chapter 2.

Chapter 3 discusses the memory cell technologies focused on scaling. To scale down memory cell size, memory cell structure has been evolved from LOCOS isolation cells to self-aligned STI cells, along with reducing the feature size (design rule).

Chapter 4 introduces the advanced operations for multilevel cells. Tight V_t distribution width is very important for multilevel cells because of enough read window margin. Advanced operations have been mainly developed for this point.

By scaling down memory cell size below 20 nm, several physical limitation phenomena are exaggerated. Chapter 5 discusses the details of physical limitations for scaling. The floating-gate capacitive coupling interference has the worst impact on scaling, even when advanced operations are used, as shown in Chapter 4. And other physical limitation factors are also discussed, including an electron injection spread, RTN, structure limitations, high field problems, and so on.

Chapter 6 describes the reliability of NAND flash memory. A program/erase cycling degrades tunnel oxide quality by generating electron/hole traps and stress-induced leakage current (SILC). Thus all reliability aspects of the cycling endurance, data retention, read disturb, program disturb, and erratic over-programming are degraded by increasing the amount of cycling. In Chapter 6, the mechanism and impact on device reliabilities are discussed.

Chapter 7 shows three-dimensional (3D) NAND flash memory cells. Many types of 3D cells have been proposed. These 3D cells are introduced, and pros and cons in structure, process, operations, scalability, performance, and so on, are discussed.

The mass production of 3D NAND flash memory was started in 2013. Full-scale production will begin in 2016. However, for future 3D NAND flash memory, many problems still remain and have to be solved. In Chapter 8, challenges of 3D NAND flash memory are discussed. Increasing the number of stacked cells is essential for reducing the effective cell size in a 3D cell. High aspect ratio process and small cell current issues will be of utmost importance, as discussed in this chapter. I tried to show some possible solutions for these problems. Other challenges, such as new program disturbance issues, data retention, power consumption, and so on, are discussed.

In Chapter 9, I summarize and describe the prospect of technologies and market for the future of NAND flash memory.

I am convinced that this book is a significant contribution to the industry of NAND flash memory and related products. I sincerely hope you find this book useful in your future work.

SEIICHI ARITOME

Kawasaki, Japan

ACKNOWLEDGMENTS

The author would like to express special thanks to Professor Fujio Masuoka. The author is proud of their collaboration on NAND flash memory, since Professor Masuoka assigned the author to the development of NAND flash memory in VLSI Research Center, Toshiba Corporation, in the early stage of development in 1988.

The author would like to thank Mr. Kiyoshi Kobayashi, Mr. Shinichi Tanaka, Mr. Masaki Momodomi, Professor Riichiro Shirota, Professor Shigeyoshi Watanabe, Dr. Koji, Sakui, Professor Fumio Horiguchi, Mr. Kazunori Ohuchi, Dr. Junichi Matsunaga, Dr. Akimichi Hojo, and Dr. Hisakazu Iizuka for their continuous encouragement ever since I joined the Research & Development Center at the Toshiba Corporation.

The author is grateful to colleagues at the Toshiba Corporation for their contributions. Without their help, this work could not have been successful. I especially thank Mr. Ryouhei Kirisawa, Dr. Kazuhiro Shimizu, Mr. Yuji Takeuchi, Mr. Hiroshi Watanabe, Dr. Gertjan Hemink, Mr. Shinji Sato, Dr. Tooru Maruyama, Mr. Kazuo Hatakeyama, Professor Hiroshi Watanabe, Professor Ken Takeuchi, and Mr. Tomoharu Tanaka. I appreciate Mr. Ryozo Nakayama, Mr. Akira Goda, Mr. K. Narita, Mr. E. Kamiya, Mr. T. Yaegashi, Ms. K. Amemiya, Mr. Toshiharu Watanabe, Dr. Fumitaka Arai, Dr. Tetsuya Yamaguchi, Ms. Hideko Oodaira, Dr. Tetsuo Endoh, Mr. Susumu Shuto, Mr. Hirohisa Iizuka, Mr. Hiroshi Nakamura, Dr. Toru Tanzawa, Dr. Yasuo Itoh, Mr. Yoshihisa Iwata, Mr. Kenichi Imamiya, Mr. Kazunori Kanebako, Mr. Kazuhisa Kanazawa, Mr. Hiroto Nakai, Mr. Takehiro Hasegawa, Dr. Katsuhiko Hieda, Dr. Akihiro Nitayama, Mr. Koichi Fukuda, and Mr. Seiichi Mori for their fruitful discussion.

The author would like to thank Mr. Eli Harari, Mr. Sanjay Mehrotra, Dr. George Samachisa, Dr. Jian Chen, Mr. Tuan D. Pham, Mr. Ken Oowada, Dr. Hao Fang, and

Dr. Khandker Quader for their continuous encouragement and fruitful discussions since SanDisk-Toshiba joint development started in 1999.

The author would like to thank Dr. Kirk Prall, the late Mr. Andrei Mihnea, Mr. Frankie Roohparvar, Dr. Luan Tran, and Mr. Mark Durcan for their continuous encouragement since I joined Micron Technology, Boise, Idaho, USA, in 2003.

The author would like to thank Mr. Krishna Parat, Dr. Pranav Kalavade, Dr. Mark Bauer, Dr. Nile Mielke, and Dr. Stefan K. Lai for their continuous encouragement and fruitful discussions since Intel-Micron joint development started.

The author would like to thank Dr. T.-J. Brian Shieh, Dr. Alex Wang, Dr. Travis C.-C.Cho, Ms. Saysamone Pittikoun, Mr. Yoshikazu Miyawaki, Mr. Hideki Arakawa, and Mr. Stephen C. K. Chen for their continuous encouragement and support since I joined Powerchip Semiconductor Corp, Hsinchu, Taiwan.

The author would like to thank Dr. Sungwook Park, Dr. Sungjoo Hong, Dr. Seok Hee Lee, Dr. Seokkiu Lee, Dr. Seaung Suk Lee, Dr. Sungkye Park, Mr. Gyuseog Cho, Mr. Jongmoo Choi, Mr. Yoohyun Noh, Mr. Hyunseung Yoo, Dr. EunSeok Choi, Mr. HanSoo Joo, Mr. Youngsoo Ahn, Mr. Byeongil Han, Mr. Sungjae Chung, Mr. Keonsoo Shim, Mr. Keunwoo Lee, Mr. Sanghyon kwak, Mr. Sungchul Shin, Mr. Iksoo Choi, Mr. Sanghyuk Nam, Mr. Dongsun Sheen, Mr. Seungho Pyi, Mr. Jinwoong Kim, Mr. KiHong Lee, Mr. DaeGyu Shin, Mr. BeomYong Kim, Mr. MinSoo Kim, Mr. JinHo Bin, Mr. JiHye Han, Mr. SungJun Kim, Mr. BoMi Lee, Mr. YoungKyun Jung, Mr. SungYoon Cho, Mr. ChangHee Shin, Mr. HyunSeung Yoo, Mr. SangMoo Choi, Mr. Kwon Hong, Mr. SungKi Park, Ms. Soonok Seo, and Mr. Hyungseok Kim for their warm consideration ever since I joined the Research & Development division at the SK Hynix Inc.

The author would like to thank Mr. Angelo Visconti, Ms. Silvia Beltrami, Ms. Gabriella Ghidini, Dr. Emilio Camerlenghi, Mr. Roberto Bez, Mr. Giuseppe Crisenza, and Mr. Paolo Cappelletti for their continuous encouragement and fruitful discussion while we performed Numonyx–Hynix joint development.

The author is profoundly grateful to Professor Masataka Hirose, Professor Mizuho Morita, Professor Seiichi Miyazaki, and Professor Yukio Osaka for their warm consideration and encouragement since I joined the laboratory of Professor Masataka Hirose in Hiroshima University on 1982.

The author would like to thank Professor Takamaro Kikkawa, Professor Shin Yokoyama, and Professor Seiichiro Higashi, of Hiroshima University for their invaluable guidance and continuous encouragement.

Finally the author would like to express his heartfelt thanks to his wife, Miho Aritome, and his son, Santa Aritome, who have continuously supported him with their love.

SEIICHI ARITOME

Kawasaki, Japan

ABOUT THE AUTHOR

Seiichi Aritome received his B.E., M.E., and Ph.D degrees in electronic engineering from Hiroshima University, Japan, in 1983, 1985, and 2013, respectively.

He joined the Toshiba Research and Development Center, Kawasaki, Japan, in 1985. Since then, he has been engaged in the development of high-density DRAM. In 1988, he joined the EEPROM development group at the same research center. At that time, the EEPROM development group started to develop NAND-type flash memory for the first time in the world. His major work is the NAND flash memory device technology, process integration, characterization, and reliability.

He has contributed to NAND flash memory technologies over 25 years (1988 to the present) in several companies and nations. He has developed over 12 generations of NAND flash memories. Many technologies which he developed had become a standard of NAND flash memories.

He had contributed to the crucial decision of the proper uniform program/erase scheme for NAND flash memory by analyzing phenomena of program/erase cycling degradation. He clarified that the uniform program/erase scheme was appropriate reliability in comparison with other schemes. As a result, the uniform program/erase scheme was decided upon for NAND flash operation. The uniform program/erase scheme had another important advantage, namely, fast programming speed ($\sim$100 Mbyte/s) due to low power consumption during the program in the uniform program/erase scheme. Because of high reliability and fast programming, the uniform program/erase scheme became the de facto standard technology. All NAND suppliers (Toshiba, Samsung, Micron/Intel, SK Hynix, etc.) have used the uniform program/erase scheme for all NAND flash products over the past 20 years.

He proposed and developed the self-aligned shallow trench isolation cell (SA-STI cell) for the first time. The cell size of NAND Flash memory could be drastically shrunk to 66% (from $6F^2$ to $4F^2$; F stands for feature size), in comparison with that of a conventional LOCOS cell. This technology could realize a 256-Mbit NAND flash for the first time in the world. Also, the SA-STI cell has an excellent reliability because of no STI corner in tunnel oxide. Therefore, the SA-STI cell has been widely used in the NAND flash product over the past 17 years. All NAND flash memory suppliers are using this cell technology.

Many of the NAND flash technologies he developed became de facto standard because of low cost, high reliability, and fast programming speed. Therefore, the NAND flash enabled us to launch the new market of a smartphone, tablet PC, and SSD (solid-state drive) and have had a large volume production, estimated to be $35 billion in 2015.

In 1998, he moved to the flash device engineering group of memory division at the Toshiba semiconductor company. While working for Toshiba (1988–2003), he had engaged many generations of NAND flash (0.7 μm, 0.4 μm, 0.2 μm, 0.16 μm, 0.12 μm, 90 nm, 70 nm). Also, he was the first technical coordinator of the Toshiba–SanDisk joint development of NAND flash memory in 1999.

He started to work at Micron Technology Inc., Boise, Idaho, USA, in December 2003. He engaged the NAND flash process and device development over several generations of 90 nm, 72 nm, 50 nm, and 34 nm. He was transferred to Powerchip Semiconductor Corporation, Hsinchu, Taiwan, on April 2007. He worked for the development of NAND flash of 70-nm and 42-nm generations as program manager. And then he moved to SK Hynix Inc., Icheon, Korea, on April 2009. He worked for NAND flash development of 26-nm, 20-nm, and mid-1X-nm generations and three-dimensional cell.

He holds 251 US patents and 76 Japanese patents on NAND flash memories, such as cell structure, process, operation scheme, and high reliability device technologies. He has authored or co-authored over 50 papers. Most of them are for NAND flash memory technologies.

He is an IEEE Fellow and a member of the IEEE Electron Device Society.

1

INTRODUCTION

1.1 BACKGROUND

Recent progress in computers and mobile equipment requires further efforts in developing higher-density nonvolatile semiconductor memories. A breakthrough in the field of nonvolatile memories was the invention of the flash memory [1], which is a new type of EEPROM (electrically erasable and programmable read-only memory), as shown in Fig. 1.1a. The first paper discussing the flash memory was presented in 1984 IEDM (International Electron Device Meeting). The flash memory has many advantages in comparison with other nonvolatile memories. Therefore, the flash memory explosively accelerated the development of higher-density EEPROMs.

In 1987, a NAND structured cell was proposed by Masuoka et al. [2]. This structure can reduce the memory cell size without scaling of device dimension. The NAND structure cell arranges a number of bits in series, as shown in Fig. 1.1b [2]. The conventional EPROM cell has one contact area per two bits. However, for a NAND structure cell, only one contact hole is required per two NAND structure cells (NAND string). As a result, the NAND cell can realize a smaller cell area per bit than the conventional EPROM.

Applications of flash memory became quite wide due to nonvolatility, fast access, and robustness. Flash memory application can be classified into two major markets (Fig. 1.1). One is for code storage applications, such as PC BIOS, cellular phones, and DVDs. The NOR-type cell is best suitable for this market due to its fast random access speed. The other is for file storage applications, such as the digital still camera

Nand Flash Memory Technologies, First Edition. Seiichi Aritome.
© 2016 The Institute of Electrical and Electronics Engineers, Inc. Published 2016 by John Wiley & Sons, Inc.

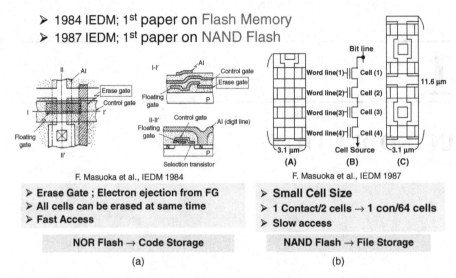

> 1984 IEDM; 1st paper on Flash Memory
> 1987 IEDM; 1st paper on NAND Flash

F. Masuoka et al., IEDM 1984

> Erase Gate ; Electron ejection from FG
> All cells can be erased at same time
> Fast Access

NOR Flash → Code Storage

(a)

F. Masuoka et al., IEDM 1987

> Small Cell Size
> 1 Contact/2 cells → 1 con/64 cells
> Slow access

NAND Flash → File Storage

(b)

FIGURE 1.1 Invention of flash memory and NAND flash memory. (a) Flash memory. All cells in the memory chip can be erased at the same time by applying erase voltage to the erase gate [1]. (b) NAND flash memory [2]. Memory cells are connected in series to share contact area. Comparison between (A) NAND cell and (C) conventional EPROM (NOR flash cell). (B) shows the equivalent circuit of the NAND structure cell having 4 cells.

(DSC), silicon audio, the smartphone, and the tablet PC. The NAND-type cell is suitable for file storage market.

Figure 1.2 shows the memory hierarchy of computer system before mass production of NAND Flash. SRAM and DRAM had been used as cash memory and main memory, respectively. And magnetic memories, such as HDD, had been used as a nonvolatile mass-storage device. NAND flash memory had been targeted to replace magnetic memory [54]. Actually, from the production start of NAND flash memory in 1992, the NAND flash memory has been widely applied to new emerging applications and has replaced magnetic memory, as shown in Fig. 1.3. At first, a photo film had been completely replaced by the memory cards of NAND flash memory. Next, the floppy disk was replaced by USB drive memory. The mobile music equipment with cassette tape was replaced by the MP3 player using flash memory storage. Also, NAND flash memory had created new market of smartphones and tablet PCs. And now, the application is extending to the SSD (solid-state drive) market, not only for the consumer but also for the enterprise server. Therefore, over 20 years, NAND flash memory has created new large-volume markets and industries of consumer, computer, mass storage, and enterprise server. NAND flash production volume was tremendously increased. The overall NAND market is expected to reach $40 billion in 2016 [55]. NAND flash has become an explosive innovation and has greatly contributed to the improvement of our lives with the advent of convenient mobile equipment such as smartphones and tablet PCs.

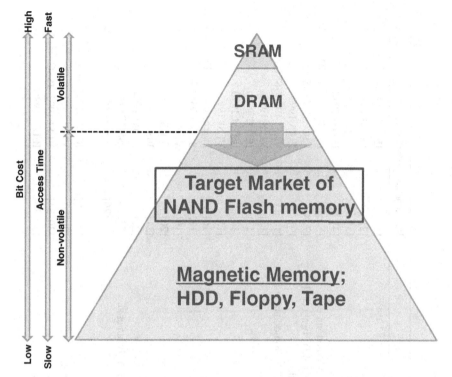

FIGURE 1.2   Target market of NAND flash memory.

Table 1.1 shows the history of NAND flash memory development, based on technical papers from 1987 to 1997. During the 10 years from the first NAND flash paper in 1987, all of the fundamental and important NAND flash technologies were established, such as page programming [7, 8], block erase, the uniform program and uniform well erase scheme [9, 12, 13], bit-by-bit verify [15, 21], the ISPP (incremental step pulse program) [25, 26, 29], the self-aligned STI cell [22, 51, 56], the shield

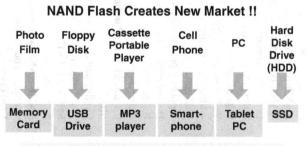

FIGURE 1.3   NAND flash memory creates a new market.

TABLE 1.1 History of the NAND Flash Memory (~1997)

Year		Authors	References	Conference/Journal
1984	Flash Memory, first paper	F. Masuoka et al.	[1]	IEDM 1984
1987	NAND-type flash memory, 1st paper	F. Masuoka et al.	[2]	IEDM 1987
1988	NAND-type flash memory	R. Shirota et al.	[3]	VLSI 1988
	Drain-FN program	M. Momodomi et al.	[4]	IEDM 1988
1989	4-Mb NAND-type flash memory	Y. Itoh et al./M. Momodomi et al.	[5, 6]	ISSCC1989/ JSSC
	Page program	M. Momodomi et al./Y. Iwata et al.	[7, 8]	CICC1989/ JSSC
1990	Well erase reliability	S. Aritome, et al	[9]	IRPS 1990
	4-Mbit tight V_t distribution	T. Tanaka et al./M. Momodomi et al.	[10, 11]	VLSI 1990/JSSC1991
	Well erase	R. Kirisawa et al.	[12]	VLSI 1990
	Bipolarity program/erase	S. Aritome et al.	[13]	IEDM 1990
	Double patterning	R. Shirota et al.	[14]	IEDM 1990
1992	Bit-by-bit verify	T. Tanaka et al.	[15]	VLSI 1992
1993	Reliability of flash	S. Aritome et al.	[16]	Proceedings of IEEE
	0.4-μm 64-Mb-cell technology	S. Aritome et al.	[17,18]	SSDM 93/ JJAP
1994	SILC	H. Watanabe et al.	[19]	VLSI 1994
	Cycling and data retention reliability	S. Aritome et al.	[20]	IEICE
	Intelligent program, shield BL scheme	T. Tanaka et al.	[21]	JSSC
	Self-aligned STI cell (SA-STI cell)	S. Aritome et al.	[22]	IEDM 1994
1995	32-Mb NAND	K Imamiya et al./Y. Iwata et al.	[23, 24]	ISSCC1995/ JSSC
	32-Mb NAND with increment step program pulse (ISPP), self-boost	K. D.Suh et al.	[25, 26]	ISSCC1995/ JSSC
	Read disturb, SILC	S. Satoh, et al	[27, 28]	ICMTS1995/ED
	Increment step program pulse (ISPP)	G. J. Hemink et al.	[29]	VLSI 1995
	Double V_t select gate	K Takeuchi et al.	[30, 31]	VLSI 1995/ JSSC
	SWATT cell	S. Aritome et al.	[32, 33]	IEDM 1995/ ED

Year	Description	Author	Ref.	Conference/Journal
1996	128-Mb MLC	T. S. Jung et al.	[34, 35]	ISSCC 1996/ JSSC
	64 Mb	J. K. Kim et al.	[36, 37]	VLSI1996/ JSSC
	SILC	G. J. Hemink et al.	[38]	VLSI 1996
	On-chip ECC	T. Tanzawa et al.	[39, 40]	VLSI1996/ JSSC
	Booster plate	J. D. Choi et al.	[41]	VLSI 1996
	High- speed NAND	D. J. Kim et al.	[42]	VLSI 1996
	SILC in STI	H. Watanabe et al.	[43]	IEDM 1996
	Shared bit line	W. C. Shin et al.	[44]	IEDM 1996
1997	Nonvolatile virtual DRAM using NAND	T. S. Jung et al.	[45, 46]	ISSCC 1997/ JSSC
	Three- level cell (1.5 bits/cell)	T. Tanaka et al,	[47]	VLSI 1997
	Multi-page cell	K Takeuchi et al.	[48, 49]	VLSI 1997/ JSSC
	Parallel program	H. S. Kim et al	[50]	VLSI 1997
	0.25 μm SA-STI cell	K. Shimizu et al.	[51]	IEDM 1997
	Program disturb	S. Satoh et al.	[52]	IEDM 1997
	Triple poly-booster gate	J. D. Choi et al.	[53]	IEDM 1997

Requirements for NAND Flash

➢ Low Bit Cost

    Small cell size & Scalability → Self-Aligned STI (SA-STI)

                                           Multi-bit cell (MLC)

➢ High-Speed Program

    Parallel (Low power) Program → Page program

                                            Bit-by-bit verify

                                            V_{pgm} **step up (ISPP)**

➢ High Reliability

    Less degradation on tunnel oxide

                                    → Uniform P/E scheme

FIGURE 1.4   Requirements for NAND flash memory of the file storage market.

bit-line scheme [21], and so on. These technologies could satisfy the requirements of file storage memory.

Requirements for file storage memory are low bit cost, high-speed programming, and high reliability, as shown in Fig 1.4 [56].

The most important requirement for file storage applications is the low bit cost. The cost of a memory device is mainly determined by the die size of the memory chip and by the fabrication process cost, which is mainly dependent on depreciation of investment on factory. Then it is very important to combine small die size with a simple and low-cost fabrication process. In order to reduce the die size, reduction of unit memory cell size is as important as scaling feature size. Ideal memory cell size is $4 * F^2$ (F stands for feature size), because both X and Y directions are determined by line (F) and space (F). However, in early 1990s, it was difficult to realize $4 * F^2$ cell size of NAND flash memory due to wide ($>2 * F$) isolation width of LOCOS (local oxidation of Si). The self-aligned shallow trench isolation cell (SA-STI cell) was proposed and implemented to the NAND flash memory product. An isolation width could be scaled down from 2–3F in the LOCOS cell to F in the SA-STI cell. Therefore, the cell size could be drastically scaled down.

The SA-STI cell has been used in mass production for a long time, from 1998 to the present, because of a lot of advantages, such as small cell size, high reliability, and excellent scalability. However, below the 20-nm feature size, it is becoming very difficult to manage physical limitations, such as the floating gate capacitive coupling effect, RTN (random telegraph noise), the high-field problem, and so on. The recent feature size for production could reach to 15–16 nm [57]. It is not still clear whether memory cell size can be scaled down further or not.

Another way to reduce the effective cell size is the "multilevel cell." The logical bits are stored in one physical memory cell; for example, 2 logical bits are stored in one physical memory cell (MLC; 2 bits/cell). And 3 bits/cell and 4 bits/cell are called TLC and QLC. The mass-production start of the MLC was in 2000 by using 0.16-µm technology. The process technology to fabricate a multilevel cell device is basically as same as the process of single bit cell (SLC); however, the operations for a multilevel cell are much different from SLC operation. It is very important to make a tight V_t distribution width in the multilevel cell, in order to have high performance and reliability.

The next requirement for file storage application is high-speed programming, as shown in Fig. 1.4. In NAND flash memory, the uniform program/erase (P/E) scheme has been used as a de facto standard over 20 years. Unlike a NOR flash, no huge hot-electron injection current is required for programming, but a uniform P/E scheme has produced very low power consumption for programs even when the number of memory cells to be programmed is increased. Therefore the NAND flash memory can be easily programmed in large pages (512-byte to 32-Kbyte cells) so that the programming speed per byte can be quite fast ($\sim$ 100 Mbytes/s). In addition, several advanced program operations, such as bit-by-bit verify, V_{pgm} step up (ISPP: incremental step pulse program), ABL (all bit line) architecture, and so on, had been developed for high-speed programming.

The other important requirement for file storage applications is "high reliability," as shown in Fig. 1.4. A high voltage (>20 V) is applied to a control gate to produce a Fowler–Nordheim (FN) tunneling current on the tunnel oxide during programming. The electric field in tunnel oxide reaches values greater than 10 MV/cm, which is normally caused by oxide breakdown in other semiconductor devices. This means that flash memory uses a breakdown-like operation in normal program and erase. Due to applying a high field, tunnel oxide has been degraded by an electron/hole trap, interface state generation, and stress-induced leakage current (SILC). Major reliability degradation aspects of flash memory are related to this tunnel oxide degradation by programming and erase cycling. Even if a tunnel oxide is degraded, stored data have to be sustained in memory cells for long time, as nonvolatile memory. Data retention time after programming and erase cycling is a key of NAND flash reliability.

In addition, read disturb and program disturb are also an important reliability phenomena in NAND flash [13,16]. During read and program operation, pass voltages are applied to unselected word lines (WLs) in the NAND string. Several kinds of disturb stress are applied to an unselected cell in a cell array. Read disturb and program disturb are caused in these unselected cells in a string in a cell array.

Reliability specifications for NAND flash memory are dependent on applications such as digital still cameras, MP3 players, SSDs (solid-state disks) for PCs, SSDs for data servers, and so on. Target specifications of a NAND flash are generally as follows. In order to guarantee the specifications of NAND, every effort has been made regarding devices, processes, operations, circuits, memory systems, and so on.

Program and erase cycles (P/E cycles): 1-K to 100-K cycles

Data retention: 1–10 years

Read cycles: 1E5 – 1E7 times

Number of page program time (NOP): 1 time for MLC,TLC,QLC, 2–8 times for SLC

In 2007, three-dimensional (3D) NAND flash device technology of BiCS (bit cost scalable) was proposed [58] in order to scale down the NAND flash memory cell further. BiCS technology has a new low-cost process concept. The vertical poly-Si channel is fabricated by through-holes in stacked multilayer word lines. After the BiCS proposal, several three-dimensional (3D) NAND flash cells have been proposed [59–63]. Due to the vertical stacked cell structure, the 3D cell has an advantage of reducing effective cell size without scaling the feature size of F. In 2013, the mass-production start of 3D NAND flash was announced. The device was a 128-Gbit MLC 3D V-NAND flash with a 24-cell stacked charge trap cell [64]. To proceed to a lower bit cost of the 3D NAND cell, a number of stacked cells are needed to increase intensively. Many technical issues, such as a high-aspect etching, data retention of a charge trap cell, a new program disturb mode, cell current fluctuation, and so on, have to be solved or managed. After overcoming these critical issues, it is expected that a 1-terabit or 2-terabit NAND flash memory device will be available around 2020.

1.2 OVERVIEW

The NAND flash memory device technologies are reviewed in this book. The chapters focus on the scaling of the NAND flash memory cell, the high-performance operation of NAND flash, the improvement of NAND flash reliability, and three-dimensional (3D) NAND flash technologies, because they are very important for present and future NAND flash memory.

After describing a background of NAND flash technology in Chapter 1, Chapter 2 presents a basic structure and operations of NAND flash memory. The structures of single-cell and NAND-cell array are described. Cell operations of read, program, and erase are introduced. And then multilevel NAND cell technology is discussed to realize low-cost NAND flash memory.

The scaling history and scenario of planar (two-dimensional) NAND flash memory cells are reviewed in Chapter 3. The layout of the NAND flash memory cell is simple: Parallel word lines (WL) are perpendicular to parallel bit lines (BL). WL pitch is normally $2 * F$, (F: feature size), which is limited by lithography technology. However, BL pitch was normally $3 * F$ or more in the case of LOCOS isolation. This is because the isolation width needed to be $2 * F$ or more to prevent a relatively high (~ 8 V) punch-through between NAND cell channels (strings) during programming. Thus, it was crucial to scale down isolation width, in order to scale down memory cell size to satisfy the requirement of low bit cost.

First, LOCOS isolation cell technologies are presented (Section 3.2). The LOCOS isolation width can be minimized with improving device performance by the field-through implantation technique (FTI) after LOCOS formation. Next, the self-aligned STI cell with the FG (floating gate) wing is discussed (Section 3.3). The FG wing is applied to reduce the aspect ratio of cell structure. And then, the self-aligned STI cell without FG wing is discussed (Section 3.4). This cell has been used from the 90-nm generation cell to the present cell (1Y-nm cell), as a defacto standard. And the planar FG cell is introduced as an alternate cell structure (Section 3.5). Also, the sidewall transfer transistor cell (SWATT cell) is described (Section 3.6). Due to sidewall transfer transistor, the V_t read window margin can be greatly improved. Then, fast programming speed can be expected. And then, recent advanced NAND flash memory cell technologies of the dummy word-line scheme and the p-type floating gate are discussed in Section 3.7.

Another important technology for low bit cost is the multilevel cell (MLC), which is a stored multilogical bit in a single memory cell. To implement MLC, smart operation schemes are crucial to produce reasonable performance and reliability. In Chapter 4, the advanced operations for a multilevel NAND flash are discussed. It is very important to make tight V_t distribution width during programming for better performance and reliability. Most of the operation schemes focus on this point.

For the scaling memory cell, it is becoming very difficult to control the V_t distribution width due to the occurrence of several physical limitations, including the floating-gate capacitive coupling effect, electron injection spread, RTN, and the high-field problem. These physical limitations make the V_t distribution width wider, and then the, cell V_t setting margin (read window margin) is degraded. The recent feature size for production could reach values below 20 nm. The read window margin is seriously degraded. Then it is important to clarify how much scaling limitation factors have an impact on the V_t margin beyond the 20-nm feature size. Thus, the scaling challenges of the self-aligned STI cell are discussed beyond 20 nm in Chapter 5.

The reliability of two-dimensional NAND flash memory cell is discussed in Chapter 6. Reliability of flash memory is attributed to data retention or read disturb after program/erase cycling endurance. Program and erase operation schemes have a serious impact on reliability of a flash memory cell. Then, many program/erase schemes were proposed to satisfy the requirement of reliability and performance. It is very important to clarify the cell degradation mechanism and the best scheme of program/erase in order to achieve the requirement of reliability and performance. Chapter 6 describes the reliability aspect of NAND flash cell. The uniform program/erase scheme has several advantages in NAND flash reliability by comparing program/erase endurance, data retention, and read disturb characteristics in several program and erase schemes.

The three-dimentional (3D) NAND flash cells are presented in Chapter 7. After describing motivation and history of 3D NAND flash, many types of three-dimensional cells are introduced. Advantages and performances are compared in several 3D cells, including BiCS cell, TCAT/V-NAND, SMArT cell, VG-NAND, and DC-SF cell.

After that, the challenges of 3D NAND cells are discussed in Chapter 8. To realize low-cost NAND flash memories, serious issues have to be solved or managed by improving process, structure, device, performance, and reliability

The future trend of NAND flash memory is discussed in Chapter 9. The perspectives on future NAND flash technologies are also discussed.

Corresponding to the above discussions, the following topics are described in this book:

1. Principle of NAND flash memory.
2. Scaling scenario of 2D NAND flash memory cell.

3. Practical framework of scaling down of 2D NAND flash memory cell.

4. The LOCOS isolation technology to scale down NAND flash memory cell.

5. The self-aligned STI technology.

6. Low-cost NAND flash process flow.

7. The planar FG cell.

8. The SWATT cell for MLC (multilevel cell).

9. Advanced operations for MLC.

10. Basic and advanced program operations for tight programmed V_t distribution width in MLC.

11. Page program sequence for MLC (2 bits/cell).

12. Page program sequence for TLC (3 bits/cell)

13. The scaling challenges of a 2D NAND flash cell.

14. The solutions to overcome the scaling limitation of a 2D NAND flash cell.

15. The factors analysis of physical scaling limitation of a 2D NAND flash cell.

16. Detail mechanism of the floating gate capacitive coupling interference, Electron injection spread, RTN, and so on., as scaling limiter.

17. Investigation on the reliability of NAND flash in several program and erase schemes, in order to clarify the dependence of program and erase scheme.

18. Investigation of the program disturb and read disturb phenomena to optimize operation of NAND flash cell.

19. Introduction of several three-dimensional (3D) NAND flash memory cells.

20. Scaling challenges of 3D NAND flash memory.

21. Detail mechanism of program disturb, cell current fluctuation, and so on., in 3D NAND flash cell.

22. Future trend of NAND flash memory technologies.

REFERENCES

[1] Masuoka, F.; Asano, M.; Iwahashi, H.; Komuro, T.; Tanaka, S. A new flash E^{2}PROM cell using triple polysilicon technology, *Electron Devices Meeting, 1984 International*, vol. 30, pp. 464–467, 1984.

[2] Masuoka, F.; Momodomi, M.; Iwata, Y.; Shirota, R. New ultra high density EPROM and flash EEPROM with NAND structure cell, *Electron Devices Meeting, 1987 International*, vol. 33, pp. 552–555, 1987.

[3] Shirota, R.; Itoh, Y.; Nakayama, R.; Momodomi, M.; Inoue, S.; Kirisawa, R.; Iwata, Y.; Chiba, M.; Masuoka, F. New NAND cell for ultra high density 5v-only EEPROMs, *Digest of Technical Papers—Symposium on VLSI Technology*, 1988, pp. 33–34.

[4] Momodomi, M.; Kirisawa, R.; Nakayama, R.; Aritome, S.; Endoh, T.; Itoh, Y.; Iwata, Y.; Oodaira, H.; Tanaka, T.; Chiba, M.; Shirota, R.; Masuoka, F. New device technologies

for 5 V-only 4 Mb EEPROM with NAND structure cell, *Electron Devices Meeting, 1988. IEDM'88. Technical Digest International*, pp. 412–415, 1988.

[5] Itoh, Y.; Momodomi, M.; Shirota, R.; Iwata, Y.; Nakayama, R.; Kirisawa, R.; Tanaka, T.; Toita, K.; Inoue, S.; Masuoka, F. An experimental 4 Mb CMOS EEPROM with a NAND structured cell, *Solid-State Circuits Conference, 1989. Digest of Technical Papers, 36th ISSCC, 1989 IEEE International*, pp. 134–135, 15–17 Feb. 1989.

[6] Momodomi, M.; Itoh, Y.; Shirota, R.; Iwata, Y.; Nakayama, R.; Kirisawa, R.; Tanaka, T.; Aritome, S.; Endoh, T.; Ohuchi, K.; Masuoka, F. An experimental 4-Mbit CMOS EEPROM with a NAND-structured cell, *Solid-State Circuits, IEEE Journal of*, vol. 24, no. 5, pp. 1238–1243, Oct. 1989.

[7] Momodomi, M.; Iwata, Y.; Tanaka, T.; Itoh, Y.; Shirota, R.; Masuoka, F. A high density NAND EEPROM with block-page programming for microcomputer applications, *Custom Integrated Circuits Conference, 1989, Proceedings of the IEEE 1989*, pp. 10.1/1–10.1/4, 15–18 May 1989.

[8] Iwata, Y.; Momodomi, M.; Tanaka, T.; Oodaira, H.; Itoh, Y.; Nakayama, R.; Kirisawa, R.; Aritome, S.; Endoh, T.; Shirota, R.; Ohuchi, K.; Masuoka, F. A high-density NAND EEPROM with block-page programming for microcomputer applications, *Solid-State Circuits, IEEE Journal of*, vol. 25, no. 2, pp. 417–424, Apr. 1990.

[9] Aritome, S.; Kirisawa, R.; Endoh, T.; Nakayama, R.; Shirota, R.; Sakui, K.; Ohuchi, K.; Masuoka, F. Extended data retention characteristics after more than 10^4 write and erase cycles in EEPROMs, *International Reliability Physics Symposium, 1990. 28th Annual Proceedings*, pages 259–264, 1990.

[10] Tanaka, T.; Momodomi, M.; Iwata, Y.; Tanaka, Y.; Oodaira, H.; Itoh, Y.; Shirota, R.; Ohuchi, K.; Masuoka, F. A 4-Mbit NAND-EEPROM with tight programmed V_t distribution, *VLSI Circuits, 1990. Digest of Technical Papers, 1990 Symposium on*, pp. 105–106, 7–9 June 1990.

[11] Momodomi, M.; Tanaka, T.; Iwata, Y.; Tanaka, Y.; Oodaira, H.; Itoh, Y.; Shirota, R.; Ohuchi, K.; Masuoka, F. A 4 Mb NAND EEPROM with tight programmed V_t distribution, *Solid-State Circuits, IEEE Journal of*, vol. 26, no. 4, pp. 492–496, Apr. 1991.

[12] Kirisawa, R.; Aritome, S.; Nakayama, R.; Endoh, T.; Shirota, R.; Masuoka, F. A NAND structured cell with a new programming technology for highly reliable 5 V-only flash EEPROM, *1990 Symposium on VLSI Technology, 1990. Digest of Technical Papers*, pages 129–130, 1990.

[13] Aritome, S.; Shirota, R.; Kirisawa, R.; Endoh, T.; Nakayama, R.; Sakui, K.; Masuoka, F. A reliable bi-polarity write/erase technology in flash EEPROMs, *International Electron Devices Meeting, 1990. IEDM'90. Technical Digest*, pages 111–114, 1990.

[14] Shirota, R.; Nakayama, R.; Kirisawa, R.; Momodomi, M.; Sakui, K.; Itoh, Y.; Aritome, S.; Endoh, T.; Hatori, F.; Masuoka, F. A 2.3 μm^2 memory cell structure for 16 Mb NAND EEPROMs, *Electron Devices Meeting, 1990. IEDM'90. Technical Digest, International*, pp. 103–106, 9–12 Dec. 1990.

[15] Tanaka, T.; Tanaka, Y.; Nakamura, H.; Oodaira, H.; Aritome, S.; Shirota, R.; Masuoka, F. A quick intelligent program architecture for 3 V-only NAND-EEPROMs, *VLSI Circuits, 1992. Digest of Technical Papers, 1992 Symposium on*, pp. 20–21, 4–6 June 1992.

[16] Aritome, S.; Shirota, R.; Hemink, G.; Endoh, T.; Masuoka, F.; Reliability issues of flash memory cells, *Proceedings of the IEEE*, vol. 81, no. 5, pages 776–788, 1993.

[17] Aritome, S.; Hatakeyama, I.; Endoh, T.; Yamaguchi, T.; Shuto, S.; Iizuka, H.; Maruyama, T.; Watanabe, H.; Hemink, G. H.; Tanaka, T.; M. Momodomi, K. Sakui, and R. Shirota, A 1.13 μm² memory cell technology for reliable 3.3 V 64 Mb EEPROMs, *1993 International Conference on Solid State Device and Material (SSDM93)*, pp. 446–448, 1993.

[18] Aritome, S.; Hatakeyama, I.; Endoh, T.; Yamaguchi, T.; Susumu, S.; Iizuka, H.; Maruyama, T.; Watanabe, H.; Hemink, G.; Koji, S.; Tanaka, T.; Momodomi, M.; and Shirota, R. An advanced NAND-structure cell technology for reliable 3.3V 64 Mb electrically erasable and programmable read only memories (EEPROMs), *Jpn. J. Appl. Phys.* vol. 33, pp. 524–528, Jan. 1994.

[19] Watanabe, H.; Aritome, S.; Hemink, G. J.; Maruyama, T.; Shirota, R. Scaling of tunnel oxide thickness for flash EEPROMs realizing stress-induced leakage current reduction, *VLSI Technology, 1994. Digest of Technical Papers. 1994 Symposium on*, pp. 47–48, 7–9 June 1994.

[20] Aritome, S.; Shirota R.; Sakui, K.; Masuoka, F. Data retention characteristics of flash memory cells after write and erase cycling, *IEICE Trans. Electron.*, vol. E77-C, no. 8, pp. 1287–1295, Aug. 1994.

[21] Tanaka, T.; Tanaka, Y.; Nakamura, H.; Sakui, K.; Oodaira, H.; Shirota, R.; Ohuchi, K.; Masuoka, F.; Hara, H. A quick intelligent page-programming architecture and a shielded bitline sensing method for 3 V-only NAND flash memory, *Solid-State Circuits, IEEE Journal of*, vol. 29, no. 11, pp. 1366–1373, Nov. 1994.

[22] Aritome, S.; Satoh, S.; Maruyama, T.; Watanabe, H.; Shuto, S.; Hemink, G. J.; Shirota, R.; Watanabe, S.; Masuoka, F. A 0.67 μm² self-aligned shallow trench isolation cell (SA-STI cell) for 3 V–only 256 Mbit NAND EEPROMs, *Electron Devices Meeting, 1994. IEDM'94. Technical Digest., International*, pp. 61–64, 11–14 Dec. 1994.

[23] Imamiya, K.; Iwata, Y.; Sugiura, Y.; Nakamura, H.; Oodaira, H.; Momodomi, M.; Ito, Y.; Watanabe, T.; Araki, H.; Narita, K.; Masuda, K.; Miyamoto, J. A 35 ns-cycle-time 3.3 V-only 32 Mb NAND flash EEPROM, *Solid-State Circuits Conference, 1995. Digest of Technical Papers. 42nd ISSCC, 1995 IEEE International*, pp. 130–131, 351, 15–17 Feb. 1995.

[24] Iwata, Y.; Imamiya, K.; Sugiura, Y.; Nakamura, H.; Oodaira, H.; Momodomi, M.; Itoh, Y.; Watanabe, T.; Araki, H.; Narita, K.; Masuda, K.; Miyamoto, J.-I. A 35 ns cycle time 3.3 V only 32 Mb NAND flash EEPROM, *Solid-State Circuits, IEEE Journal of*, vol. 30, no. 11, pp. 1157–1164, Nov. 1995.

[25] Suh, K.-D.; Suh, B.-H.; Um, Y.-H.; Kim, J.-Ki; Choi, Y.-J.; Koh, Y.-N.; Lee, S.-S.; Kwon, S.-C.; Choi, B.-S.; Yum, J.-S; Choi, J.-H.; Kim, J.-R.; Lim, H.-K. A 3.3 V 32 Mb NAND flash memory with incremental step pulse programming scheme, *Solid-State Circuits Conference, 1995. Digest of Technical Papers. 42nd ISSCC, 1995 IEEE International*, pp.128–129, 350, 15–17 Feb. 1995.

[26] Suh, K.-D.; Suh, B.-H.; Lim, Y.-H.; Kim, J.-K.; Choi, Y.-J.; Koh, Y.-N.; Lee, S.-S.; Kwon, S.-C.; Choi, B.-S.; Yum, J.-S.; Choi, J.-H.; Kim, J.-R.; Lim, H.-K. A 3.3 V 32 Mb NAND flash memory with incremental step pulse programming scheme, *Solid-State Circuits, IEEE Journal of*, vol. 30, no. 11, pp. 1149–1156, Nov. 1995.

[27] Satoh, S.; Hemink, G. J.; Hatakeyama, F.; Aritome, S. Stress induced leakage current of tunnel oxide derived from flash memory read-disturb characteristics, *Microelectronic Test Structures, 1995. ICMTS 1995. Proceedings of the 1995 International Conference on*, pp. 97–101, 22–25 Mar. 1995.

[28] Satoh, S.; Hemink, G.; Hatakeyama, K.; Aritome, S.; Stress-induced leakage current of tunnel oxide derived from flash memory read-disturb characteristics, *IEEE Transactions on Electron Devices*, vol. 45, no. 2, pp. 482–486 1998.

[29] Hemink, G. J.; Tanaka, T.; Endoh, T.; Aritome, S.; Shirota, R. Fast and accurate programming method for multi-level NAND EEPROMs, *VLSI Technology, 1995. Digest of Technical Papers. 1995 Symposium on*, pp. 129–130, 6–8 June 1995.

[30] Takeuchi, K.; Tanaka, T.; Nakamura, H. A double-level-V_{th} select gate array architecture for multi-level NAND flash memories, *VLSI Circuits, 1995. Digest of Technical Papers., 1995 Symposium on*, pp. 69–70, 8–10 June 1995.

[31] Takeuchi, K.; Tanaka, T.; Nakamura, H. A double-level-V_{th} select gate array architecture for multilevel NAND flash memories, *Solid-State Circuits, IEEE Journal of*, vol. 31, no. 4, pp. 602–609, Apr. 1996.

[32] Aritome, S.; Takeuchi, Y.; Sato, S.; Watanabe, H.; Shimizu, K.; Hemink, G.; Shirota, R. A novel side-wall transfer-transistor cell (SWATT cell) for multi-level NAND EEPROMs, *Electron Devices Meeting, 1995. International*, pp. 275–278, 10–13 Dec. 1995.

[33] Aritome, S.; Takeuchi, Y.; Sato, S.; Watanabe, I.; Shimizu, K.; Hemink, G.; Shirota, R. A side-wall transfer-transistor cell (SWATT cell) for highly reliable multi-level NAND EEPROMs, *Electron Devices, IEEE Transactions on*, vol. 44, no. 1, pp. 145–152, Jan. 1997.

[34] Jung, T.-S.; Choi, Y.-J.; Suh, K.-D.; Suh, B.-H.; Kim, J.-K.; Lim, Y.-H.; Koh, Y.-N.; Park, J.-W.; Lee, K.-J.; Park, J.-H.; Park, K.-T.; Kim, J.-R.; Lee, J.-H.; Lim, H.-K. A 3.3 V 128 Mb multi-level NAND flash memory for mass storage applications, in Solid-State Circuits Conference, 1996. Digest of Technical Papers. 42nd ISSCC., 1996 IEEE International, pp. 32–33, 10-10 Feb. 1996.

[35] Jung, T.-S.; Choi, Y.-J.; Suh, K.-D.; Suh, B.-H.; Kim, J.-K.; Lim, Y.-H.; Koh, Y.-N.; Park, J.-W.; Lee, K.-J.; Park, J.-H.; Park, K.-T.; Kim, J.-R.; Yi, J.-H.; Lim, H.-K. A 117-mm^2 3.3-V only 128-Mb multilevel NAND flash memory for mass storage applications, *Solid-State Circuits, IEEE Journal of*, vol. 31, no. 11, pp. 1575–1583, Nov. 1996.

[36] Kim, J.-K.; Sakui, K.; Lee, S.-S.; Itoh, J.; Kwon, S.-C.; Kanazawa, K.; Lee, J.-J.; Nakamura, H.; Kim, K.-Y.; Himeno, T.; Jang-Rae, Kim; Kanda, K.; Tae-Sung, Jung; Oshima, Y.; Kang-Deog, Suh; Hashimoto, K.; Sung-Tae Ahn; Miyamoto, J. A 120 mm^2 64 Mb NAND flash memory achieving 180 ns/byte effective program speed, *VLSI Circuits, 1996. Digest of Technical Papers 1996 Symposium on*, pp. 168–169, 13–15 June 1996.

[37] Kim, J.-K.; Sakui, K.; Lee, S.-S.; Itoh, Y.; Kwon, S.-C.; Kanazawa, K.; Lee, K.-J.; Nakamura, H.; Kim, K.-Y.; Himeno, T.; Kim, J.-R.; Kanda, K.; Jung, T.-S.; Oshima, Y.; Suh, K.-D.; Hashimoto, K.; Ahn, S.-T.; Miyamoto, J. A 120-mm^2 64-Mb NAND flash memory achieving 180 ns/Byte effective program speed, *Solid-State Circuits, IEEE Journal of*, vol. 32, no. 5, pp. 670–680, May. 1997.

[38] Hemink, G. J.; Shimizu, K.; Aritome, S.; Shirota, R. Trapped hole enhanced stress induced leakage currents in NAND EEPROM tunnel oxides, *IEEE International Reliability Physics Symposium, 1996. 34th Annual Proceedings*, pp. 117–121, 1996.

[39] Tanzawa, T.; Tanaka, T.; Takeuchi, K.; Shirota, R.; Aritome, S.; Watanabe, H.; Hemink, G.; Shimizu, K.; Sato, S.; Takeuchi, Y.; Ohuchi, K. A compact on-chip ECC for low cost flash memories, *VLSI Circuits, 1996. Digest of Technical Papers, 1996 Symposium on*, pp. 74–75, 13–15 June 1996.

[40] Tanzawa, T.; Tanaka, T.; Takeuchi, K.; Shirota, R.; Aritome, S.; Watanabe, H.; Hemink, G.; Shimizu, K.; Sato, S.; Takeuchi, Y.; Ohuchi, K. A compact on-chip ECC for low cost flash memories, *Solid-State Circuits, IEEE Journal of*, vol. 32, no. 5, pp. 662–669, May 1997.

[41] Choi, J. D.; Kim, D. J.; Tang, D. S.; Kim, J.; Kim, H. S.; Shin, W. C.; Ahn, S. T.; Kwon, O.H. A novel booster plate technology in high density NAND flash memories for voltage scaling-down and zero program disturbance, *VLSI Technology, 1996. Digest of Technical Papers. 1996 Symposium on*, pp. 238–239, 11–13 June 1996.

[42] Kim, D.J.; Choi, J. D.; Kim, J.; Oh, H. K.; Ahn, S. T.; Kwon, O. H. Process integration for the high speed NAND flash memory cell, *VLSI Technology, 1996. Digest of Technical Papers. 1996 Symposium on*, pp. 236–237, 11–13 June 1996.

[43] Watanabe, H.; Shimizu, K.; Takeuchi, Y.; Aritome, S. Corner-rounded shallow trench isolation technology to reduce the stress-induced tunnel oxide leakage current for highly reliable flash memories, *Electron Devices Meeting, 1996. IEDM'96., International*, pp. 833–836, 8–11 Dec. 1996.

[44] Shin, W. C.; Choi, J. D.; Kim, D. J.; Kim, H. S.; Mang, K. M.; Chung, C. H.; Ahn, S. T.; and Kwon, O. H.. A new shared bit line NAND Cell technology for the 256 Mb flash memory with 12V programming, *Electron Devices Meeting, 1996. IEDM'96, International*, Dec. 1996.

[45] Jung, T.-S.; Choi, D.-C.; Cho, S.-H.; Kim, M.-J.; Lee, S.-K.; Choi, B.-S.; Yum, J.-S.; Kim, S.-H.; Lee, D.-G.; Son, J.-C.; Yong, M.-S.; Oh, H.-K.; Jun, S.-B.; Lee, W.-M.; Haq, E.; Suh, K.-D.; Ali, S.; Lim, H.-K. A 3.3 V 16 Mb nonvolatile virtual DRAM using a NAND flash memory technology, *Solid-State Circuits Conference, 1997. Digest of Technical Papers. 43rd ISSCC, 1997 IEEE International*, pp. 398–399, 493, 6–8 Feb. 1997.

[46] Jung, T.-S.; Choi, D.-C.; Cho, S.-H.; Kim, M.-J.; Lee, S.-K.; Choi, B.-S.; Yum, Jin-Sun; Kim, S.-H.; Lee, D.-G.; Son, J.-C.; Yong, M.-S.; Oh, H.-K.; Jun, S.-B.; Lee, W.-M.; Haq, E.; Suh, K.-D.; Ali, S. B.; Lim, H.-K.; A 3.3-V single power supply 16-Mb nonvolatile virtual DRAM using a NAND flash memory technology, *Solid-State Circuits, IEEE Journal of*, vol. 32, no. 11, pp. 1748–1757, Nov. 1997.

[47] Tanaka, T.; Tanzawa, T.; Takeuchi, K.; , A 3.4-Mbyte/sec programming 3-level NAND flash memory saving 40% die size per bit, *VLSI Circuits, 1997. Digest of Technical Papers., 1997 Symposium on*, pp. 65–66, 12–14 June 1997.

[48] Takeuchi, K.; Tanaka, T.; Tanzawa, T. A Multi-page Cell Architecture for high-speed programming multi-level NAND flash memories, *VLSI Circuits, 1997. Digest of Technical Papers, 1997 Symposium on*, pp. 67–68, 12–14 June 1997.

[49] Takeuchi, K.; Tanaka, T.; Tanzawa, T. A multipage cell architecture for high-speed programming multilevel NAND flash memories, *Solid-State Circuits, IEEE Journal of*, vol. 33, no. 8, pp. 1228–1238, Aug. 1998.

[50] Kim, H. S.; Choi, J. D.; Kim, J.; Shin, W. C.; Kim, D. J.; Mang, K. M.; Ahn, S. T. Fast parallel programming of multi-level NAND flash memory cells using the booster-line technology, *VLSI Technology, 1997. Digest of Technical Papers, 1997 Symposium on*, pp. 65–66, 10–12 June 1997.

[51] Shimizu, K.; Narita, K.; Watanabe, H.; Kamiya, E.; Takeuchi, Y.; Yaegashi, T.; Aritome, S.; Watanabe, T. A novel high-density $5F^2$ NAND STI cell technology suitable for 256 Mbit and 1 Gbit flash memories, *Electron Devices Meeting, 1997. IEDM'97. Technical Digest, International*, pp. 271–274, 7–10 Dec. 1997.

[52] Satoh, S.; Hagiwara, H.; Tanzawa, T.; Takeuchi, K.; Shirota, R. A novel isolation-scaling technology for NAND EEPROMs with the minimized program disturbance, *Electron Devices Meeting, 1997. IEDM'97. Technical Digest, International*, pp. 291–294, 7–10 Dec. 1997.

[53] Choi, J. D.; Lee, D. G.; Kim, D. J.; Cho, S. S.; Kim, H. S.; Shin, C. H.; Ahn, S. T. A triple polysilicon stacked flash memory cell with wordline self-boosting programming, *Electron Devices Meeting, 1997. IEDM'97. Technical Digest, International*, pp. 283–286, 7–10 Dec. 1997.

[54] F. Masuoka, flash memory makes a big leap, *Kogyo Chosakai*, vol. 1, pp. 1–172, 1992. (in Japanese).

[55] Aritome, S., NAND flash innovations, *Solid-State Circuits Magazine, IEEE*, vol. 5, no. 4, pp. 21,29, Fall 2013.

[56] Aritome, S. Advanced flash memory technology and trends for file storage application *Electron Devices Meeting, 2000. IEDM Technical Digest International*, pp. 763–766, 2000.

[57] Hwang, J.; Seo, J.; Lee, Y.; Park, S.; Leem, J.; Kim, J.; Hong, T.; Jeong, S.; Lee, K.; Heo, H.; Lee, H.; Jang, P.; Park, K.; Lee, M., Baik, S.; Kim, J.; Kkang, H.; Jang, M.; Lee, J.; Cho, G.; Lee, J.; Lee, B.; Jang, H.; Park, S.; Kim, J.; Lee, S.; Aritome, S.; Hong, S.; and Park, S. A middle-1X nm NAND flash memory cell (M1X-NAND) with highly manufacturable integration technologies, *Electron Devices Meeting (IEDM), 2011 IEEE International*, pp. 199–202, Dec. 2011.

[58] Tanaka, H.; Kido, M.; Yahashi, K.; Oomura, M.; Katsumata, R.; Kito, M.; Fukuzumi, Y.; Sato, M.; Nagata, Y.; Matsuoka, Y.; Iwata, Y.; Aochi, H.; Nitayama, A. Bit cost scalable technology with punch and plug process for ultra high density flash memory, *VLSI Symposium Technical. Digest., 2007*, pp. 14–15.

[59] Katsumata, R.; Kito, M.; Fukuzumi, Y.; Kido, M.; Tanaka, H.; Komori, Y.; Ishiduki, M.; Matsunami, J.; Fujiwara, T.; Nagata, Y.; Zhang, L.; Iwata, Y.; Kirisawa, R.; Aochi, H.; Nitayama, A. Pipe-shaped BiCS flash memory with 16 stacked layers and multi-level-cell operation for ultra high density storage devices, *VLSI Symposium Technology. Digest.*, pp. 136–137, 2009.

[60] Kim, J.; Hong, A. J.; Ogawa, M.; Ma, S.; Song, E. B.; Lin, Y.-S.; Han, J.; Chung, U.-I.; Wang, K. L. Novel 3-D structure for ultra high density flash memory with VRAT (vertical-recess-array-transistor) and PIPE (planarized integration on the same plane), *VLSI Symposium Technology. Digest., 2008*, pp. 122–123.

[61] Kim, W. J.; Choi, S.; Sung, J.; Lee, T.; Park, C.; Ko, H.; Jung, J.; Yoo, I.; Park, Y. Multi-layered vertical gate NAND flash overcoming stacking limit for terabit density storage, *VLSI Symposium Technology. Digest., 2009*, pp. 188–189.

[62] Jang, J.; Kim, H.-S.; Cho, W.; Cho, H.; Kim, J.; Shim, S.I.; Jang, Y.; Jeong, J.-H.; Son, B.-K.; Kim, D. W.; Kim, K.; Shim, J.-J.; Lim, J. S; Kim, K.-H.; Yi, S. Y.; Lim, J.-Y.; Chung, D.; Moon, H.-C.; Hwang, S.; Lee, J.-W.; Son, Y.-H.; U-In Chung and Lee, W.-S. Vertical cell array using TCAT (terabit cell array transistor) technology for ultra high density NAND flash memory", *VLSI Symposium Technology. Digest, 2009*, pp. 192–193.

[63] Lue, H.-T.; Hsu, T.-H.; Hsiao, Y.-H.; Hong, S. P.; Wu, M. T.; Hsu, F. H.; Lien, N. Z.; Wang, S.-Y.; Hsieh, J.-Y.; Yang, L.-W.; Yang, T.; Chen, K.-C.; Hsieh, K.-Y.; Lu, C.-Y.; A highly scalable 8-layer 3D vertical-gate (VG) TFT NAND flash using junction-free buried channel BE-SONOS device, *VLSI Technology (VLSIT), 2010 Symposium on*, pp. 131–132, 15–17 June 2010.

[64] Park, K.-T.; Nam, S.; Kim, D.; Kwak, P.; Lee, D.; Choi, Y.-H.; Choi, M.-H.; Kwak, D.-H.; Kim, D.-H.; Kim M.-S.; Park, H.-W.; Shim, S.-W.; Kang, K.-M.; Park, S.-W.; Lee, K.; Yoon, H.-J.; Ko, K.; Shim, D.-K.; Ahn, Y.-L.; Ryu, J.; Kim, D.; Yun, K.; Kwon, J.; Shin, S.; Byeon, D.-S.; Choi, K.; Han, J.-M.; Kyung, K.-H.; Choi, J.-H.; Kim, K. Three-dimensional 128 Gb MLC vertical NAND flash memory with 24-WL stacked layers and 50 MB/s high-speed programming, *Solid-State Circuits, IEEE Journal of*, vol. 50, no. 1, pp. 204, 213, Jan. 2015.

2

PRINCIPLE OF NAND FLASH MEMORY

2.1 NAND FLASH DEVICE AND ARCHITECTURE

2.1.1 NAND Flash Memory Cell Architecture

Figure 2.1 shows the single-cell architecture of flash memory. The single cell has an Nch MOS transistor with poly-silicon floating gate (FG). FG is electrically isolated by tunnel oxide and interpoly dielectric (IPD). Charge is stored in FG, and potential of FG is controlled by control gate (CG) voltage with capacitive coupling between CG and FG. Figure 2.2 shows the NAND string structure [1]. The NAND string consists of 32 cells (typical) and two select transistors (SGD, SGS). Thirty-two cells are connected in series. SGD connects at a drain to isolate it from a bit line (BL), and SGS connects at a source to isolate it from a source line (SL). The number of cells in one NAND string has been increased with steps of 8 cells → 16 cells → 32 cells (0.12-μm generation~) → 64 cells (~43-nm generation).

Figure 2.3 shows (a) a top view and (b) a cross-sectional view of NAND flash memory cells. One NAND string consists of 32 series-connected stacked gate memory transistors and two select gate transistors. The entire memory array is formed by straight active area lines (horizontal lines) and straight gate lines (vertical lines). This simple cell structure enables easy fabrication of memory cells, which have small feature size. The memory cell is located at an intersection of the two cross lines. Thirty-two memory transistors are arranged between two select transistors, so that both the bit-line and source-line contacts are arranged every 34 gate lines (vertical lines).

Nand Flash Memory Technologies, First Edition. Seiichi Aritome.
© 2016 The Institute of Electrical and Electronics Engineers, Inc. Published 2016 by John Wiley & Sons, Inc.

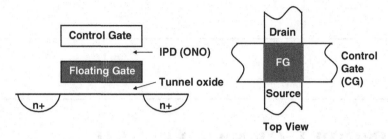

FIGURE 2.1   Single-cell architecture of floating-gate flash memory. Nch MOS transistor with poly-silicon floating gate (FG). The FG is electrically isolated by tunnel oxide and inter-poly dielectric (IPD). Charge is stored in FG.

Figure 2.4 shows the array architecture of a NAND flash memory cell. Page size is typically 2 K to 16 KByte (2- to 16-KByte cells + ECC code). Page size is increasing for enhancing the performance of read and program operations. The original page size was 512-Byte cells + ECC code [2–5]. And the page size was increased with steps of 512 B → 2 KB (0.12-μm generation~) → 4 KB (~56-nm generation) → 8 KB (~43-nm generation) → 16 KB (~2X- to 1X-nm generation, with an all bit-line scheme). The page size will be more increased for future devices because much higher performance will be required. The block size is basically page × 2 × 32 cells = 128-K to 256-KByte cells (physical) in this case. The block size was also increased by increasing the number of cells in the NAND string and increasing the page size. The read and program operations are performed per page. And the erase operation is performed per block.

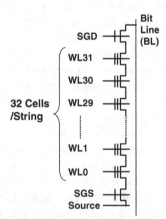

FIGURE 2.2   NAND cell string architecture. NAND cell string consists of 32 cells (for example) and two select transistors (SGD, SGS). Thirty-two cells are connected in series. SGD and SGS connect at drain and source to isolate from bit line and source line, respectively.

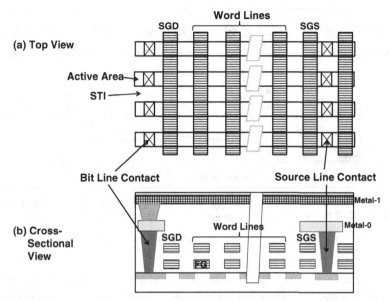

FIGURE 2.3 (a) Top schematic view and (b) cross-sectional view of NAND flash memory cells.

2.1.2 Peripheral Device

Figure 2.5 shows the cross-sectional view of a typical NAND flash memory device [6]. A memory cell array is fabricated on a double well consisting of a *p*-well and an *n*-well. Low-voltage transistors of N-ch (N-channel) and P-ch (P-channel) are on the *p*-well and the *n*-well, respectively. A high-voltage transistor of N-ch is fabricated

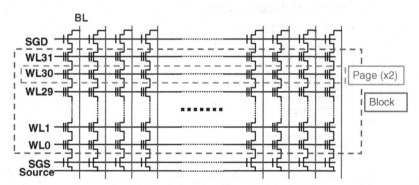

FIGURE 2.4 Array architecture of NAND flash memory cells. Page size is typically 2-K to 16-KByte (2- to 16-KByte cells + ECC code). Block size is typically page × 2 × 32 cells = 128-K to 256-KByte cell (physical). Read and program are performed in per page. Erase is performed in per block.

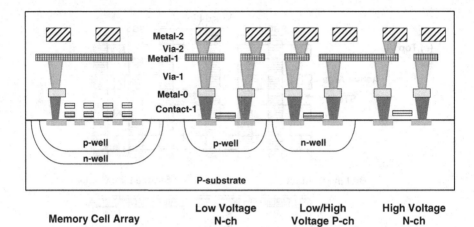

FIGURE 2.5   The cross-sectional view of NAND flash memory.

on a P-substrate. A high-voltage transistor of P-ch is fabricated on an *n*-well, which is basically the same *n*-well as for a low-voltage P-ch transistor. A NAND flash memory device has normally three metal layers of Metal-0, Metal-1, and Metal-2. Cu metal layers with a damascene process are used for Metal-1 and Metal-2 in several companies' products.

The peripheral transistors for a NAND flash memory device are listed in Fig. 2.6. The lineup of peripheral transistor types is dependent on circuit requirement. Figure 2.6 shows a typical case of a transistors list. A thick gate oxide of 25–40 nm is used for a high-voltage transistor. A low-voltage transistor has a thin gate oxide (~9 nm) which is the same as tunnel oxide in a memory cell. In order to realize low process cost, it is important that peripheral Tr is fabricated without increasing process steps.

NAND Flash Peripheral transistors

| Low-voltage (LV) transistor |
| N-ch E-type |
| P-ch E-type |
| High-voltage (HV) transistor |
| N-ch E-type |
| N-ch I-type |
| N-ch D-type |
| P-ch E(+)-type |

Where E-type = Enhancement type
       I-type = Intrinsic type ($V_t \sim 0$ V)
       D-type = Depletion type
       E(+)-type = deep Enhancement type ($V_t \sim -3$ V)

FIGURE 2.6   List of peripheral transistors in NAND flash memory. V_t; threshold voltage.

The scaling of a high-voltage transistor (HV Tr) is one of the important challenges for NAND flash memories. The size scaling of a high-voltage transistor (HV Tr) is required for placing HV Tr and isolation on a certain determined cell pitch. There are two critical locations of HV Tr scaling. One is the row decoder/word-line driver, and HV Tr + isolation width have to be in one string pitch basically. The other location is the bit-line selector area. The size of an HV transistor has to be as small as possible to make a small area of bit line selector.

2.2   CELL OPERATION

2.2.1   Read Operation

Figure 2.7 shows basic read operation of single cell. Data (0 or 1) is judged by cell current (I_{cell}) flow "ON" or not flow "OFF" during $V_{gate} = 0$ V applied (for SLC (1 bit/cell) case). The erased state has a positive charge in an FG. The programmed state has a negative charge in FG.

Read operations are performed in page units, as shown in Fig. 2.8. A page corresponds to a row of cells and is accessed by a single word line. V_{passR} ($\sim$6 V) is applied to an unselected word line to be as a pass transistor. Random read speed (tR) is normally 25 μs for SLC and 60 μs for MLC (2 bits/cell). However, tR is becoming longer as a cell scaling because of decreasing cell current, I_{cell}.

A read disturb problem is mainly caused in unselected cells where V_{passR} is applied to a control gate (CG) and channel is 0 V. It is the weak electron injection mode, and then cell V_t (threshold voltage) is gradually increased. Detail read disturb characteristics are discussed in Chapter 6.

2.2.2   Program and Erase Operation

Several program and erase schemes were considered to use a NAND flash memory product in early stage of development, as shown in Fig. 2.9. There are only two

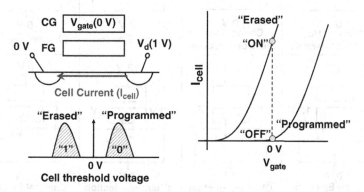

FIGURE 2.7   Principle of read of single cell. Erased: Positive charge in FG. I_{cell} "ON". Programmed: Negative charge in FG. I_{cell} "OFF".

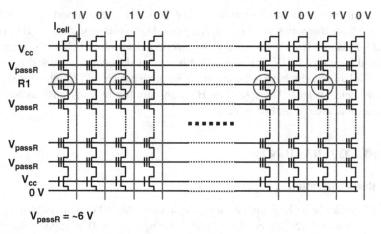

FIGURE 2.8   Read operation of NAND cell array. Page read: 2–16 KByte at the same time. Typical random read speed, tR: 25 μs for SLC, 60 μs for MLC.

ways of giving an electron injection to a floating gate (FG) of a channel hot electron (CHE) injection and a Fowler–Nordheim (FN) channel injection. And also, there are only two ways of electron ejection from an FG of a drain/source FN ejection and a channel FN ejection. Four schemes of programming and erase operations

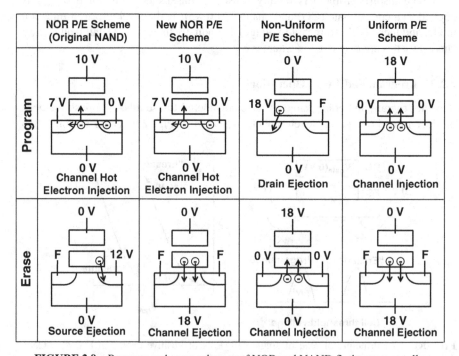

FIGURE 2.9   Program and erase schemes of NOR and NAND flash memory cells.

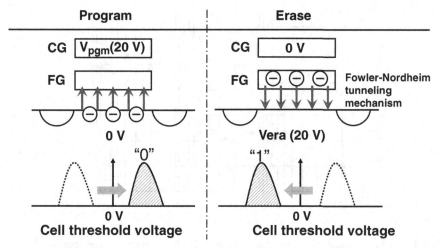

FIGURE 2.10   Principle of single-cell program and erase. Program: Injected electron to the FG through tunnel oxide. Erase: Ejected electron from the FG.

are possible by combination of these ways of injection and ejection, as shown in Fig. 2.9. The first one is the NOR-type flash program/erase scheme. CHE injection is used for programming, and source FN ejection is used for erase. The second one is the new NOR-type flash program/erase scheme [7]. CHE injection is used for programming, and channel FN ejection is used for erase. The third one is the old program/erase scheme for NAND flash [8, 9]. Drain FN ejection is used for programming, and channel FN injection is used for erasing. The fourth one is the current NAND-type flash program/erase scheme [7, 10, 11]. Channel FN injection is used for programming, and channel FN ejection is used for erasing.

Power consumption is worse in both CHE injection and drain/source FN ejection schemes due to large current flow. The memory cell scalability is also worse in CHE injection and drain/source FN ejection schemes due to applying large voltage to a drain/source. And reliability is worse in drain/source FN ejection schemes due to degradation by a generated hot hole, as described in Chapter 6. Therefore, the program/erase schemes for a NAND flash memory product use a uniform program/erase (P/E) scheme, as shown in Fig. 2.9 [7, 10, 11]. A detailed program and an erase operation are described in the following.

The program operation of a NAND flash cell is performed by applying high-voltage V_{pgm} to control gate (CG) while keeping substrate/source/drain 0 V, as shown in Fig. 2.10. Electrons are injected to the floating gate (FG) by a Fowler–Nordheim (FN) tunneling mechanism through the tunnel oxide. V_t of cell has a positive shift. An erase operation is performed by applying high-voltage V_{era} to substrate (p-well), while keeping CG at 0 V. Electrons in an FG are ejected to a substrate through a tunnel oxide. V_t has a negative shift.

Figure 2.11 shows a typical program characteristic. Higher program voltage (V_{pgm}) or longer program pulse width can cause faster programming.

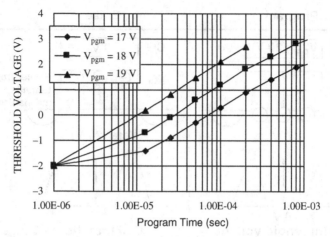

FIGURE 2.11 Typical program characteristics. High program voltage can be faster operation.

Figure 2.12 shows the cell array program scheme. At first, the programming starts at the source side cells. V_{pgm} is applied to the control gate (CG0; WL0) of the selected cells while V_{pass} is applied to the control gates of the unselected cells. These unselected cells act as pass transistors and make the boosting voltage in a channel to prevent boosting mode program disturb, as described in Section 2.2.4. A 0-V charge is applied to the bit line, and then electrons are injected from the bit line (channel) to a floating gate by the electric field between the bit line and the floating gate of the selected cell. The threshold voltage of the selected cell is pushed up into the

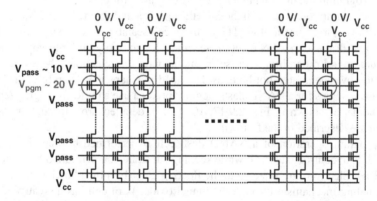

FIGURE 2.12 Program operation of NAND flash cell array. Page program: 2- to 16-KByte program at the same time. All cells in page should be programmed in one program sequence (prohibit partial page program). Program speed, t_{prog}: 200 μs for SLC, 800–1600 μs for MLC. Order of page programming is from source side page to drain side page (sequential page program in block). Random order of page programming is normally prohibited.

enhancement mode of approximately 2 V. In the case of a program inhibit mode, the bit line is raised to V_{cc}. The voltage of $V_{cc} - V_t$ (V_t of select Tr) is transferred to the channel of cell strings before applying V_{pass} to unselected CGs. Then V_{pass} are applied to the control gate of the unselected CGs to make the boosting voltage in the channel of unselected string. No electrons are injected from the channel (V_{boost}) to the floating gate, because the electric field between the bit line and the floating gate is insufficient to initiate tunneling. The threshold voltage of the selected cell remains at erase state of -3 V. After programming the cells connected with CG0 (WL0), the programming of the cells connected with CG1 (WL1) starts. V_{pgm} is applied to the control gate of the selected cell (CG1), and V_{pass} is applied to the control gates of the unselected cells (CG0, CG2–G31). Either 0 V or V_{cc} is applied to a bit line as corresponding with data. Subsequently, the programming continues from the source side cell to the bit-line side cell successively. Typical programming time per page, including the data load sequence, is 200 μs for SLC, 800–1600 μs for MLC.

Incremental step pulse programming (ISPP) achieves fast program performance under process and environmental variations while keeping a tight programmed cell V_t distribution [12–14]. Figure 2.13 shows the V_{pgm} waveform of an ISPP scheme. V_{pgm} has stepped up by each pulse. An ISPP scheme suppresses process variation issues effectively by allowing fast programmed cells to be programmed with a lower program voltage and slow program cells to be programmed with a higher program voltage. After an initial 15.5-V program pulse, each subsequent pulse (if required) is incremented in 0.5-V (ΔV_{pgm}) steps up to 20 V, for example. Since sufficiently programmed cells are automatically switched to the program inhibit state in the verification step, easily programmed cells are not affected by the higher program voltages. A 1-V program pulse increment is approximately as effective as five pulses without the increment. Thus, ISPP scheme has the effect of increasing pulse width

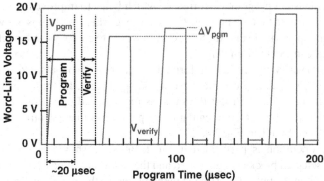

Incremental Step Pulse Programming (ISSP)

FIGURE 2.13 Incremental step pulse programming (ISSP) waveform. ISPP is used for page program to make a tight cell V_t distribution. Degradation of tunnel oxide can be mitigated due to relaxed maximum electric field in tunnel oxide during program.

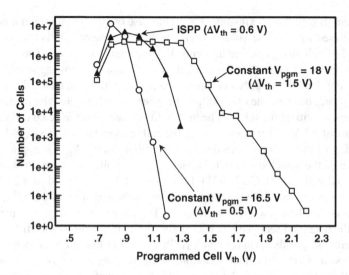

FIGURE 2.14 Comparison of programmed cell threshold voltage distribution in devices with ISPP and without ISPP (constant voltage programming with 16.5 and 18 V).

[15] without actually increasing the program time by dynamically optimizing program voltage to cell characteristics. Through program pulse/verify cycles, programmed cell V_t is maintained to within 0.6 V by using a 0.5-V ISPP step as shown in Fig. 2.14 [12]. With an ISPP scheme, a page is typically programmed within 2–6 program pulses/verify cycles for SLC (~300 μs). The constant 16.5-V program voltage device of Fig. 2.14 has the tightest V_t distribution (~0.5 V), however, program verify cycles are larger, consisting of 11–37 cycles. This means 2–16 times slower programming speed. Then an ISPP scheme provides an optimum combination of both a tight V_t distribution and a fast program time.

By effectively adjusting to process and environment variations, an ISPP scheme maintains a consistent program performance which helps to improve the yield of the device. Marginal cells that are previously out-of-spec when conditions are varied are brought within-spec with ISPP. An ISPP scheme is able to compensate for cell-by-cell variations that can exist within a die.

The program verify operation is performed just after program pulse (V_{pgm}) to check whether each cell V_t reach to verify level of V_{verify} or not, as shown in Fig. 2.13. The operation condition is almost the same as regular read operation, as shown in Fig. 2.8. The different point from regular read operation is that a control gate voltage is V_{verify}, replaced by R1 in Fig. 2.8.

The erase operation is performed in block units, as shown in Fig. 2.15. The wordlines of selected blocks are grounded and the wordlines of unselected blocks are floating. A high erase pulse (~ 20 V) is then applied to the p-well and n-well in the memory cell area (Fig. 2.5). In the selected blocks, the erase voltage creates a large (~20 V) potential difference between the p-well and the control gates. This causes FN tunneling of electrons from the floating gate into the p-well, resulting in a typical cell

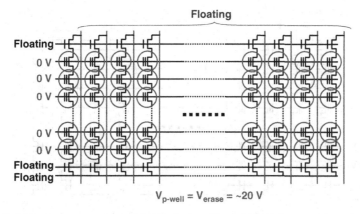

FIGURE 2.15 Erase operation of NAND flash cell array. Block erase: 128- to 256-KByte (2-KByte × 2 × 32) cells are erased at the same time. Erase speed, t_{erase}: 2–5 ms with erase verify.

threshold voltage that shifts negative. Since over-erasure is not a concern in NAND flash, cells are normally erased to −3 V. Also, the low erased cell threshold voltage provides an additional margin against upward threshold voltage shifts that arise from program/erase cycling, program disturb, read disturb, and V_t shift by floating gate capacitive coupling interference (see Chapter 5).

Figure 2.16 shows the typical erase characteristics. Cells can be erased to −3 V by applying a 17- to 18-V, 1-ms erase pulse width.

The erase verify operation is performed just after an erase pulse (V_{erase}) to check whether all cell V_t values in string become less than a certain V_t (for example, 0 V)

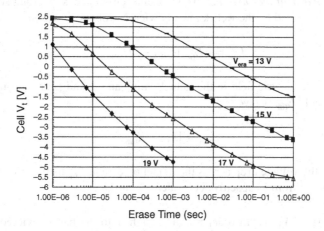

FIGURE 2.16 Erase characteristics. Higher erase voltage can be faster operation. Erase speed, t_{era}: 2–5 ms.

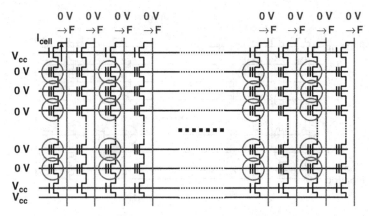

FIGURE 2.17   Erase verify operation.

or not. Figure 2.17 shows the operation condition of erase verify. A 0-V charge is applied to all control gates in a block, while V_{cc} and 0 V are applied to a source line and a bit line, respectively. The bit line is initially set to 0 V and then set to floating (F). During erase verify read, bit-line voltage is increased by cell current flow through a string, and then it judges whether all cells in the string are erased or not. If some cells are not be erased, an additional erase operation is performed in the same manner as that of the program/verify operation.

2.2.3   Program and Erase Dynamics

The device model is described for program and erase dynamics [16].

A. Calculation of Tunnel Current.   The tunneling current density through the tunnel oxide is approximated by the well-known Fowler–Nordheim equation [17, 18].

$$J_{tun} = \alpha E_{tun}^2 \left(\exp(-\beta / E_{tun}) \right) \tag{2.1}$$

where E_{tun} is the electric field in the tunnel oxide, and α and β are constants. The tunnel oxide field E_{tun} is given by

$$E_{tun} = |V_{tun}| / X_{tun} \tag{2.2}$$

where V_{tun} is the voltage drop across the tunnel oxide and X_{tun} is the thickness. V_{tun} can be calculated from a capacitive equivalent circuit of the cell.

B. Calculation of V_{tun}.   In order to gain insight into the basic device operation, a simplified equivalent circuit is used, shown in Fig. 2.18. In Fig. 2.18, C_{pp} is the interpoly capacitance, C_{tun} is the thin oxide capacitance between the floating gate

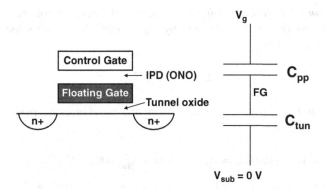

FIGURE 2.18 A simplified capacitive equivalent circuit of the NAND flash memory cell.

and the substrate. Q_{fg} is the stored charge on the floating gate. V_{tun} can be expressed for an electrically neutral floating gate in terms of simple coupling ratios:

$$|V_{tun}|_{write} = V_g * K_w \tag{2.3}$$

where

$$K_w = C_{pp}/(C_{pp} + C_{tun}) \tag{2.4}$$

and

$$|V_{tun}|_{erase} = V_{well} * K_e \tag{2.5}$$

where

$$K_e = 1 - C_{tun}/(C_{pp} + C_{tun}) \tag{2.6}$$

The coupling ratios K_w and K_e denote the fraction of the applied voltage that appears across the tunnel oxide. Note that (2.3) and (2.5) are applicable only when $Q_{fg} = 0$. During program operation, buildup of negative stored charge of the floating gate will reduce the tunnel-oxide voltage according to

$$|V_{tun}|_{write} = V_g * K_w + Q_{fg}/(C_{pp} + C_{tun}) \tag{2.3'}$$

In the ERASE operation, the initial negative stored charge on the floating gate will increase the tunnel-oxide voltage according to

$$|V_{tun}|_{erase} = V_{well} * K_e - Q_{fg}/(C_{pp} + C_{tun}) \tag{2.5'}$$

at the end of the erase operation when positive charge is built up on the floating gate, and the last term in (2.5′) will reduce the tunnel-oxide voltage.

C. Calculation of Threshold Voltages. The initial threshold voltage of the cell, corresponding to $Q_{fg} = 0$, is denoted by V_{ti}. Stored charge shifts the threshold according to the relation

$$\Delta Vt = -Q_{fg}/C_{pp} \qquad (2.7)$$

Using (2.3′) and (2.5′) for Q_{fg} at the end of the program/erase pulse, the cell's threshold voltages are

$$V_{tw} = V_{ti} - Q_{fg}/C_{pp} = V_{ti} + V_g * (1 - V'_{tun}/(K_w * V_g)) \qquad (2.8)$$

$$V_{te} = V_{ti} - Q_{fg}/C_{pp} = V_{ti} - V_{well} * (K_e/K_w - V'_{tun}/(K_w * V_{well})) \qquad (2.9)$$

Here V_{tw} is the threshold of a programmed cell, and V_{te} is the threshold of an erased cell. V_g and V_{well} are the program/erase pulse amplitudes, respectively, and V'_{tun} is the tunnel-oxide voltage at the end of the pulse. Assuming that the program/erase pulse is sufficiently long, the thin-oxide field will be reduced to below about 1×10^7 V/cm, when tunneling practically "stops." An approximation of V'_{tun} can be calculated from (2.2), and it can be substituted in (2.8) and (2.9) to give the approximate programming window of the cell and its dependence on cell parameters and programming voltage. Typical results are shown in Fig. 2.19 [16].

In order to maximize the cell window at a given tunnel-oxide thickness and program/erase voltage, the coupling ratios should approach unity. Both coupling

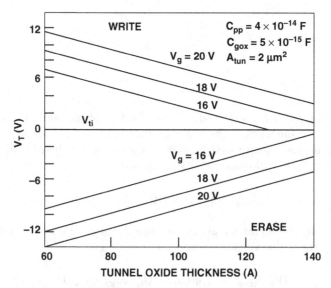

FIGURE 2.19 Program/erase threshold window versus tunnel-oxide thickness, calculated with the approximation of (2.8) and (2.9), assuming that $V_{tun}' = 1 \times 10^7 * X_{tun}$ at the end of the operation.

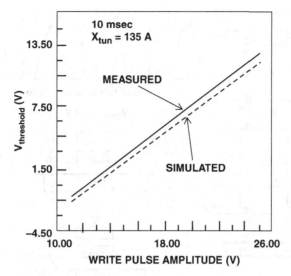

FIGURE 2.20   Measured and simulated threshold voltage as a function of program pulse amplitude, for a fixed program time of 10 ms.

ratios can be increased by reducing C_{tun} and increasing C_{pp}. At a given tunnel-oxide thickness, this is usually achieved by minimizing the thin-oxide area and adding extra poly–poly overlap area on the sides of the cell transistor. Typical coupling ratios are about 0.6.

Figure 2.20 shows the calculated and measured results for the program operation [16]. The threshold voltages as a function of program pulse amplitude are shown. The simulation results fit the measured data closely. And we can see write (program) pulse amplitude and V_t has linear relationship $\Delta V_t = \Delta V_{pgm}$ in same pulse width. This is important in the case of considering ISPP and V_t window setting.

2.2.4   Program Boosting Operation

Program disturb is a phenomenon in which the threshold voltage (V_t) of unselected cell is increased during program operation. The basic program disturb modes are shown in Fig. 2.21. There are two modes of program disturb. One is "V_{pass} mode" in selected NAND string, where bit-line (BL) and cell channel are 0 V. V_{pass} (~10 V) is applied to control gate (wordline), while a 0-V charge is applied to source and drain (channel). This condition is a weak electron injection mode to FG, and then V_t is increased, especially in the case of higher V_{pass}. The other is "boosting mode" in unselected NAND string. The program voltage V_{pgm} is applied to control gate, while channel is the boosting voltage. The boosting voltage (~8 V) is mainly generated by V_{pass} voltage of unselected wordlines in string. The "boosting mode" is also in weak electron injection mode. V_t of a memory cell is increased, especially in the case of high V_{pgm} and lower V_{pass} (lower boosing voltage). As indicated in Fig. 2.21, the V_t

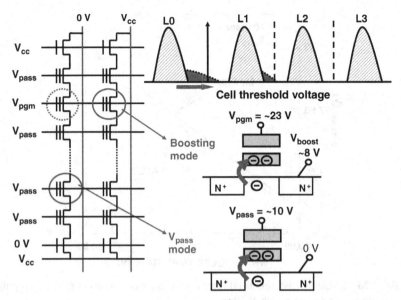

FIGURE 2.21   Program disturb of NAND flash memory cells. Weak electron injection mode is caused in an unselected cell. There are two modes of boosting mode and V_{pass} mode. The "self-boosting" scheme is used for unselected string. The channel voltage in unselected string has boosted up to $V_{boost} \sim 8$ V.

distribution tail is mainly increased in L0 or L1 states in MLC V_t setting due to higher tunnel-oxide field.

Figure 2.22 [19] shows program disturb data dependent on V_{pass}. In the case of higher V_{pass} (14–18 V), the V_t shift of a V_{pass} mode is increased. On the other hand, in the case of lower V_{pass} ($\sim$10 V), the V_t shift of boosting mode is increased. The middle of V_{pass} voltage (10–14 V) has to be used due to small V_t shift (small program disturb). Normally the range of available V_{pass} voltage is called "V_{pass} window", indeed, the V_{pass} region where V_{th} shifts by disturbs does not cause incorrect operation.

There are several program boosting schemes for multilevel NAND flash cells. Figure 2.23 shows three basic boosting schemes. The first one is a conventional self-boosting (SB) scheme. The SB scheme is mainly used for an SLC (1 bit/cell) device. The second one is a local self-boosting scheme [13]. The boosting voltage (V_{boost}) can be increased because an adjacent WL cell is cut off due to applied 0 V. Then a program disturb of boosting mode can be improved due to reducing voltage difference between CG (V_{pgm}) and channel (V_{boost}). The third one is the erase-area self-boosting scheme (EASB). EASB was widely used for MLC (multibit cell) to obtain a higher boosting voltage. The reason why higher boosting voltage can be obtained is separated erased cells and programmed cells in string. Due to page program order, source-side cells from selected WL have already been programmed. Boosting efficiency is lower in source side cells. However, drain-side cells from selected WL are still in an erased state. Boosting efficiency of drain side cells is higher than source-side due to lower cell V_t. Then it is very effective to obtain high V_{boost} by cutting off

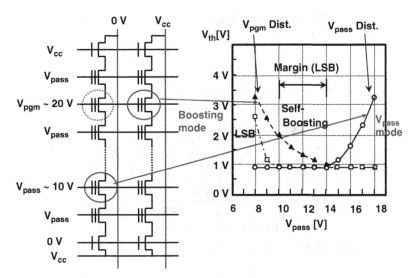

FIGURE 2.22 V_t shift of program disturb in the self-booting scheme. V_{pass} mode dominates in higher V_{pass}. Boosting mode dominates in lower V_{pass} due to low V_{boost}. V_{pass} window is between V_{pass} mode and boosting mode.

boosting node between drain-side and source-side cells. As NAND flash cell scaling, the boosting voltages produce a higher electric field in the source/drain area. A higher electric field generates hot electron and hot hole, and they are unexpectedly injected to FG. In order to relax high electric field, the self-boosting scheme is becoming a more advanced and complicated scheme for each generation of NAND flash memory.

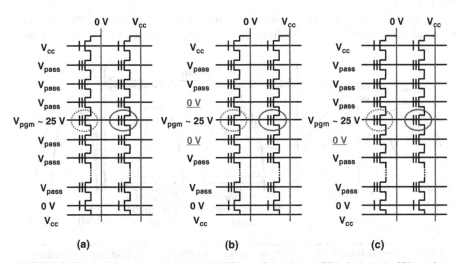

FIGURE 2.23   Self-boosting schemes. (a) Self-boosting scheme (SB). (b) Local self-boosting scheme (LSB). (c) Erase area self-boosting scheme (EASB).

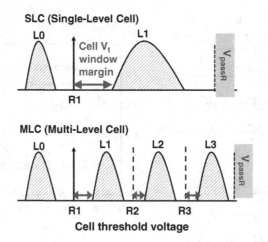

FIGURE 2.24   Cell V_t distribution of an SLC and an MLC.

2.3   MULTILEVEL CELL (MLC)

2.3.1   Cell V_t Setting

Figure 2.24 shows an image of cell V_t distribution setting of an SLC (single-level cell, 1 bit/cell) and MLC (Multilevel Cell, 2 bits/cell). An SLC has a wider cell V_t window margin, and thus SLC has a better program and read performance and also has a better reliability than MLC. As described in Section 2.2.1, in a read operation the unselected cells have to be pass a transistor during reading. Therefore, all cell V_t distributions have to be lower than V_{passR}, as shown in Fig. 2.24. This is one of limitation of cell V_t setting.

Figure 2.25 shows a read V_t window of MLC cells. V_t window is defined by the right side edge of erase V_t distribution and by the left side edge of L3 V_t distribution,

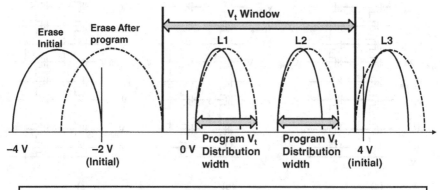

FIGURE 2.25   Read V_t window of an MLC NAND cell.

after programmed all pages in string (or block). Two programmed V_t distributions of L1 and L2 should be on the inside of a V_t window. And a read window margin (RWM) is defined as shown in Fig. 2.25 [20]. By the scaling down of memory cell size, the RWM has degraded due to an increasing impact of many inevitable physical phenomena, as discussed in Section 5.2. Major reliability issues might occur due to a narrow cell V_t window margin.

REFERENCES

[1] Masuoka, F.; Momodomi, M.; Iwata, Y.; Shirota, R. New ultra high density EPROM and flash EEPROM with NAND structure cell, *Electron Devices Meeting, 1987 International*, vol. 33, pp. 552–555, 1987.

[2] Itoh, Y.; Momodomi, M.; Shirota, R.; Iwata, Y.; Nakayama, R.; Kirisawa, R.; Tanaka, T.; Toita, K.; Inoue, S.; Masuoka, F. An experimental 4 Mb CMOS EEPROM with a NAND structured cell, *Solid-State Circuits Conference, 1989. Digest of Technical Papers. 36th ISSCC, 1989 IEEE International*, pp. 134–135, 15–17 Feb. 1989.

[3] Momodomi, M.; Itoh, Y.; Shirota, R.; Iwata, Y.; Nakayama, R.; Kirisawa, R.; Tanaka, T.; Aritome, S.; Endoh, T.; Ohuchi, K.; Masuoka, F. An experimental 4-Mbit CMOS EEPROM with a NAND-structured cell, *Solid-State Circuits, IEEE Journal of*, vol. 24, no. 5, pp. 1238–1243, Oct. 1989.

[4] Momodomi, M.; Iwata, Y.; Tanaka, T.; Itoh, Y.; Shirota, R.; Masuoka, F. A high density NAND EEPROM with block-page programming for microcomputer applications, *Custom Integrated Circuits Conference, 1989, Proceedings of the IEEE 1989*, pp. 10.1/1–10.1/4, 15–18 May 1989.

[5] Iwata, Y.; Momodomi, M.; Tanaka, T.; Oodaira, H.; Itoh, Y.; Nakayama, R.; Kirisawa, R.; Aritome, S.; Endoh, T.; Shirota, R.; Ohuchi, K.; Masuoka, F. A high-density NAND EEPROM with block-page programming for microcomputer applications, *Solid-State Circuits, IEEE Journal of*, vol. 25, no. 2, pp. 417–424, Apr. 1990.

[6] Takeuchi, Y.; Shimizu, K.; Narita, K.; Kamiya, E.; Yaegashi, T.; Amemiya, K.; Aritome, S. A self-aligned STI process integration for low cost and highly reliable 1 Gbit flash memories, *VLSI Technology, 1998. Digest of Technical Papers. 1998 Symposium on*, pp. 102–103, 9–11 June 1998.

[7] Aritome, S.; Shirota, R.; Kirisawa, R.; Endoh, T.; Nakayama, R.; Sakui, K.; Masuoka, F. A reliable bi-polarity write/erase technology in flash EEPROMs, *International electron devices meeting, 1990. IEDM'90. Technical Digest*, pp. 111–114, 1990.

[8] Shirota, R., Itoh, Y., Nakayama, R., Momodomi, M., Inoue, S., Kirisawa, R., Iwata, Y., Chiba, M., Masuoka, F. New NAND cell for ultra high density 5V-only EEPROMs, *Digest of Technical Papers—Symposium on VLSI Technology*, pp. 33–34, 1988.

[9] Momodomi, M.; Kirisawa, R.; Nakayama, R.; Aritome, S.; Endoh, T.; Itoh, Y.; Iwata, Y.; Oodaira, H.; Tanaka, T.; Chiba, M.; Shirota, R.; Masuoka, F. New device technologies for 5 V-only 4 Mb EEPROM with NAND structure cell, *Electron Devices Meeting, 1988. IEDM'88. Technical Digest, International*, pp. 412–415, 1988.

[10] Aritome, S.; Kirisawa, R.; Endoh, T.; Nakayama, R.; Shirota, R.; Sakui, K.; Ohuchi, K.; Masuoka, F. Extended data retention characteristics after more than 10^4 write and erase

cycles in EEPROMs, *International Reliability Physics Symposium, 1990. 28th Annual Proceedings.*, pp. 259–264, 1990.

[11] Kirisawa, R.; Aritome, S.; Nakayama, R.; Endoh, T.; Shirota, R.; Masuoka, F. A NAND structured cell with a new programming technology for highly reliable 5 V-only flash EEPROM, *1990 Symposium on VLSI Technology*. Digest of Technical Papers, pp. 129–130, 1990.

[12] Suh, K.-D.; Suh, B.-H.; Um, Y.-H.; Kim, J.-K.; Choi, Y.-J.; Koh, Y.-N.; Lee, S.-S.; Kwon, S.-C.; Choi, B.-S.; Yum, J.-S.; Choi, J.-H.; Kim, J.-R.; Lim, H.-K. A 3.3 V 32 Mb NAND flash memory with incremental step pulse programming scheme, *Solid-State Circuits Conference, 1995. Digest of Technical Papers. 42nd ISSCC, 1995 IEEE International*, pp. 128–129, 350, 15–17 Feb. 1995.

[13] Suh, K.-D.; Suh, B.-H.; Lim, Y.-H.; Kim, J.-K.; Choi, Y.-J.; Koh, Y.-N.; Lee, S.-S.; Kwon, S.-C.; Choi, B.-S.; Yum, J.-S.; Choi, J.-H.; Kim, J.-R; Lim, H.-K. A 3.3 V 32 Mb NAND flash memory with incremental step pulse programming scheme, *Solid-State Circuits, IEEE Journal of*, vol. 30, no. 11, pp. 1149–1156, Nov. 1995.

[14] Hemink, G.J.; Tanaka, T.; Endoh, T.; Aritome, S.; Shirota, R. Fast and accurate programming method for multi-level NAND EEPROMs, *VLSI Technology, 1995. Digest of Technical Papers. 1995 Symposium on*, pp. 129–130, 6–8 June 1995.

[15] Tanaka, T.; Tanaka, Y.; Nakamura, H.; Sakui, K.; Oodaira, H.; Shirota, R.; Ohuchi, K.; Masuoka, F.; Hara, H. A quick intelligent page-programming architecture and a shielded bitline sensing method for 3 V-only NAND flash memory, *Solid-State Circuits, IEEE Journal of*, vol. 29, no. 11, pp. 1366–1373, Nov. 1994.

[16] Kolodny, A.; Nieh, S.T.K.; Eitan, B.; Shappir, J. Analysis and modeling of floating-gate EEPROM cells, *Electron Devices, IEEE Transactions on*, vol. 33, no. 6, pp. 835–844, June 1986.

[17] Lenzlinger, M.; Snow, E. H. Fowler–Nordheim tunneling into thermally grown SiO, *Journal of Applied Physics*, vol. 40. p. 278, 1969.

[18] Weinberg, Z. A. On tunneling in metal-oxide–silicon structures, *J. Appl. Phys.*, vol. 53, p. 5052, 1982.

[19] Jung, T.-S.; Choi, D.-C.; Cho, S.-H.; Kim, M.-J.; Lee, S.-K.; Choi, B.-S.; Yum, J.-S.; Kim, S.-H.; Lee, D.-G.; Son, J.-C.; Yong, M.-S.; Oh, H.-K.; Jun, S.-B.; Lee, W.-M.; Haq, E.; Suh, K.-D.; Ali, S.B.; Lim, H.-K. A 3.3-V single power supply 16-Mb nonvolatile virtual DRAM using a NAND flash memory technology, *Solid-State Circuits, IEEE Journal of*, vol. 32, no. 11, pp. 1748–1757, Nov. 1997.

[20] Aritome, S.; Kikkawa, T. Scaling challenge of self-aligned STI cell (SA-STI cell) for NAND flash memories, *Solid-State Electronics* vol. 82, pp. 54–62, 2013.

3

NAND FLASH MEMORY DEVICES

3.1 INTRODUCTION

The most important requirement for NAND flash memory [1] is a low bit cost. In order to realize a low bit cost, the scaling down of memory cell size is essential. In this chapter, the NAND flash memory cell and its scaling technologies are discussed.

A NAND flash technology road map and structure scaling of a two-dimensional NAND flash memory cell are shown in Fig. 3.1 and Fig. 3.2, respectively.

Production of NAND flash memory was started in 0.7-μm technology in 1992. The line/space pitch of word line (WL) was ideal 2*F (F: feature size); however, the line/space pitch of bit line (BL) was 3–4*F because of limitation of the LOCOS (LOCal Oxidation of Silicon) isolation. Requirements for isolation in NAND flash memory cell are more severe than other devices due to high-voltage operation during programming. Therefore, it was difficult to scale down of LOCOS isolation width beyond 1.5-μm width due to boron diffusion from isolation bottom by LOCOS oxidation process. Then a new FTI process (field through implantation process) was developed [2, 3], as described in Section 3.2. Due to the FTI process, a LOCOS isolation width could be scaled down to 0.8 μm (2*F of 0.4-μm rule) and the technology node could be scaled down to 0.35-μm technology, as indicated by "1) LOCOS cell scaling" in Fig. 3.1.

Next, the self-aligned shallow trench isolation cell (SA-STI cell) with floating gate (FG) wing had been developed [4, 5], as described in Section 3.3. Due to STI, isolation width could be drastically scaled down to 50% (0.8 μm to 0.4 μm) and then cell size could be scaled down to 67% (bit-line pitch: 1.2 μm to 0.8 μm) in the same design rule,

Nand Flash Memory Technologies, First Edition. Seiichi Aritome.
© 2016 The Institute of Electrical and Electronics Engineers, Inc. Published 2016 by John Wiley & Sons, Inc.

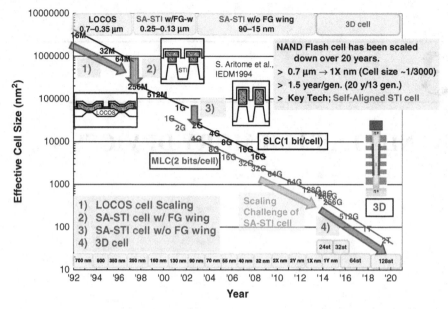

FIGURE 3.1 NAND flash memory technology road map.

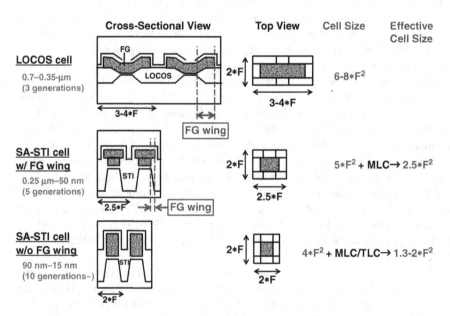

FIGURE 3.2 Structure scaling of two-dimensional NAND flash memory cell.

as indicated by "2) SA-STI cell w/ FG wing" in Fig. 3.1. Also, the isolation capability was much improved because of deep isolation of STI. Furthermore, reliability of tunnel oxide was improved due to no STI edge corner in tunnel oxide.

After that, SA-STI cell without FG wing had been developed [6], as described in Section 3.4. By using this cell structure, a high coupling ratio could be obtained due to large capacitance of IPD using an FG sidewall, as indicated by "3) SA-STI cell w/o FG wing" in Fig. 3.1. The SA-STI cell has a very simple structure, and layout allows for the formation of a very small cell size with bit-line and word-line pitch of $2*F$. Then the cell size becomes ideal $4*F^2$, as shown in Fig. 3.2. The SA-STI technology has also demonstrated an excellent reliability, because the FG does not overlap the STI corner because FG over the tunnel oxide is patterned as the shape of the active area. Therefore, the SA-STI cell realizes very low bit cost and high reliability [4–9]. The SA-STI cell has been extensively used for more than 15 years (since 1998), over 10 generations (0.25 μm to 1Y nm) of NAND flash memory product.

In Section 3.5, the planar FG cell [10–12] is presented. The planar FG cell has a very thin floating gate of around 10-nm thickness and the high-k block dielectric as IPD. The aspect ratio of stacked gate and control gate (CG) fill are much improved. The control gate (CG) fill issue, which is one of the serious problems in the SA-STI cell as described in Section 5.6, can be eliminated by planar FG structure. Also, the planar FG cell has a very small FG capacitive coupling interference due to thin FG. The planar FG cell also started to be used from a 2X-nm technology node in one supplier of NAND flash memory.

In Section 3.6, the side wall transfer-transistor (SWATT) cell [13, 14] is discussed as alternate memory cell technology for a multilevel NAND flash memory cell. By using a SWATT cell, a wide threshold voltage (V_{th}) distribution width could be allowed. The key technology that allows this wide V_{th} distribution width is the transfer transistor which is located at the side wall of the shallow trench isolation (STI) region and is connected in parallel with the floating-gate transistor. During read, the transfer transistors of the unselected cells (connected in series with the selected cell) work as pass transistors. So, even if the V_{th} of the unselected floating-gate transistor is higher than the control gate voltage, the unselected cell can be in the ON state. As a result, the V_{th} distribution of the floating-gate transistor can be wider and the programming speed can be faster because the number of program/verify cycles can be reduced.

Other advanced NAND flash device technologies are presented in Section 3.7.

First, a dummy word-line scheme in NAND flash memory [15–17] is discussed. Dummy word line (dummy cell) is located between edge word lines (edge memory cells) of NAND string and select transistors (GSL or SSL). The program disturb of a GIDL-generated hot electron injection mechanism can be suppressed. In addition, capacitive coupling noise between select gate and edge word line can be reduced. Then the program disturbance failure, read failure, and erase distribution width can be greatly improved by reducing coupling noise. The dummy word-line scheme was started to be used from 40-nm technology node due to stable operations in edge cells.

Second, the p-type doping floating gate (FG) [18–21] is introduced. The p-type FG has an advantage of better cycling endurance and data retention characteristics than an n-type floating gate. The p-type FG started to be used from a 2X-nm technology node due to better reliabilities.

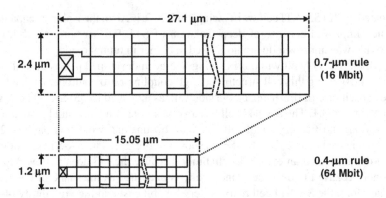

FIGURE 3.3   Top view of a 0.4-μm-rule NAND-structure cell in comparison with that of a 0.7-μm rule cell. This NAND-structured cell has 16 memory transistors arranged in series between two select transistors. Word-line and bit-line pitches are 0.8-μm (L/S = 0.4 μm/0.4 μm) and 1.2 μm, respectively. The cell size is 1.13 μm², including the select transistors and the drain contact area. Copyright 1994, The Japan Society of Applied Physics.

3.2   LOCOS CELL

3.2.1   Conventional LOCOS Cell

The mass production of the NAND flash memory [1] started in 1992. A 16-Mega-bit device was first produced by using 0.7-μm technology with a conventional LOCOS isolation process. In order to scale down memory cell size beyond 0.7-μm technology, the scaling down of LOCOS isolation width was a key issue. In a conventional LOCOS process in 0.7-μm generation, an isolation width was 1.7-μm and a bit-line pitch was 2.4 μm, as shown in Fig. 3.3 [2, 3]. It was hard to scale down isolation width beyond 1.5 μm, because of required high-voltage junction breakdown (>8 V) and high inverting voltage (V_t) of parasitic field transistor. Isolation stopper doping of boron is implanted before oxidation of LOCOS isolation. A dopant of boron is easy to diffuse during LOCOS oxidation process. Due to diffusion of boron, it became difficult to satisfy requirements of both a high-voltage junction breakdown (>8 V) and high inverting voltage (V_t) of a parasitic field transistor.

3.2.2   Advanced LOCOS Cell

A small NAND-structure cell (1.13 μm² per bit) had been developed in 0.4-μm technology [2, 3]. The chip size of a 64-Mb NAND flash memory using this cell was estimated to be 120 mm², which was 60% of a 64-Mb DRAM die size. In order to realize the small cell size, 0.8-μm width field isolation was developed with the field-through implantation (FTI) technique. A negative bias of -0.5 V to the p-well of the memory cell is applied during programming. Read disturb could be also ensured for more than 10 years even after 1 million write/erase cycles.

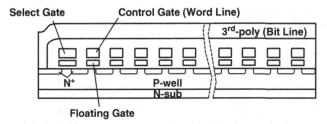

FIGURE 3.4   Cross-sectional view of a 0.4-μm-rule NAND-structure cell. Copyright 1994, The Japan Society of Applied Physics.

A. Scaling of NAND Cell   Figure 3.3 compares the top view of the advanced 0.4-μm NAND cell with that of the 0.7-μm conventional one. Figure 3.4 shows the cross-sectional view of the 0.4-μm NAND cell [2, 3]. This NAND-structure cell has 16 memory transistors arranged between two select transistors in series. The word-line pitch is 0.8 μm (line/space = 0.4 μm/0.4 μm). The bit-line pitch could be reduced to 1.2 μm by using 0.8-μm field isolation technology. Figure 3.5 shows the cross-sectional SEM photograph of the 0.4-μm NAND-structure cell after the self-aligned stacked gate etching process. The floating gates are made of first-layer polysilicon (phosphorus-doped). The control gates are made of second-layer polysilicon (phosphorus-doped polysilicon/tungsten-silicide). The process technology is summarized in Table 3.1.

As the design rule of the word line is scaled, it becomes apparent that the NAND cell has an advantage of scaling down the gate length of the memory cell. This is because the NAND cell has punch-through free operation since there is no voltage difference between the drain and source during programming and erasing. And also, the NAND cell has a very small gate–drain overlap region because the impurity concentration of the diffusion layer can be less than 10^{18} cm^{-3} due to no hot electron injection programming and no source erase. As a result, the gate length of the NAND cell can be smaller as feature size is scaled down.

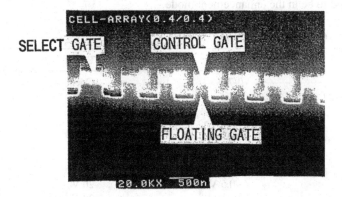

FIGURE 3.5   Cross-sectional SEM photograph of a NAND-structure cell after the self-aligned stacked etching process. Copyright 1994, The Japan Society of Applied Physics.

TABLE 3.1   Process Technology

Process	Double-well CMOS Triple-poly-silicon technology 1 layer-metal	
Cell	Cell size	1.13 μm^2
	W/L	0.4/0.4 μm
	Tunnel oxide	80 A
	Interpoly dielectric	ONO 200 A (effective)
Peripheral	L-poly N-ch	0.8 μm
	P-ch	0.8 μm

Source: Copyright 1994, The Japan Society of Applied Physics.

With respect to scaling in the bit-line direction (bit-line pitch), isolation technology is very important, as discussed in the Section 3.2.3.

B. Operation of NAND Cell   The operating conditions are shown in Table 3.2. During writing, 18 V is applied to the selected control gate while the bit lines are grounded; electrons tunnel from the substrate to the floating gate, resulting in a positive threshold voltage shift. If a voltage of 7 V is applied to the bit line (not self-boosting operation), tunneling is inhibited, and the threshold voltage remains the same. The negative bias to the p-well is effective in preventing the parasitic field transistor from turning on. During erasing, 20 V is applied to both the p-well and the (N-type) substrate while keeping the bit lines floating and all the selected control gates grounded. Electrons tunnel from the floating gate to the substrate, and the threshold voltage of the memory cells becomes negative.

The reading method is also shown in Table 3.2. Zero volt is applied to the gate of the selected memory cell, while 3.3 V (V_{cc}) is applied to the gates of the other cells. Therefore, all of the other memory transistors, except for the selected transistor, serve as transfer gates. A cell current flows if a selected memory transistor is in the depletion mode. On the other hand, cell current does not flow if the memory cell is programmed to be in the enhancement mode.

TABLE 3.2   Operation Conditions

	Write 1	Erase	Read
Bit line	0/7 V	Open	1.5 V
Select gate	7 V	20 V	3.3 V
Control gate 1	7 V	0 V	3.3 V
Control gate 2	18 V(selected)	0 V	0 V
Control gate 16	7 V	0 V	3.3 V
Select gate	0 V	20 V	3.3 V
Source	0 V	Open	0 V
p-well	−0.5 V	20 V	0 V
n-sub	3.3 V	20 V	3.3 V

Source: Copyright 1994, The Japan Society of Applied Physics.

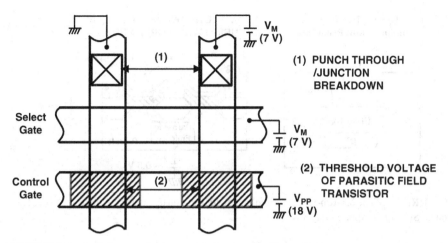

FIGURE 3.6 Isolation between two neighboring bit lines: (1) Punch-through or junction breakdown in the bit-line contact area. (2) The threshold voltage of the parasitic field transistor. Copyright 1994, The Japan Society of Applied Physics.

3.2.3 Isolation Technology

For the NAND cell, high-voltage field isolation technology is important to reduce the bit-line pitch. The isolation between the bit lines must satisfy two requirements, as shown in Fig. 3.6 [2, 3]. One is the punch-through or the junction breakdown voltage of the bit-line junction area. During programming, 7 V is applied to the bit line to prevent electron injection in cells that should remain in the erased state. Zero volt is applied to a bit line which is connected to a cell that should be programmed. The punch-through voltage between neighboring bit-line junctions must be higher than 7 V, as must the bit-line junction breakdown voltage. Another requirement is a high threshold voltage of the parasitic field transistor. During programming, the selected control gate is biased with a high voltage of 18 V, which may easily turn on the field transistor between neighboring bits (Table 3.2).

In order to avoid bit-line junction breakdown/punch-through and to prevent the field transistor from turning on, the field-through implantation process (FTI process) and p-well negative bias method have been developed [2, 3], as shown in Fig. 3.7. In the FTI process, the boron ions (160 keV, 1.13 cm^{-2}) are implanted to form a field stopper after LOCOS fabrication. The field-oxide thickness at field-through implantation is 420 nm. The punch-through voltage and the threshold voltage of the parasitic transistor increase without decreasing the junction breakdown voltage, because the lateral diffusion of the boron stopper impurity is decreased in comparison with the conventional field stopper implantation before LOCOS fabrication. A negative p-well bias prevents punch-through and increases the threshold voltage of the parasitic field transistor.

Figure 3.8 shows the breakdown voltage between two neighboring bit lines as a function of the bit-line contact distance. The breakdown voltage is ensured to be

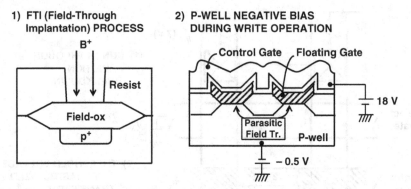

FIGURE 3.7   The 0.8-μm-wide high-voltage field isolation technology. Copyright 1994, The Japan Society of Applied Physics.

higher than 7 V, in the case of a negative-biased p-well. Figure 3.9 shows the threshold voltage of the parasitic field transistor. By using the FTI process, the threshold voltage of the 0.8-μm field transistor becomes more than 28 V. Moreover, a negative bias of −0.5 V is applied to the p-well to increase the threshold voltage of the parasitic field transistor (V_{tf}). In comparison with a zero-biased p-well, V_{tf} is increased to more than 30 V. Therefore, the field width margin is increased from 0.05 μm to 0.15 μm. As a result, a very small bit-line pitch of 1.2 μm could be realized.

The FTI process also reduces the body effect of the cell transistors, because the boron impurity concentration under the channel region of the cell transistors and the

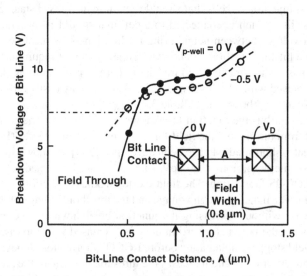

FIGURE 3.8   The breakdown voltage of the bit-line junction. The punch-through voltage and junction breakdown voltage are higher than 7 V at a 0.8-μm bit-line contact distance. Copyright 1994, The Japan Society of Applied Physics.

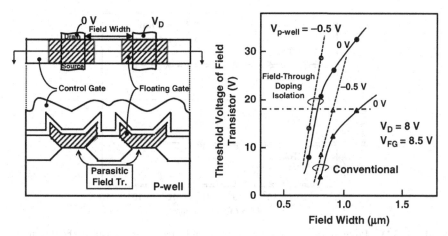

FIGURE 3.9   The threshold voltage of the parasitic field transistor as a function of the field width. By using a field-through implantation process (FTI process), the threshold voltage of a 0.8-μm field transistor becomes higher than 28 V when −0.5 V is applied to the *p*-well. Copyright 1994, The Japan Society of Applied Physics.

select transistor become lower. The decrease of the body effect results in an increased cell current. Figure 3.10 shows the cell current of a NAND cell as a function of the cell gate length. The cell currents were measured at the nearest cell from the bit-line contact side, which has the smallest cell current among all cells because that cell should be strongly influenced by the body effect due to the series resistance of the other

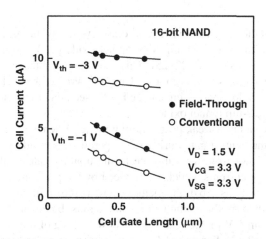

FIGURE 3.10   The cell current during a read operation using both the field-through implantation (FTI) process and the conventional process. The cell current of the FTI process is larger than that of the conventional process due to the suppression of boron diffusion to the channel region. Copyright 1994, The Japan Society of Applied Physics.

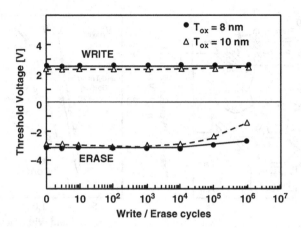

FIGURE 3.11   Program (write) and erase cycling endurance characteristics. The threshold voltage is defined as the control gate voltage which flows a drain current of 1 μA with a drain-source voltage of 1.5 V. For an 8-nm tunnel oxide, we have the following: Program (write): $V_{cg} = 18$ V, 0.1 ms. Erase: $V_{p\text{-well}} = 20$ V, 1 ms. For a 10-nm tunnel oxide, we have the following: Program (write): $V_{cg} = 20.4$ V, 0.1 ms. Erase: $V_{p\text{-well}} = 22.7$ V, 1 ms. Window narrowing is almost not observed up to 1 million program (write)/erase cycles. Copyright 1994, The Japan Society of Applied Physics.

cells. The threshold voltage of the selected cell is −1 V and −3 V, the threshold voltage of the unselected cells, which, connected in series, is from 0.5 to 1.5 V. In the case of the FTI process, the cell current is larger than the current in the conventional case.

3.2.4   Reliability

Figure 3.11 shows the program (write) and erase cycling endurance characteristics of a NAND cell with 8- and 10-nm-thick tunnel oxide [2, 3] using the bipolarity uniform program/erase scheme [22–25]. This scheme guarantees a wide cell threshold window of as large as 4 V, even after 1 million write/erase cycles. In the 8-nm tunnel-oxide cell, window narrowing can be hardly seen due to the small number of electron traps in the 8-nm tunnel oxide.

Read disturb occurs as a weak programming mode. When a certain positive voltage is applied to the control gate during read operation, a small Fowler–Nordheim tunneling current flows from the substrate to the floating gate. Unfortunately, the tunnel-oxide leakage currents, which are induced by the program and erase cycling stress, degrade the read disturb of the memory cell, as shown in Fig. 3.12. In order to suppress the read disturb, the applied gate voltage must be lowered. A reduction of the gate voltage from 5 V to 3.3 V allows the downscaling of the tunnel oxide from 10 nm to 8 nm (Fig. 3.13). Sensing at 3.3 V can be performed by a bit-by-bit verified programming scheme [26], which results in a written cell threshold voltage between 0.5 V and 1.8 V.

Even for an 8-nm tunnel-oxide thickness, read disturb suppression can be ensured for more than 10 years even after 10^6 W/E cycles, as presented in Fig. 3.13. By scaling

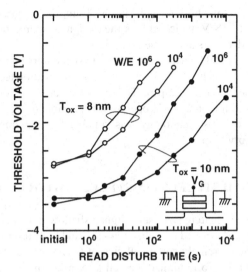

FIGURE 3.12 The read disturb characteristics of a NAND flash cell with 8- and 10-nm tunnel-oxide thickness. The voltage of the control gate (V_G) is 9 V as an accelerated condition. The threshold voltage is measured under the same condition as in Fig. 3.11. Copyright 1994, The Japan Society of Applied Physics.

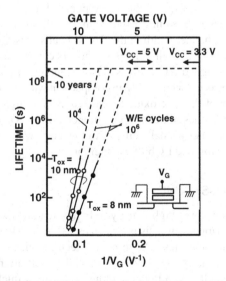

FIGURE 3.13 The read disturb lifetime is defined as the time at which V_{th} reaches -1.0 V during the applied gate voltage stress. Even if 8-nm tunnel oxide thickness is used, the read disturb time is far more than 10 years when a $+3.3/-0.3$ V supplied voltage (V_{cc}) is used. Copyright 1994, The Japan Society of Applied Physics.

down the tunnel-oxide thickness from 10 nm to 8 nm, the program voltage can be reduced from 21 V to 18 V, which allows the design of more compact peripheral circuits such as row decoders and sense amplifiers.

A 1.13-μm^2 memory cell for a 64-Mbit NAND flash memory had been successfully developed using 0.4-μm technology. High-voltage field isolation technology realizes a very small bit-line pitch of 1.2-μm. The tunnel-oxide thickness can be scaled down from 10 nm to 8 nm, and 3.3-V operation is possible using a bit-by-bit verify programming method. This technology was suitable for realizing a low-cost and highly reliable memory chip.

3.3   SELF-ALIGNED STI CELL (SA-STI CELL) WITH FG WING

A high-density 5*F^2 (F: feature size) self-aligned shallow trench isolation cell (SA-STI cell) technology is described to realize low-cost and high-reliability NAND flash memories [4, 5]. The extremely small cell size of 0.31 μm^2 has been obtained for the 0.25 μm design rule. To minimize the cell size, a floating gate is isolated with a shallow trench isolation (STI) and a slit formation by a novel SiN spacer process, which has made it possible to realize a 0.55 μm-pitch isolation at a 0.25 μm design rule. Another structural feature to the cell and its small size is the borderless bit-line and source-line contacts which are self-aligned with the select gate. The proposed NAND cell with the gate length of 0.2 μm and the isolation space of 0.25 μm shows a normal operation as a transistor without any punch-through. A tight distribution of the threshold voltages (2.0 V) in 2-Mbit memory cell array is achieved due to a good uniformity of the channel width in the SA-STI cells. Also, the peripheral low-voltage CMOS transistors and high-voltage transistors can be fabricated at the same time by using the self-aligned STI process. The advantages are as follows: (1) The number of process steps is reduced to 60% in comparison with a conventional process, and (2) high reliability of the gate oxide is realized even at high-voltage transistors because a gate electrode does not overlap the trench corner. Therefore this SA-STI process integration combines a small cell size (a low cost) with a high reliability for a manufacturable 256-Mbit and l-Gbit flash memory.

3.3.1   Structure of SA-STI Cell

This section describes a novel high density 5*F^2 (F: feature size) NAND STI cell technology [4] and peripheral transistors devices [5] which have been developed for a low bit-cost flash memories. The three key technologies to minimize the cell size have been introduced, as shown in Fig. 3.14. The self-aligned shallow trench isolation cell (SA-STI cell) has a high coupling ratio with a thick floating gate [6]. However, its high aspect ratio of the gate space has made it difficult to control the planarization process of the trench isolation by chemical mechanical polishing (CMP). To overcome this problem, a stacked floating gate structure is applied. A first thin poly-Si gate is self-aligned with the active area of the cells to control the

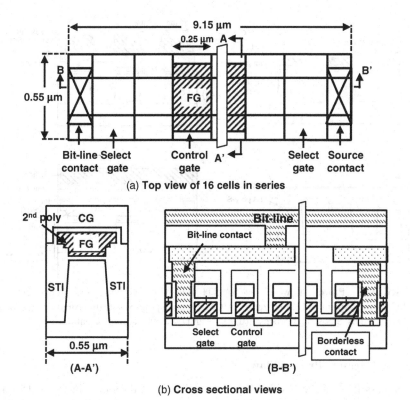

(a) **Top view of 16 cells in series**

(b) **Cross sectional views**

FIGURE 3.14 Schematic view of the SA-STI cell with floating-gate (FG) wing. (a) Top view of 16 cells in series. (b) Cross-sectional views of A–A' and B–B' in (a). Three key technologies to achieve $5*F^2$ (F: feature size) cell size has been introduced. (1) A stacked floating-gate structure has been applied in order to reduce the gate space aspect. A first poly-Si gate is self-aligned with the active area. A second poly-Si gate is formed over the exposed first gate to achieve a high coupling ratio (>0.6) with floating-gate wing. (2) The second poly-Si gate has been patterned with spacing of 0.15 μm by a novel SiN spacer process. This process has made it possible to realize 0.55-μm-pitch isolation. (3) The borderless bit-line and source-line contacts which are self-aligned with the select gate can eliminate a space between the contacts and the gate.

channel width precisely. A global planarization by CMP process is very controllable due to the reduction of the gate space aspect. A second poly-Si gate is formed on the first poly-Si gate to achieve a high coupling ratio (>0.6) of the cells. The second poly-Si gate is patterned with spacing of 0.15 μm by a novel SiN spacer process. This process has made it possible to realize 0.55-μm-pitch isolation. Another feature of integration to the cell and its small size is the borderless bit-line and source-line contacts which are self-aligned with the select gate. By the above technologies, an extremely small cell size of 0.31 μm^2 has been obtained for the 0.25 μm design rule.

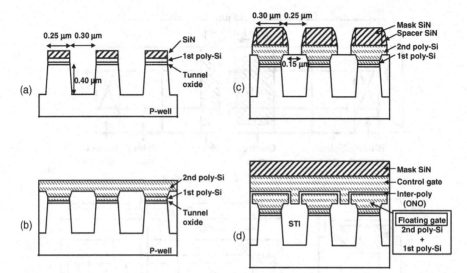

FIGURE 3.15 Process flow of the SA-STI cell with FG wing. (a) Trench etching. (b) LPCVD SiO_2 fill-in and planarization by CMP, second poly-Si gate deposited. (c) Floating-gate formation by SiN spacer process. (d) ONO and the control-gate formation.

3.3.2 Fabrication Process Flow

The process flow of the SA-STI cell with FG wing is described in Fig. 3.15. [4]. The active area is isolated by the STI formation using a self-aligned mask of a first thin poly-Si gate (a). After CVD SiO_2 deposition and planarization by CMP, the second poly-Si gate is deposited on the exposed first poly-Si layer, resulting in the stacked floating gate structure (b). The second poly-Si layer is striped with spacing of 0.15 μm, which is less than a design rule by a novel SiN spacer process as follows. A SiN mask is patterned at spacing of 0.25 μm, and a 50-nm-thick spacer SiN is then deposited. By etching the SiN mask back, a stripe mask pattern with 0.15-μm space is obtained (c). After removal of the SiN mask, an inter-poly dielectric (ONO) and the control gate are successively deposited (d). The control gate and the floating gate are continuously patterned, followed by deposition of a barrier SiN layer and an interlayer. The SiN layer covering the control gate prevents a short circuit between the gate and the borderless contacts. Finally, a doped poly-Si is filled within the bit-line contact and source-line contact and is etched back, followed by the metallization. Figures 3.16 and Fig. 3.17 show the cross-sectional SEM micrographs of the $5*F^2$ memory cell array with a cell size of 0.31 $μm^2$ using a 0.25-μm design rule. The key process parameters are listed in Table 3.3.

Figure 3.18 shows the schematic view of the SA-STI cell and peripheral transistors. The comparison between a novel process and a conventional process is schematically shown in Fig. 3.19. The number of fabrication steps of the process is reduced to about 60%. The memory cells and the peripheral transistors can be formed simultaneously without any additional process steps.

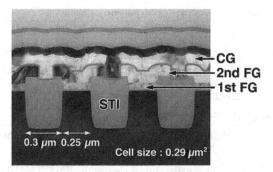

FIGURE 3.16   Cross-sectional SEM micrographs of the cell, parallel to the control gate.

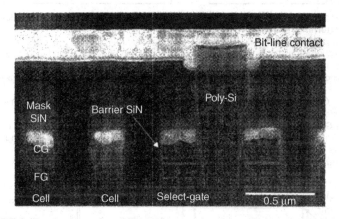

FIGURE 3.17   Cross-sectional SEM micrographs of the cell, parallel to the bit line.

TABLE 3.3   The Main Device Parameters

Technology	0.25-μm Double-well CMOS Self-aligned shallow trench isolation	
Cell	Tunnel oxide	9.0 nm
	Gate length	0.25 μm
	Channel width	0.25 μm
	Cell size	0.31 μm²
Peripheral	High-Voltage gate oxide	40.0 nm
	Low-Voltage gate oxide	9.0 nm
	NMOS effective gate length	0.28 μm
	PMOS effective gate length	0.38 μm

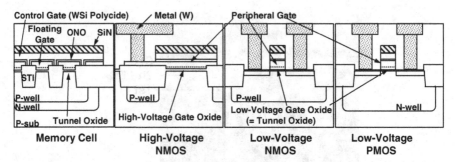

FIGURE 3.18   The schematic cross-sectional view of the SA-STI NAND flash memory.

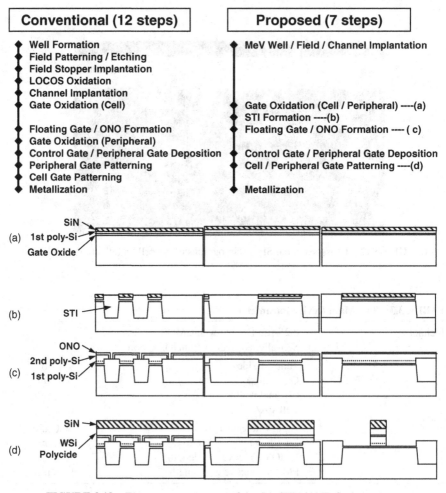

FIGURE 3.19   The process sequence of the SA-STI NAND flash memory.

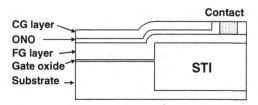

FIGURE 3.20   Cross-sectional view of peripheral transistor. The gate electrode (FG layer) is self-aligned with STI.

First, a retrograde well profile is formed by a high-energy ion implantation. Each implantation is carried out for a well formation, a field punch-through stopper, and a channel threshold adjustment. Next, 40-nm-thick gate oxides for the high-voltage transistors and 9-nm-thick gate oxides for the cells and the low-voltage transistors are formed, and then the first poly-Si layer for the floating gate and a SiN layer are successively deposited, as shown in Fig. 3.19a. The shallow trench, which is self-aligned with the first poly-Si layer, is etched, followed by SiO_2 filled-in planarization by CMP, as shown in Fig. 3.19b. After the SiN mask removal and second poly-Si layer deposition, the second poly-Si layers are patterned with a 0.15-μm space by a SiN spacer process. The stacked poly-Si structure acts as the floating gate of the cells and the gate electrode of the peripheral transistors. Then, ONO of an inter-poly dielectric is deposited, as shown in Fig. 3.19c.

Next, a WSi polycide layer for the control gate of the cell is deposited, and then the polycide layer, ONO, and the stacked poly-Si layer are continuously patterned, as shown in Fig. 3.19d. The peripheral gate electrode is also patterned with the cell. The polycide layer in the peripheral transistors is partially removed for a gate contact formation.

Finally, interconnections and peripheral contacts are formed by a dual damascene process. Figure 3.20. shows the cross-sectional view of the peripheral transistor [8]. The main device parameters of peripheral devices are also summarized in Table 3.3.

3.3.3   Characteristics of SA-STI with FG Wing Cell

Figure 3.21 shows a bit-line junction breakdown characteristics as a function of the isolation width. There occurs no field punch-through between the bit-line contacts at the STI width of up to 0.25 μm with an implantation of boron for the field stopper. Moreover, the 0.4-μm-thick STI field oxide results in a high threshold voltage (>30 V) of the parasitic field transistor between the neighboring bits. Figure 3.22. shows that the threshold voltage of the cell transistors shows a weak dependence on the channel width because no boron atoms implanted for the field stopper diffuses into the channel region from the trench bottom. From these results, the STI cell is very suitable for scaling of the isolation pitch.

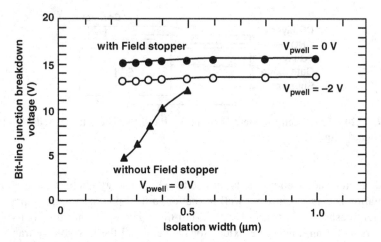

FIGURE 3.21   A breakdown voltage of the bit-line junctions, which is isolated by STI. There is no punch-through at less than 15 V, which is the junction breakdown voltage using boron field stopper implantation.

Figure 3.23 shows I_d–V_g characteristics of the SA-STI cell transistors. No anomalous hump is observed in the subthreshold characteristics of the wide channel transistors with W = 10 μm since the floating gate never overlaps the STI corners. In the case of the NAND cell transistors, the maximum drain voltage of around 1 V (less than V_{cc}) is applied only in the read operation. Figure 3.24 shows the I_d–V_g characteristics of the cell transistors with various gate lengths at $V_d = V_{cc}$ (2.5 V).

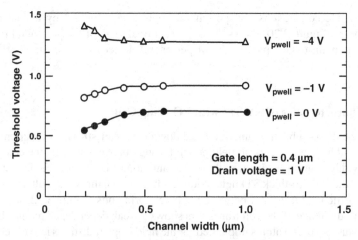

FIGURE 3.22   A threshold voltage of the SA-STI cell as a function of the channel width. The SA-STI cells show a weak dependence of the threshold voltage on the channel width. Therefore, the STI cell is suitable for scaling of the isolation pitch.

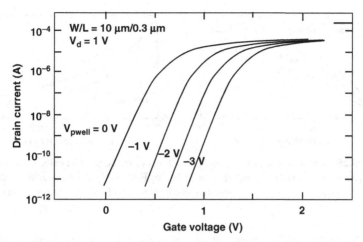

FIGURE 3.23 I_d–V_g characteristics of the wide channel width in cell transistors with various voltages of *p*-well. No anomalous hump is seen in the subthreshold characteristics since the floating gate does not overlap the STI corners.

There is a sufficient margin at the gate length of 0.2 μm for device operation. These results enable a 0.2-μm-rule SA-STI cell with the cell area of 0.31 μm². In the SA-STI cells, Fowler–Nordheim tunneling can achieve a fast programming (20 μs) by applying 17 V to the control gate and achieve a fast erasing (2 ms) by applying 18 V to a *p*-well, as shown in Fig. 3.25. Figure 3.26 shows the TDDB characteristics of the tunnel oxides. Since Q_{BD} in the stripe capacitors with trench edges is almost the

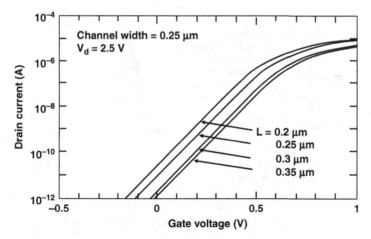

FIGURE 3.24 I_d–V_g characteristics of the short channel cell transistors at $V_d = 2.5$ V, which is applied in the read operation. There is a sufficient margin at the gate length of 0.2 μm for device operation.

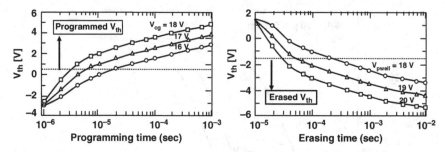

FIGURE 3.25   Programming and erasing characteristics of the SA-STI cells. Fast programming (20 µs) and erasing (2 ms) can be accomplished by Fowler–Nordheim tunneling, applying 17 V to the control gate during programming and 18 V to *p*-well during erasing, respectively.

same as that in the flat capacitors without trench edges, the process damages into the tunnel oxides during the SA-STI fabrication steps are negligibly small. Therefore, the endurance characteristics of the SA-STI cells are excellent, as shown in Fig. 3.27. The threshold voltage window narrowing has not been observed up to 1 million cycles.

A threshold voltage distribution of programmed and erased cells is evaluated by measuring a 2-Mbit cell array, as shown in Fig. 3.28. Both the programming and the erasing are performed by Fowler–Nordheim tunneling of electrons. A tight distribution of about 2.0 V is realized though the programming and the erasing are carried out by one pulse without verification, because of a good uniformity of the channel width in the memory cell by using the self-aligned STI structure.

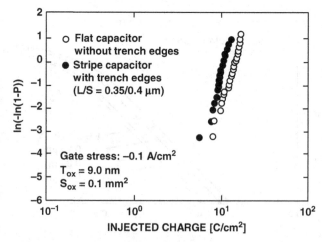

FIGURE 3.26   TDDB (Time Dependent Dielectric Breakdown) characteristics of the tunnel oxide in the STI stripe capacitor and a flat capacitor. The process damages into the tunnel oxides during the SA-STI fabrication steps are negligibly small.

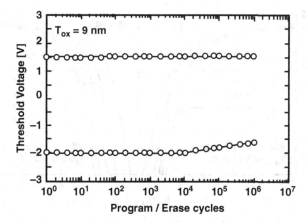

FIGURE 3.27 The program and erase cycling endurance characteristics of the SA-STI cell. The threshold voltage window narrowing has not been observed up to 1 million cycles.

3.3.4 Characteristics of Peripheral Devices

Figure 3.29 shows subthreshold characteristics of low voltage peripheral transistor as a function of the well voltage. No hump is observed in the subthreshold region as a result of avoiding to overlap the gate electrodes with the STI corners.

Figure 3.30 shows a junction breakdown voltage and a threshold voltage of a parasitic field transistor. The isolating ability of the STI is greatly higher than that of LOCOS. Furthermore, the breakdown voltage is higher than a demand (>22.5 V)

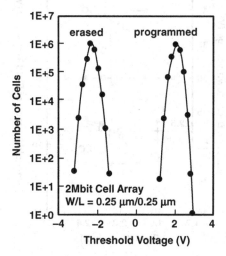

FIGURE 3.28 A cell threshold voltage distribution in one program and one erase pulse (no verify). Programming and erasing are carried out by 17 V, 10 μs and 18 V, 1 ms, respectively.

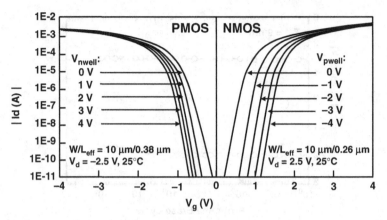

FIGURE 3.29 A subthreshold characteristics of the low-voltage peripheral transistors (NMOS and PMOS) as a function of well voltage.

with a sufficient margin. Therefore, the self-aligned STI process is suitable to the peripheral transistor as well as to the memory cell.

Figure 3.31 shows the TDDB characteristics of the gate oxides in the high-voltage transistors. The evaluated lifetime of the gate oxide is sufficiently long. The result implies that the process damages into the gate oxides are negligibly small.

A $0.31\text{-}\mu m^2$ SA-STI cell with FG wing and peripheral integration process have been successfully developed using $0.25\text{-}\mu m$ design rules. This technology makes it

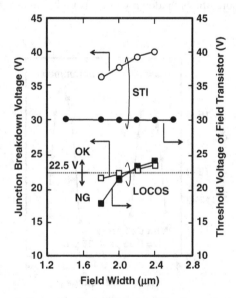

FIGURE 3.30 A punch-through voltage and threshold voltage of a parasitic field transistor for high-voltage isolation.

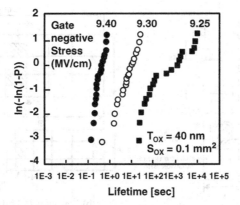

FIGURE 3.31 TDDB characteristics of the high-voltage gate oxide as a function of applied-gate electric field at negative gate voltage.

possible to realize reliable memory cell and peripheral devices with a simple process. Therefore, the SA-STI cell with FG wing is suitable for a low-cost flash memories of 256 Mbit and 1 Gbit for mass storage applications [7, 8]. The SA-STI cell with FG wing had been successfully adopted to a NAND flash memory product of 0.25-μm rule [7, 8], 0.15 to 0.16-μm rule [27], 0.12- to 0.13-μm rule [28, 29], and 90-nm rule [30], to achieve low-cost and reliable flash memory, as shown in Figs. 3.1 and 3.2.

3.4   SELF-ALIGNED STI CELL (SA-STI CELL) WITHOUT FG WING

An ultra-high-density NAND flash memory cell, using a self-aligned shallow trench isolation (SA-ST1) technology, had been developed for a high-performance and low-bit cost flash memory [6]. The SA-STI technology results in an extremely small cell size of ideal $4*F^2$ (F:feature size). The key technologies to realize a small cell size are (1) 0.4-μm (F) width shallow trench isolation (STI) to isolate neighboring bits and (2) a floating gate that is self-aligned with the STI, eliminating the floating-gate wings. Even though the floating-gate wings are eliminated, a high coupling ratio of 0.65 can be obtained by using the side walls of the floating gate to increase the coupling ratio. Using this self-aligned structure, a reliable tunnel oxide can be obtained because the floating gate does not overlap the trench corners, so enhanced tunneling at the trench corner is avoided. Therefore, the SA-STI cell combines a low bit cost with a high performance and a high reliability.

3.4.1   SA-STI Cell Structure

A self-aligned shallow trench isolation (SA-ST1) cell without FG wing is described for a high-performance and low-bit-cost NAND flash memory cell [6]. A small cell size of 0.67 μm^2, including the select transistor and drain contact area, was obtained

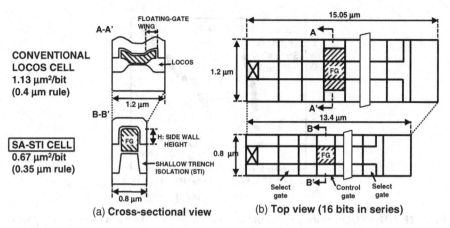

(a) Cross-sectional view   (b) Top view (16 bits in series)

FIGURE 3.32   (a) The cross-sectional view and (b) top view of the self-aligned shallow trench isolation cell (SA-STI cell) without FG wing in comparison with that of the conventional LOCOS cell.

under a 0.35-μm design rule, in comparison with a 1.13-μm^2 LOCOS cell [3]. The key technology in obtaining small cell size is the bit-line isolation technology, which uses the shallow trench isolation (ST1) process. This technology also realizes a high reliability and a high performance.

Figure 3.32 compares the cross-sectional and top view of the SA-STI cell with that of the conventional LOCOS cell. This NAND structure cell has 16 memory transistors arranged between two select transistors in series. The word-line pitch is 0.7 μm. The bit-line pitch can be reduced to 0.8 μm by using 0.4-μm STI technology. As a result, the cell size of the SA-STI cell is about 60% of that of the conventional LOCOS cell [3].

In general, as the isolation width between the memory cells is reduced, the coupling ratio is reduced due to the decreased floating-gate wing area. However, in the SA-STI cell without FG wing, even if very tight 0.4-μm-width isolation is used, a high coupling ratio of 0.65 can be obtained because the 0.3-μm high side wall (H) of the floating gate is used to increase the coupling ratio, as shown in Fig. 3.33. Table 3.4 shows major cell parameters.

3.4.2 Fabrication Process

The fabrication of the SA-STI cell is simple and uses only conventional techniques, as shown in Fig. 3.34. First, a stacked layer of the gate oxide, the floating-gate poly-silicon, and the cap oxide is formed. Next, the trench isolation region is defined by patterning these three layers, followed by the trench etching, as shown in Fig. 3.34a and filling with LP-CVD SiO$_2$, as shown in Fig. 3.34b. Subsequently, the LP-CVD SiO$_2$ is etched back until the sidewall of the floating gate poly-silicon is exposed. After

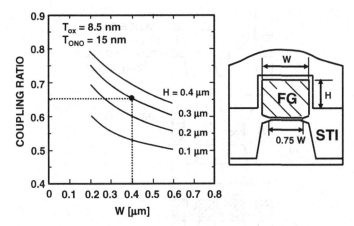

FIGURE 3.33 The coupling ratio of the SA-STI cell as a function of the gate width (*W*). A high coupling ratio of 0.65 can be obtained because the 0.3-μm-high sidewall (H) of the floating gate is used.

that, the inter-poly dielectric (ONO) (Fig. 3.34c) and the control-gate poly-silicon are formed (Fig. 3.34d), followed by the stacked-gate patterning. In this process, the floating-gate patterning and STI patterning are carried out by the same mask, so the number of fabrication steps for the SA-STI process can be decreased with about 10% in comparison with that for a conventional LOCOS process. Figure 3.35 shows cross-sectional SEM photograph of an SA-STI cell.

3.4.3 Shallow Trench Isolation (STI)

In the case of LOCOS isolation, the punch-through of the bit-line junctions occurs at a 0.5-μm isolation width, as shown in Fig. 3.36. However, in the case of STI, the punch-through and junction breakdown voltage are higher than 15 V even at a

TABLE 3.4 Memory Cell Parameters of the SA-STI Cell without FG Wing in 0.4-μm Technology

Cell size	0.67 μm^2 (including select Tr, etc.)
Gate length	0.35 μm
Gate width	0.4 μm
Trench isolation width	0.4 μm
Tunnel oxide	8.5 nm
Interpoly dielectric	ONO 15 nm (effective)
Programming time	0.195 μs/byte
Erase time	2 ms/sector
	2 ms/chip

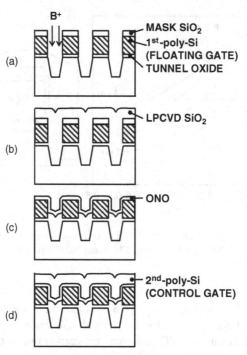

FIGURE 3.34 The process sequence of the SA-STI process without FG wing. (a) Trench etching, B+ implantation. (b) LP-CVD SiO_2 fill-in. (c) Oxide etch-back and ONO formation. (d) Control-gate formation. The floating-gate and STI patterning are carried out by the same mask, so the number of fabrication steps for the SA-STI process can be decreased with about 10% in comparison with that for a conventional LOCOS process.

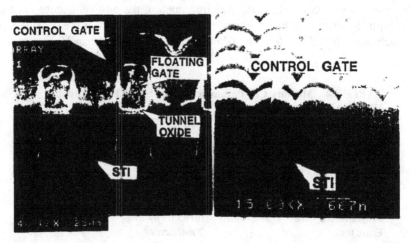

FIGURE 3.35 Cross-sectional SEM photograph of the SA-STI cell without FG wing.

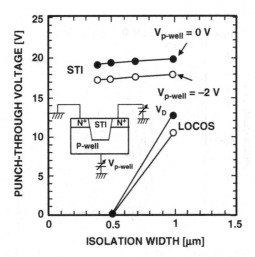

FIGURE 3.36 The punch-through voltage of the bit-line junction, which is isolated by shallow trench isolation (STI), in comparison with LOCOS isolation. The punch-through is higher than 15 V, which is high enough to realize 0.4-μm trench isolation.

0.4-μm isolation width, as shown in Fig. 3.36. Furthermore, the 0.7-μm-thick STI field oxide results in a high threshold voltage (>30 V) of the parasitic field transistor between the neighboring bits. As a result, a very tight 2∗F (0.8 μm) bit-line pitch can be realized by using STI.

Figure 3.37 illustrates the leakage of $n+$−p junction for STI and LOCOS. The leakage current of STI is comparable to that of LOCOS.

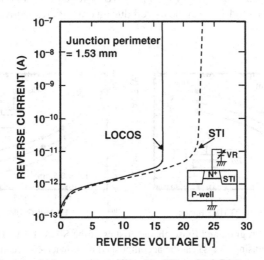

FIGURE 3.37 Junction leakage current of the SA-STI and LOCOS processes. The junction leakage current of the SA-STI process is comparable to that of LOCOS process.

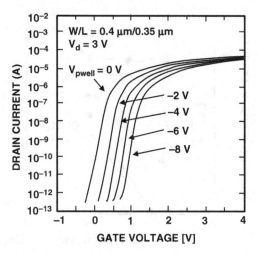

FIGURE 3.38   The subthreshold characteristics of the SA-STI cell with various substrate (*p*-well) bias conditions. Anomalous subthreshold characteristics (hump) cannot be seen because the floating-gate dose not overlap the trench comer.

3.4.4   SA-STI Cell Characteristics

Figure 3.38 shows the subthreshold characteristics of the SA-STI cell with a 0.4-µm channel width. Anomalous subthreshold characteristics (hump) cannot be seen as a result of the SA-STI structure.

Figure 3.39 shows the program and erase characteristics of an SA-STI cell. The fast programming (100 µs/512 byte) and fast erasing (2 ms) can be accomplished by Fowler–Nordheim tunneling, applying a positive voltage of 17 V to the control gate during programming and 17 V to the *p*-well during erasing, respectively, as shown in Fig. 3.39.

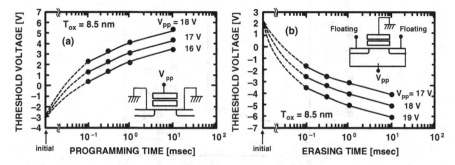

FIGURE 3.39   (a) Program and Erase characteristics of the SA-STI cell. The fast programming (100 µs) and erasing (2 ms) can be accomplished by Fowler–Nordheim tunneling over the channel area, applying a positive voltage of 17 V to the control gate during programming and 17 V to the *p*-well during erasing, respectively.

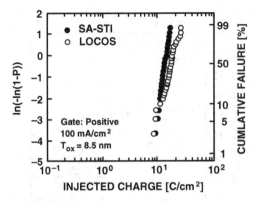

FIGURE 3.40 The TDDB characteristics of the 8.5-nm-thick tunnel oxide. The TDDB characteristics in the SA-STI process without FG wing are almost the same as that in the LOCOS process.

The TDDB characteristics of the tunnel oxide in the SA-STI process are almost the same as that in the LOCOS process, as shown in Fig. 3.40., because the floating gate does not overlap the trench edges. Therefore, the endurance characteristics of the SA-STI cell are comparable with that of the conventional LOCOS cell, as shown in Fig. 3.41. The SA-STI cell guarantees a wide cell threshold window as large as 3 V, even after 1 million write/erase cycles. Furthermore, read disturb characteristics can be ensured for more than 10 years even after 1 million write/erase cycles, as shown in Fig. 3.42.

The SA-STI cell without FG wing has a very simple cell structure and has a very small cell size of ideal $4*F^2$ with bit-line and word-line pitch of $2*F$ (F; feature size),

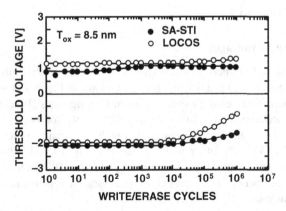

FIGURE 3.41 The program(write)/erase endurance characteristics of the SA-STI cell and the LOCOS cell. In the SA-STI cell, window narrowing has not been observed up to 1 million program (write)/erase cycles.

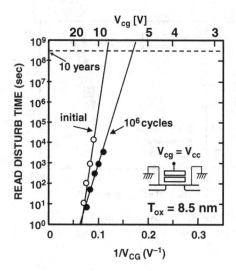

FIGURE 3.42 The read disturb characteristics of the SA-STI cell without FG wing. The read disturb time is more than 10 years when a V_{cc} of 3.0 V is used, even after 1 million program/erase cycles .

as shown in Fig. 3.2. The SA-STI technology has also demonstrated an excellent reliability and performance. Therefore, as shown in Fig. 3.1, the SA-STI cell without FG wing has been extensively used for more than 12 years since around (2002), over eight generations (90 nm to 1X nm) of NAND flash memory product, such as a 90 nm cell [31], 70- nm cell, a 50-nm cell, a 43-nm cell [32], a 30-nm cell [33], a 27-nm cell [34], a 20-nm cell, and a mid-1X-nm cell [21, 35].

3.5 PLANAR FG CELL

3.5.1 Structure Advantages

The conventional self-aligned STI cell (SA-STI cell) has a structure problem of control gate formation between floating gates, as shown in Fig. 3.43 (as described in Section 5.6). There is not enough space between floating gates (FGs) to fabricate a control gate in a scaled cell [36]. To solve this problem, two solutions were proposed, as shown in Fig. 3.43. One is the slimming FG width to obtain an enough space for a control gate [21, 35]. The other is the planar FG cell [10–12], which has a thin ($\sim$10 nm) floating gate with a high-k inter-poly dielectric (IPD). Thanks to a high-k IPD, the capacitance between CG and FG becomes large enough to operate a memory cell. Then FG thickness can be very thin.

Figure 3.44 shows cross sections of (a) a conventional SA-STI cell, (b) a planar FG cell [10], and (c) the stacked structure of a planar FG cell [11]. Thickness of

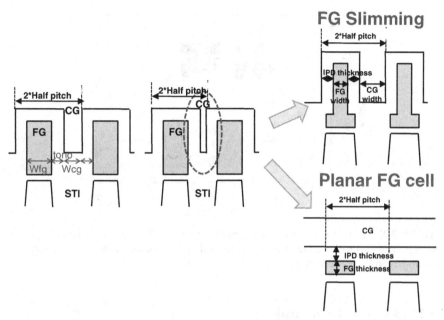

FIGURE 3.43 Structure problem of the self-aligned STI cell and two solutions of floating-gate (FG) slimming and the planar FG cell.

floating gate is very thin, around 10 nm. And the high-k block dielectric (BD) is stacked on thin FG as IPD. The aspect ratio of stacked gate and control gate (CG) fill are compared between conventional SA-STI cell (wrap cell) and a planar FG cell, as shown in Fig. 3.45 [12]. In the SA-STI cell, the aspect ratio becomes more than 10 for both the stacked gate and CG fill in a sub-20-nm cell. The planar FG cell can much mitigate this limitation.

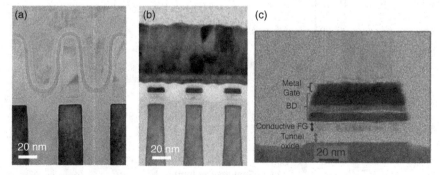

FIGURE 3.44 Cross sections of (a) a conventional SA-STI cell without FG cell, (b) a 20-nm planar FG cell, and (c) a stacked structure of planar FG cell.

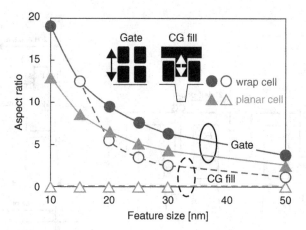

FIGURE 3.45 Aspect ratio (A.R.) of a conventional SA-STI cell (wrap cell) and a planar cell. A.R. increases with scaling. A.R. is >10 for both the word-line and the bit-line directions in a sub-20-nm SA-STI cell (solid: Gate, open: CG fill).

3.5.2 Electrical Characteristics

The planar FG cell can drastically reduce the floating-gate capacitive coupling interference (cell-to-cell interference), as shown in Fig. 3.46. Due to small floating-gate capacitive coupling interference, the read window margin (RWM) (see Section 5.2) can be much improved. Also, the erase V_t setting can be shallower V_t (higher V_t). Then program/erase cycling endurance and data retention can be expected to improve

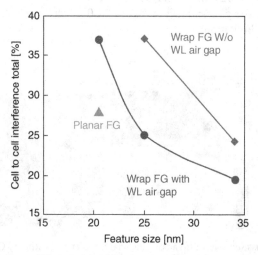

FIGURE 3.46 Cell-to-cell interference (floating-gate capacitive coupling interference) scaling. An ~30% total interference reduction is achieved with the planar FG cell.

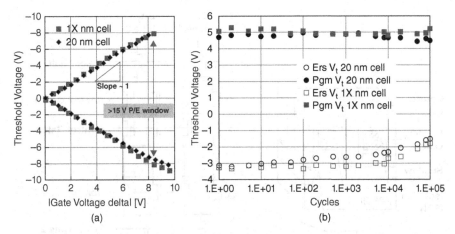

FIGURE 3.47   (a) Program/erase characteristics and (b) program/erase cycling endurance characteristics of a 20-nm and a 1X-nm planar FG cell.

because oxide stress is reduced due to smaller amount of charge through oxide during program/erase operations.

Figure 3.47a shows program/erase characteristics. Excellent program/erase window and program slope ($\sim$1) are demonstrated in the planar FG cell of a 20-nm cell and 1X-nm cell. Both of these characteristics are important for enabling a highly reliable MLC NAND flash memory. Figure 3.47b shows the program/erase cycling endurance characteristics. Excellent cycling endurance characteristics are also demonstrated.

The planar FG cell has a potential to extend NAND cell scaling very effectively by removing the structure problem and by small floating-gate capacitive coupling interference.

3.6   SIDEWALL TRANSFER TRANSISTOR CELL (SWATT CELL)

A multilevel NAND flash memory cell, using a sidewall transfer-transistor (SWATT) structure, had been developed for a high-performance and low-bit-cost flash memory [13, 14]. With the SWATT cell, a relatively wide threshold voltage (V_{th}) distribution width of about 1.1 V can be obtained for MLC (2 bits/cell) in contrast to a narrow 0.6-V distribution width that is required for a conventional cell. The key technology that allows this wide V_{th} distribution width is the transfer transistor, which is located at the side wall of the shallow trench isolation (STI) region and is connected in parallel with the floating-gate transistor. During read, the transfer transistors of the unselected cells (connected in series with the selected cell) work as pass transistors. So, even if the V_{th} of the unselected floating-gate transistor is higher than the control-gate voltage, the unselected cell will be in the ON state. As a result, the V_{th} distribution of

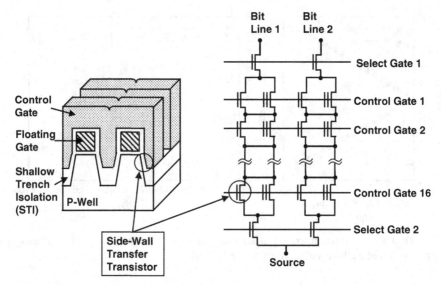

FIGURE 3.48   The schematic view and equivalent circuit of the sidewall transfer transistor cell (SWATT cell). A transfer transistor is located at the sidewall of the shallow trench isolation (STI) region and is connected in parallel with the floating-gate transistor.

the floating-gate transistor can be wider and the programming can be faster because the number of program/verify cycles can be reduced.

3.6.1   Concept of the SWATT Cell

The concept of a sidewall transfer-transistor cell (SWATT cell) for multilevel NAND flash memory [13, 14] is described. The schematic view and equivalent circuit of the SWATT cell are shown in Fig. 3.48. One cell consists of both a floating-gate transistor and a transfer transistor, which is located at the sidewall of the shallow trench isolation (STI) region. These two transistors are connected in parallel. Sixteen cells are connected in series between two select transistors to form a NAND cell string. The read conditions of a conventional NAND cell and a SWATT cell for the two-level scheme (SLC) are shown in Fig. 3.49. In a conventional cell, zero volt is applied to the gate of the selected memory cell, while 5.0 V is applied to the gates of the unselected cells in the NAND string. All the memory cells, except for the selected cell, serve as transfer gates. Therefore, for the conventional NAND cell, the threshold voltage of the in-series connected cells must be lower than the unselected control-gate (CG) voltage of 4.5–5.5 V. Thus, the V_t distribution of the cells in the programmed state must be narrow with a width of less than 3.0 V for two-level operation, as shown in Fig. 3.49a.

On the other hand, the sidewall transfer transistor in the SWATT cell works as a pass transistor instead of a floating-gate transistor, as shown in Fig. 3.49b. So the

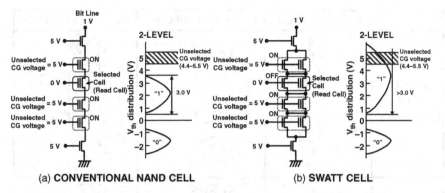

FIGURE 3.49 Read condition of (a) a conventional NAND cell and (b) the SWATT cell for a two-level scheme. In the conventional NAND cell, the V_{th} distribution of the cells in programmed state must be narrow with a width of 3.0 V or less, because the unselected cells must work as pass transistors for a control-gate voltage of 5 V. However, in the SWATT cell, the sidewall transfer transistor works as a pass transistor. Therefore, the V_{th} distribution of the floating-gate transistor in the programmed state is allowed to be very wide with a width of >3.0 V for two-level operation.

threshold voltage of the floatin-gate transistor does not have to be lower than the unselected CG voltage of 4.5–5.5 V. Therefore, the V_t distribution of the floating-gate transistor in the programmed state is allowed to be very wide with a width >3 V for two-level operation. As a result, the V_t distribution can be wider in comparison with the conventional NAND cell.

The threshold voltage distributions of the two-level (SLC) and four-level scheme (MLC) are compared in Fig. 3.50. In a conventional NAND cell, the V_t distribution of the cells in the programmed states ("1," "2," and "3") must be very narrow (0.6 V), because the in-series connected cells must work as pass transistors. However, in the SWATT cell, the V_t distribution of the floating-gate transistor in the programmed "1" and "2" states is allowed to be very wide (1.1 V). The V_t distribution in the programmed "3" state is allowed to be even wider than 1.1 V.

This wide threshold voltage distribution results in a high programming speed because of reducing the number of program/verify cycles and good data retention characteristics.

3.6.2 Fabrication Process

The developed SWATT cell has 16 memory transistors connected in series between two select transistors. The word-line pitch is 0.7 μm. A very narrow bit-line pitch of 0.8 μm can be realized by using 0.4-μm-width shallow trench isolation (STI) technology. As a result, a small cell size of $5.5*F^2$ (0.67 μm²), including the select transistor and drain contact area, can be obtained under a 0.35-μm design rule. The thickness of sidewall dielectric (ONO) is 40 nm effective.

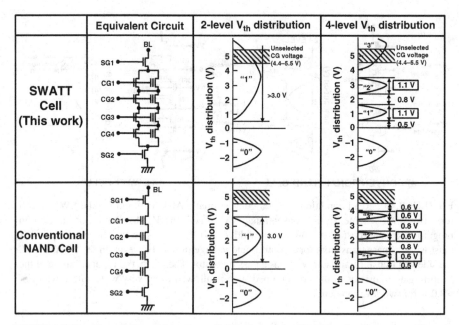

FIGURE 3.50 The cell threshold voltage distribution of the SWATT cell and the conventional NAND cell for two-level (SLC) and four-level (MLC) operation. In the SWATT cell, a wide threshold distribution of 1.1 V is allowed for four-level operation, in comparison with a 0.6-V distribution that is required for the conventional NAND cell. The range of the unselected CG (control gate) voltage is limited because of read disturb.

The fabrication process of the SWATT cell is similar to that of the SA-STI cell. The process sequence of the SWATT cell is shown in Fig. 3.51. First, a stacked layer of the gate oxide, the floating-gate poly-silicon, and the cap oxide is formed. Next, the trench isolation region is defined by patterning these three layers, followed by the trench etching, trench bottom boron implantation, and filling with LP-CVD SiO_2, as shown in Fig. 3.51a. Subsequently, the LP-CVD SiO_2 is etched back until the sidewall of the STI is exposed (Fig. 3.51b). Boron (B+) ion implantation (60 KeV, $2E12/cm^2$) is carried out for V_{th} adjustment of the sidewall transfer transistor. After that, the interpoly dielectric (ONO) and transfer-transistor gate oxide are formed at the same time, as shown in Fig. 3.51c. Then the control-gate poly-silicon is deposited, followed by the stacked gate patterning (Fig. 3.51d). In this process, the thermal oxide of the STI sidewall is about two times thicker than that on the poly-silicon due to oxidation enhancement at the STI sidewalls. As a result, breakdown of the control gate does not occur even if a high voltage of about 20 V is applied to the control gate during the program operation.

A cross-sectional TEM photograph is shown in Fig. 3.52. Both the trench isolation and channel width (gate width) of the floating gate transistor are 0.4 μm. The vertical channel width of the sidewall transfer transistor is about 0.2 μm.

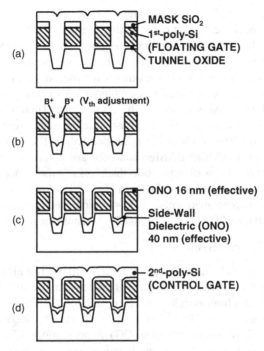

FIGURE 3.51 The process sequence of the SWATT process. (a) Trench etching, LP-CVD SiO$_2$ fill-in. (b) Oxide etch-back and B+ implantation of the V_{th} adjustment of the side wall transfer transistor. (c) ONO formation. (d) Control gate formation. The thermal oxide of the STI sidewall is about two times thicker than that on the poly-silicon due to oxidation enhancement at the STI sidewalls.

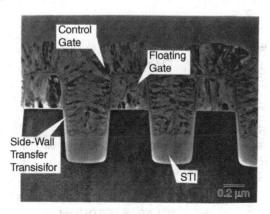

FIGURE 3.52 Cross-sectional TEM photograph of the SWATT cell along the word-line (Control Gate) direction.

An accurate control of the threshold voltage of the sidewall transfer transistor is important for the SWATT cell. The range of the threshold voltage is determined as follows. The sidewall transfer transistor must be in the ON state when the unselected CG voltage (4.5–5.5 V) is applied to the control gate. So, the upper limit of the V_{th} of the sidewall transistor is 4.5 V. On the other hand, the sidewall transfer transistor must be in the OFF state when the read voltage (about 3.9 V) between the "2" and "3" state for four-level operation is applied to the control gate. Therefore, the threshold voltage of the side-wall transfer transistor must be in the range from 3.9 V to 4.5 V for four-level operation (from 0 V to 4.5 V for two-level). The important statistical parameters of V_{th} of a sidewall transfer transistor are boron concentration in the channel region and the sidewall gate-oxide thickness. Boron concentration is well controlled by boron implantation, as shown in Fig. 3.51b. Also, the oxide thickness of the STI side wall is controlled within 10% variation. Therefore, the narrow range of the threshold voltage of the sidewall transfer transistor can be adjusted.

3.6.3 Electrical Characteristics

A. Isolation   For the NAND flash cell, the high-voltage isolation technology is important to reduce the bitline pitch. The isolation between the bit lines must satisfy two demands. One is a high punch-through or junction breakdown voltage of the bit-line junction area (>10 V). The other is a high threshold voltage of the parasitic field transistor (>25 V) of the control gate (CG) in the memory cell.

The breakdown voltage of the bit-line junction occurs at about 19 V while no punch-through is observed. The breakdown voltage is higher than the required 10 V, which is high enough to apply NAND flash cell.

Figure 3.53 shows the threshold voltage of the parasitic field transistor in the SWATT cell. The 0.3-μm-thick STI field oxide results in a high threshold voltage (>30 V) of the parasitic field transistor between the neighboring bits.

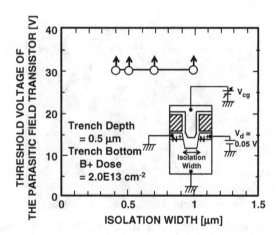

FIGURE 3.53   The threshold voltage of the parasitic field transistor in the SWATT cell, which is isolated by shallow trench isolation (STI). The threshold voltage of the field transistor is higher than 30 V, which is high enough to realize 0.4-μm-width trench isolation.

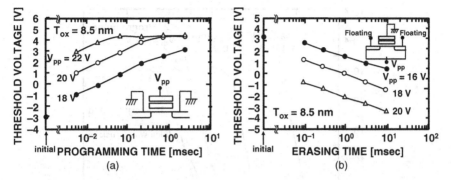

(a)   (b)

FIGURE 3.54   (a) The program and (b) erase characteristics of the SWATT cell. A short programming time of 200 μs and short erase time of 2 ms can be accomplished by Fowler–Nordheim tunneling over the channel area, applying a positive voltage of 21 V to the control gate during programming and 19 V to the p-well during erasing, respectively.

B. Cell Characteristics   The program and erase characteristics of the SWATT cell are shown in Fig. 3.54a and Fig. 3.54b, respectively. The threshold voltage of programming saturates at about 4.2 V. This is explained as followed. In this memory cell (observed V_{th} = 4.2 V), a floating-gate transistor is programmed to high threshold voltage (V_{th} > 4.2 V), so a floating-gate transistor is in the OFF state for a measurement condition. On the other hand, the sidewall transfer transistor is in the ON state for V_{cg} > 4.2 V, because the V_{th} of the sidewall transfer transistor is about 4.2 V. Therefore, the V_{th} of sidewall transfer transistor is observed. Then, V_{th} saturates at about 4.2 V even after long programming time (>0.1 ms at 22 V). It can be seen that a fast programming (200 μs/512 byte) and erase operation (2 ms) can be obtained.

Figure 3.55 shows the subthreshold I_d–V_g characteristics of the SWATT cell at the erased "0" and programmed "1", "2", "3" states. In the programmed "3" state, the V_{th} of the floating-gate transistor is higher than 4.5 V, so only the I_d of the sidewall transfer transistor can be observed.

Figure 3.56 shows the coupling ratio of the SWATT cell as a function of the gate width (W). In general, as the isolation width between the memory cells is reduced, the coupling ratio is reduced due to the decreased floating-gate wing area. However, in the SWATT cell, even if very tight 0.4-μm-width isolation is used, a high coupling ratio of 0.65 can be obtained because the 0.3-μm-high sidewall (H) of the floating gate is used to increase the coupling ratio. Moreover, the coupling ratio increases as the gate width W is scaled down. This means that the programming voltage and erasing voltage (V_{pp}) can be reduced as the memory cell is scaled down, which allows the design of more compact peripheral circuits such as row decoders and sense amplifiers. Furthermore, the variation of the coupling ratio of the SWATT cell can be very small because the sidewall (H) of the floating gate is determined by the thickness of the floating-gate poly-silicon. Therefore, a very tight V_t distribution of the SWATT cell is expected.

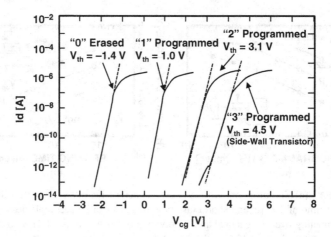

FIGURE 3.55 The subthreshold characteristics of the SWATT cell for the erased "0" and programmed "1," "2," "3" states. In the programmed "3" state, the side-wall transfer transistor is in the ON state.

C. Reliability   Figure 3.57 shows the program/erase cycling endurance characteristics of a SWATT cell using the uniform program/erase scheme [22–25]. This scheme guarantees a wide cell threshold window of as large as 3 V, even after one million write/erase cycles. These endurance characteristics of the SWATT cell are comparable to that of the conventional NAND cell [4–6].

Read disturb occurs as a weak programming mode. The tunnel-oxide leakage currents, which are induced by the program and erase cycling stress, degrade the read disturb of the memory cell, as shown in Fig. 3.58. However, even after one million program/erase cycles, the read disturb time is more than 10 years when a V_{cg} of 5.0 V is used.

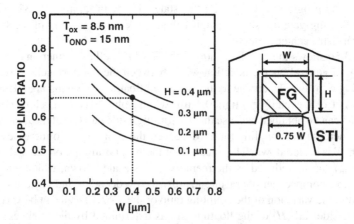

FIGURE 3.56 The coupling ratio of the SWATT cell as a function of the channel width (*W*). A high coupling ratio of 0.65 can be obtained because the 0.3-μm-high side-wall (*H*) of the floating gate is used to increase the coupling ratio.

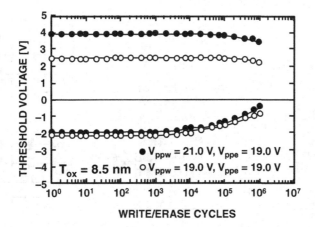

FIGURE 3.57   The program (write)/erase cycling endurance characteristics of the SWATT cell. Window narrowing has not been observed up to one million program/erase cycles.

3.7   ADVANCED NAND FLASH DEVICE TECHNOLOGIES

3.7.1   Dummy Word Line

A dummy word-line (dummy cell) scheme in NAND flash memory was proposed to eliminate abnormal program disturb of edge memory cell [15–17]. Dummy word line (dummy cell) is located between edge word lines (edge memory cells) of NAND

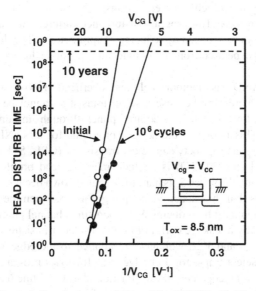

FIGURE 3.58   The read disturb characteristics of the SWATT cell. The read disturb time is more than 10 years when a V_{cc} of 5.0 V is used, even after 1 million program/erase cycles.

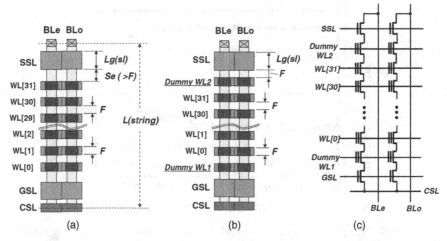

FIGURE 3.59   (a) Layout view of conventional NAND flash memory. (b) Layout view of dummy word lines in NAND flash memory. (c) Schematic diagram of dummy word lines in NAND flash memory. Copyright 2007, The Japan Society of Applied Physics.

string and select gate transistors (GSL or SSL). The program disturb of a GIDL generated hot electron injection mechanism [37] can be suppressed by increasing the distance between an edge cell and a select transistor. Also, the program boosting potential drop from edge cell to select transistor can be well controlled by using the proper dummy word-line voltage and proper dummy cell V_t. Therefore, abnormal program disturbance of an edge memory cell can be greatly suppressed. In addition, capacitive coupling noise between select transistor and edge memory cell can be reduced to less than 50%. The program disturbance failure, read failure, and erase distribution width can be reduced by reducing coupling noise. The dummy word-line scheme was started to be used from a 40-nm technology node due to stable operations in edge cells [17].

By scaling a NAND flash memory cell, area overhead of two select transistors in a NAND string is increasing because a select transistor cannot be scaled down as memory cell scale down due to the required punch-through immunity for program boosting voltage. This is one of the scaling problems for a NAND flash memory cell. Also, area of space (S_e) between select transistors (GSL, SSL) and edge word lines (WLs: WL[0] and WL[31]) is another area overhead problem, as shown in Fig. 3.59(a). Reducing the space S_e is hard to be scaled down because of the following two reasons. One is that the capacitive coupling noise between the select transistor and edge WLs is increased by reducing S_e. A boosting channel potential of program inhibit is decreased by a leakage current through a select transistor which is slightly turned on when V_{pass} and V_{pgm} voltages are applied to the edge WL due to large coupling between select transistor and edge WL. It causes program inhibit failure. Also, during read for an edge cell, the voltage of edge word line has a bump due to coupling with a select gate, which is ramped up after ramping up of voltage of edge

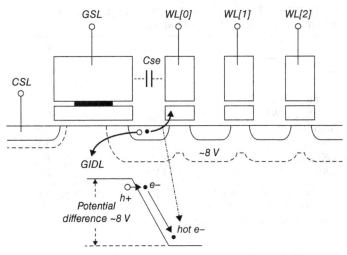

FIGURE 3.60 Enhanced program disturbance of edge memory cell by hot carrier induced by GIDL. Copyright 2007, The Japan Society of Applied Physics.

word lines [39]. It causes read failure. The other reason is that hot-carrier disturbance due to large electric field in junction between select transistor and edge WL [37], as shown in Fig. 3.60 (see Section 6.5.2). The hot carriers are mainly generated by a GIDL (gate-induced drain leakage) mechanism, and hot electrons are enhanced by an electric field between a select transistor and an edge cell. Some hot electrons are injected to the floating gate of an edge memory cell. It was reported that at least a larger than 110 nm S_e is required to avoid severe hot carrier program disturbance, as shown in Fig. 3.61 [37].

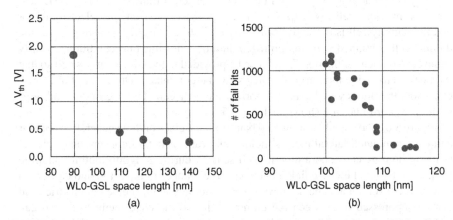

FIGURE 3.61 (a) Simulation result of the number of electrons injected to the WL0 cell for cell arrays having various WL0-GSL spaces. (b) Number of fail bits measured with 1 Mb block array at $V_{pass} = 10$ V. Copyright 2007, The Japan Society of Applied Physics.

TABLE 3.5   Read and Erase Condition for the Dummy Word-Line Scheme in NAND Flash Memory Cells

	Read	Erase
BL	V_{pc}	F
SSL	V_{cc}	F
Dummy WL2	V_{read}	0 V
Unselected WL	V_{read}	0 V
Selected WL	V_r	0 V
Dummy WL1	V_{read}	0 V
GSL	V_{cc}	F
CSL	0 V	F

V_{pc}, BL precharge voltage; V_r, read voltage for selected WL; V_{read}, read voltage for unselected WL; F, Floating).
Source: Copyright 2007, The Japan Society of Applied Physics.

Furthermore, an edge memory cell has a different condition compared with middle cells. In an edge cell, one side of a source/drain is connected to the select transistor while the other side is connected to a neighbor cell. However, in the middle cell, both sides are connected to neighbor cells. The potential of a floating gate is different between an edge cell and middle cells during operations. It causes abnormal electrical characteristics such as erase and program characteristics, as compared to middle cells. This is because the coupling ratio of the floating gate and the voltage condition applied around neighbor gates are different between an edge cell and a middle cell for each operation. It eventually results in wide V_{th} distributions of erase and program state.

To solve these scaling issues of an edge memory cell, a dummy word-line scheme and the new operation conditions were proposed [15–17]. Figures 3.59b and 3.59c show a structures of a dummy cell scheme. A dummy cell which is identical to normal memory cell is additionally placed between each select transistors (GSL, SSL) and the edge memory cell (WL[0], WL[31]). The space between the select transistor and the dummy cell is basically formed by F (feature size). By adjusting V_{th} of the dummy cell combined with an optimized dummy word-line bias condition, a nearly equal environment of the middle cell can be provided to the edge memory cell so that the unexpected edge memory cell effects can be eliminated. During read and erase operation, the dummy cell acts as a normal memory cell. The operation conditions of dummy word lines are shown in Table 3.5.

Figure 3.62 shows (a,b) a simulated band-to-band electron/hole generation contour and (c) a simulated lateral electric field in the case of both a conventional scheme (without dummy cell) and a dummy cell scheme during program inhibit condition [15]. A high lateral electric field generates large number of band-to-band carriers in case of a conventional scheme, however, in a dummy cell scheme, an electric field can be suppressed in between the dummy cell and the edge memory cell. It can be explained by that an optimized biased voltage and adjusted V_{th} of the dummy cell mitigate an electrostatic potential difference between a dummy cell and an edge memory cell, so that it results in a decreasing injection of hot electron carriers to

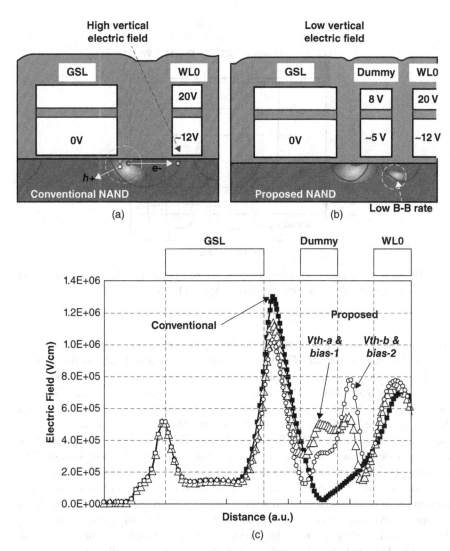

FIGURE 3.62 Simulated band-to-band electron/hole generation rate and lateral electric field during program: (a) Band-to-band electron/hole generation contour in conventional NAND (without dummy word line). (b) Band-to-band electron/hole generation contour with dummy word line scheme. (c) Lateral electric field across structure. Copyright 2007, The Japan Society of Applied Physics.

floating gate of edge cells. Depending on the dummy word-line bias voltage and adjusted V_{th} of the dummy cell, the generated lateral electric field can be reduced further, as shown in Fig. 3.62c.

The dummy cell scheme is also able to shield a memory cell from a high-voltage boosted select gate of GSL/SSL during an erase operation. Figure 3.63 shows a

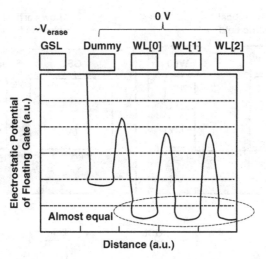

FIGURE 3.63   Simulated electrostatic potential of floating gate of memory cells during erase operation. Copyright 2007, The Japan Society of Applied Physics.

simulated electrostatic potential of a floating gate during an erase operation. The generated voltage of floated GSL during erasing is boosted up to almost the same of a high erase voltage so that the potential of a floating gate of dummy cell becomes slightly higher than that of middle cells due to capacitive coupling between GSL and the dummy cell. Thanks to the shielding effect of a dummy cell, the potentials of a floating gate of edge cells are almost equal to middle cells. Then it leads to improve the erase V_{th} distribution width. Figure 3.64 shows the measured V_{th} distributions of the erase state for each WL in the conventional and dummy word-line scheme. The erased V_{th} distribution of edge WLs in the conventional NAND is as high as 0.5–1.2 V compared to middle WLs. The erase V_{th} distribution is about 1.65 V wide. By using the dummy WL scheme, the difference of erased V_{th} distribution between edge WLs and middle WLs becomes negligible, as shown in Fig. 3.64(b). Thus, the erase V_{th} distribution of dummy word-line scheme is about 1.1 V, which is about 31% narrow width. This leads to better V_{th} distribution of programmed state cells compared to the conventional memory cells.

3.7.2   The P-Type Floating Gate

The n-type phosphorus-doped poly-Si floating gate is a legacy process from an initial production of NAND flash memory in 1992. As n-type poly-Si has several advantages of better dopant controllability, a better scalability of a surface channel nMOS cell, and a low sheet resistance for a select gate. Especially, in the NAND cell, it was important for n-type poly-Si layer to have lower sheet resistance due to a short RC (resistance and capacitance) delay of a select gate in a LOCOS cell and an SA-STI

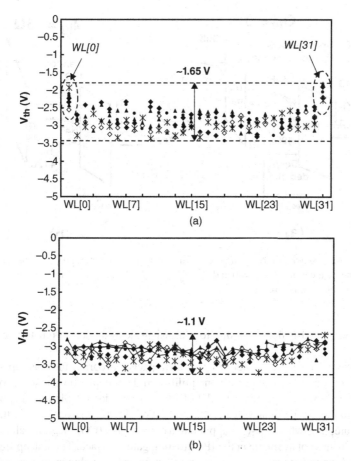

FIGURE 3.64 Measured erased V_{th} of memory cells in (a) conventional NAND flash memory cell without dummy word lines and (b) NAND flash memory cell with dummy word lines. Copyright 2007, The Japan Society of Applied Physics.

cell with FG wing (Section 3.2 and 3.3). Due to higher sheet resistance, p-type poly-Si could not be used for a NAND flash memory cell. However, in the SA-STI cell without FG wing, a high sheet resistance of a floating gate is not a problem because a floating gate and a control gate are directly connected as forming select gate transistor and peripheral transistors.

It had been reported that the p-type floating gate had an advantage to improve the data retention of flash memory cells [18]. However, the depletion effect of a p-type floating gate is not negligible because it is hard to maintain the required doping concentrations after subsequent heat budget processes because of faster inherent boron segregation compared with n-type phosphorus-doped poly-Si. Thus, the doping concentration of boron in the p-type floating gate is normally several times lower than that in an n-type floating gate. If the doping concentration is insufficient,

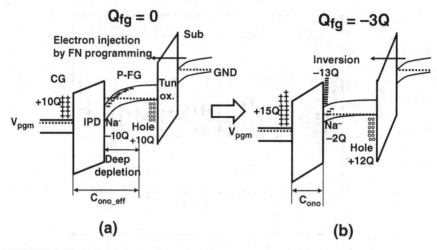

FIGURE 3.65 Schematic band diagram of a transient deep-depletion phenomenon in a p-type floating gate cell in (a) $Q_{fg} = 0$ and (b) $Q_{fg} = -3Q$. The charge values are not calibrated, just for conceptual illustrations.

the program speed is degraded due to a transient deep depletion phenomenon in a floating gate.

A transient deep depletion behavior has an impact on program and erase operations, which are based on a model of a nonequilibrium deep depletion phenomenon [38]. Figure 3.65 shows a conceptual model of the transient deep-depletion phenomenon [19, 20]. In the p-type floating gate, the amount of electrons is very low in the equilibrium state; thus when V_{PGM} is applied to a control gate (CG), negative charges in a floating gate are not available at the IPD/floating gate interface. Then, deep-depletion occurs and extends more deeply into the floating gate, as shown in Fig. 3.65a. In the nonequilibrium condition, the conduction band energy at the IPD/floating gate interface is much lower than that of the Fermi level of the floating gate, thus a large voltage drop occurs, resulting in the loss in the coupling ratio. There are several ways to break the deep-depletion conditions, such as electron injection through the tunnel oxide, electron generation by impact ionization by the injected electrons, thermal electron generation by a SRH, and BTBT electron generation by a strong electric field. All these mechanisms have a contribution toward breaking the deep-depletion.

In the case of a high enough p-type dopant concentration ($N_a > $ 1E20), program and erase operations are successfully performed with a breaking deep-depletion condition. Figure 3.66 shows experimental program and erase characteristics of an n-FG/n-CG cell and a p-FG/n-CG cell in a 42-nm generation cell [19]. The p-FG/n-CG cell appears to have a slower erase speed with ~1.5 V than the n-FG/n-CG cell as mentioned above, while its program speed appears to be faster with ~1 V.

The program/erase cycling endurance of the p-type floating gate is better, compared with that of the n-type floating gate, as shown in Fig. 3.67 [19] and Fig. 3.68 [20]. The midgap voltage shift by N_{OT} (oxide trapping charge) of the p-type floating

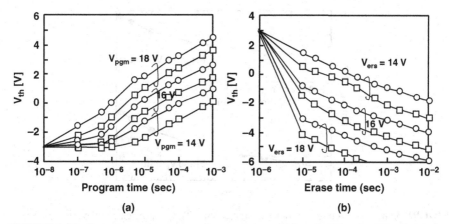

FIGURE 3.66 Experimental program and erase characteristics of *n*-FG and *p*-FG cells (circle is *p*-FG/*n*-CG and rectangle is *n*-FG/*n*-CG).

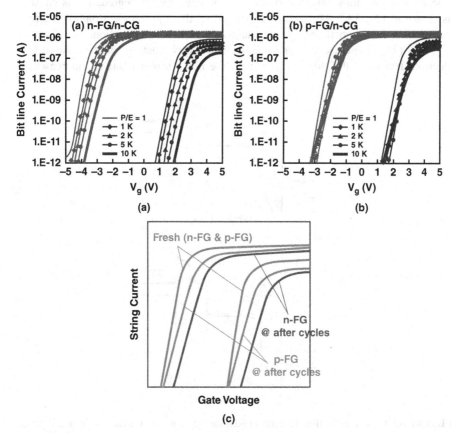

FIGURE 3.67 Experimental I_d–V_g characteristics of (a) *n*-FG and (b) *p*-FG cells ((a) *n*-FG/*n*-CG cell and (b) *p*-FG/*n*-CG cell) before and after P/E cycling. (c) Schematic I_d–V_g curve of cell transistor for *n*-type and *p*-type floating gate before/after P/E cycling.

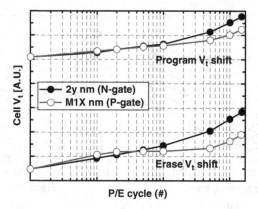

FIGURE 3.68   P/E cycling endurance characteristics. The $p+$ poly gate of the middle 1X nm cell shows an improved cycling endurance compared with 2y nm with p-type gate.

gate is very low, thus only N_{IT} (interface trapping charge) degradation is caused, as shown in Fig. 3.67. This can be explained by the injected electron/hole current ratio during erase operation, which mainly caused the degradation, as shown in Fig. 3.69 [20]. The erase voltage increases in a p-type floating gate because of low electron density at the floating gate, thus the hole current relatively increases to

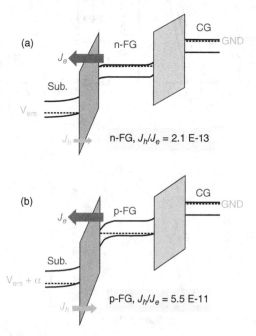

FIGURE 3.69   Schematic illustrations of endurance improvement in the p-type floating gate cell. (a) n-type FG and (b) p-type FG. The ratio of hole current in the p-type floating gate is increased with that in the p-type floating gate.

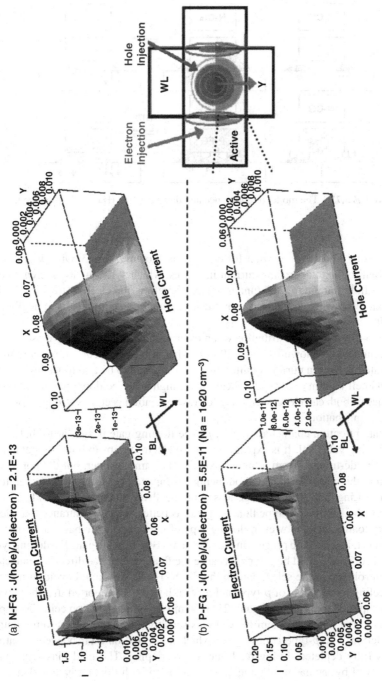

FIGURE 3.70 Distribution of hole/electron erasing tunneling currents with (a) *n*-FG cells and (b) *p*-FG cells with Na = 1e20cm^{-3}.

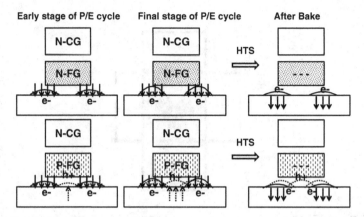

FIGURE 3.71   The model of endurance and data retention (HTS) of a p-FG cell.

have the required electron current for erasing, as shown in Fig. 3.69b. The ratio of hole current to the total erasing current increases, which subsequently increases the amount of hole trapping in the tunnel oxide. Therefore, the electron trapping in the tunnel oxide is mostly compensated by the hole trapping, and it results in negligible N_{OT} shift.

The two-dimensional distribution of an electron/hole current during erase operation is simulated, as shown in Fig. 3.70 [19]. The ratio of hole current injected from Si substrate to electron current emitted from FG is found to be 260 times higher in a p-type FG cell than in an n-type FG cell. With higher p-type doping concentration, the balance of both carriers contributing to erase operation becomes stronger for hole tunneling contribution.

The data retention characteristic of a p-type floating gate is similar to that of an n-type floating gate [19]. It is explained that the electron traps in both p-type/n-type FG cells are de-trapped from a tunnel oxide in the same manner, but the holes still remained in the hole trap sites without being de-trapped even after high temperature baking, resulting in the same charge loss in both cells, as shown in Fig. 3.71.

As described above, a p-type floating gate has better cycling endurance than does an n-type floating gate. However, doping type of the control gate has not been well discussed yet. In a realistic process in the case of a p-type floating gate, the doping type of the control gate should be p-type, because the floating gate has directly connected to the control gate in the select gate in the SA-STI cell. We should avoid to mixing and canceling out dopants of p-type and n-type. It had been reported that the p-type poly-Si is applied to the control gate [21] in a mid-1X-nm generation cell. The paper [21] pointed out a new problem of read bias sensitivity, caused by a severe control gate depletion. The p-type control gate, which is located on STI (between floating gates), is fully depleted during read due to low doping. The read bias sensitivity can be solved by increasing doping concentration in both floating gate and control gate [21].

REFERENCES

[1] Masuoka, F.; Momodomi, M.; Iwata, Y.; Shirota, R. New ultra high density EPROM and flash EEPROM with NAND structure cell, *Electron Devices Meeting, 1987 International*, vol. 33, pp. 552– 555, 1987.

[2] Aritome, S.; Hatakeyama, I.; Endoh, T.; Yamaguchi, T.; Shuto, S.; Iizuka, H.; Maruyama, T.; Watanabe, H.; Hemink, G.H.; Tanaka, T.; Momodomi, M.; Sakui, K.; and Shirota, R. A 1.13 um^2 memory cell technology for reliable 3.3 V 64 Mb EEPROMs, *1993 International Conference on Solid State Device and Material (SSDM93)*, pp. 446–448, 1993.

[3] Aritome S.; Hatakeyama I.; Endoh T.; Yamaguchi T.; Shuto S.; Iizuka H.; Maruyama T.; Watanabe H.; Hemink G.; Sakui K.; Tanaka T.; Momodomi, M.; and Shirota R. An advanced NAND-structure cell technology for reliable 3.3 V 64 Mb electrically erasable and programmable read only memories (EEPROMs), *Japanese Journal of Applied Physics*, vol. 33, part 1, no. 1B, pp. 524–528, Jan. 1994.

[4] Shimizu, K.; Narita, K.; Watanabe, H.; Kamiya, E.; Takeuchi, Y.; Yaegashi, T.; Aritome, S.; Watanabe, T. A novel high-density $5F^2$ NAND STI cell technology suitable for 256 Mbit and 1 Gbit flash memories, *Electron Devices Meeting, 1997. IEDM '97. Technical Digest, International*, pp. 271–274, 7–10 Dec. 1997.

[5] Takeuchi, Y.; Shimizu, K.; Narita, K.; Kamiya, E.; Yaegashi, T.; Amemiya, K.; Aritome, S. A self-aligned STI process integration for low cost and highly reliable 1 Gbit flash memories, *VLSI Technology, 1998. Digest of Technical Papers. 1998 Symposium on*, pp. 102–103, 9–11 June 1998.

[6] Aritome, S.; Satoh, S.; Maruyama, T.; Watanabe, H.; Shuto, S.; Hemink, G. J.; Shirota, R.; Watanabe, S.; Masuoka, F. A 0.67 μm^2 self-aligned shallow trench isolation cell (SA-STI cell) for 3 V-only 256 Mbit NAND EEPROMs, *Electron Devices Meeting, 1994. IEDM '94. Technical Digest, International*, pp. 61–64, 11–14 Dec. 1994.

[7] Imamiya, K.; Sugiura, Y.; Nakamura, H.; Himeno, T.; Takeuchi, K.; Ikehashi, T.; Kanda, K.; Hosono, K.; Shirota, R.; Aritome, S.; Shimizu, K.; Hatakeyama, K.; Sakui, K. A 130 mm^2 256 Mb NAND flash with shallow trench isolation technology, *Solid-State Circuits Conference, 1999. Digest of Technical Papers. ISSCC. 1999 IEEE International*, pp. 112–113, 1999.

[8] Imamiya, K.; Sugiura, Y.; Nakamura, H.; Himeno, T.; Takeuchi, K.; Ikehashi, T.; Kanda, K.; Hosono, K.; Shirota, R.; Aritome, S.; Shimizu, K.; Hatakeyama, K.; Sakui, K. A 130-mm^2, 256-Mbit NAND flash with shallow trench isolation technology, *Solid-State Circuits, IEEE Journal of*, vol. 34, no. 11, pp. 1536–1543, Nov. 1999.

[9] Aritome, S. Advanced flash memory technology and trends for file storage application, *Electron Devices Meeting, 2000. IEDM Technical Digest. International*, pp. 763–766, 2000.

[10] Goda, A.; Parat, K. Scaling directions for 2D and 3D NAND cells, *Electron Devices Meeting (IEDM), 2012 IEEE International*, pp. 2.1.1, 2.1.4, 10–13 Dec. 2012.

[11] Ramaswamy, N.; Graettinger, T.; Puzzilli, G.; Liu H.; Prall, K.; Gowda, S.; Furnemont, A.; Changhan K.; Parat, K. Engineering a planar NAND cell scalable to 20nm and beyond, *Memory Workshop (IMW), 2013 5th IEEE International*, pp. 5,8, 26–29 May 2013.

[12] Goda, A. Recent progress and future directions in NAND flash scaling, *Non-Volatile Memory Technology Symposium (NVMTS), 2013 13th*, pp. 1,4, 12–14 Aug. 2013.

[13] Aritome, S.; Takeuchi, Y.; Sato, S.; Watanabe, H.; Shimizu, K.; Hemink, G. J.; Shirota, R. A novel side-wall transfer-transistor cell (SWATT cell) for multi-level NAND EEPROM's, in *IEEE IEDM Technical Digest*, pp. 275–278, 1995.

[14] Aritome, S.; Takeuchi, Y.; Sato, S.; Watanabe, I.; Shimizu, K.; Hemink, G.; Shirota, R. A side-wall transfer-transistor cell (SWATT cell) for highly reliable multi-level NAND EEPROMs, *Electron Devices, IEEE Transactions on*, vol. 44, no. 1, pp.145–152, Jan 1997.

[15] Park, K.-T.; Lee, S.C.; Sel, J.-S.; Choi, J.; Kim, K. Scalable wordline shielding scheme using dummy cell beyond 40 nm NAND flash memory for eliminating abnormal disturb of edge memory cell, *SSDM*, pp. 298–299, 2006.

[16] Park, K.-T.; Lee, S.C.; Sel, J.-S.; Choi, J.; Kim, K. Scalable wordline shielding scheme using dummy cell beyond 40 nm NAND flash memory for eliminating abnormal disturb of edge memory cell, *Japanese Journal of Applied Physics*, vol. 46, no. 4B, pp. 2188–2192, 2007.

[17] Kanda, K.; Koyanagi, M.; Yamamura, T.; Hosono, K.; Yoshihara, M.; Miwa, T.; Kato, Y.; Mak, A.; Chan, S.L.; Tsai, F.; Cernea, R.; Le, B.; Makino, E.; Taira, T.; Otake, H.; Kajimura, N.; Fujimura, S.; Takeuchi, Y.; Itoh, M.; Shirakawa, M.; Nakamura, D.; Suzuki, Y.; Okukawa, Y.; Kojima, M.; Yoneya, K.; Arizono, T.; Hisada, T.; Miyamoto, S.; Noguchi, M.; Yaegashi, T.; Higashitani, M.; Ito, F.; Kamei, T.; Hemink, G.; Maruyama, T.; Ino, K.; Ohshima, S. A 120 mm^2 16 Gb 4-MLC NAND flash memory with 43 nm CMOS technology, *Solid-State Circuits Conference, 2008. ISSCC 2008. Digest of Technical Papers. IEEE International*, pp. 430–625, 3–7 Feb. 2008.

[18] Shen, C.; Pu, J.; Li, M.-F.; Cho, J. Byung, P-Type Floating Gate for Retention and P/E Window Improvement of flash memory devices, *Electron Devices, IEEE Transactions on*, vol. 54, no. 8, pp. 1910, 1917, Aug. 2007.

[19] Lee, C.H.; Fayrushin, A.; Hur, S.; Park, Y.; Choi, J.; Choi, J.; Chung, C. Physical modeling and analysis on improved endurance behavior of *p*-type floating gate NAND flash memory, *Memory Workshop (IMW), 2012 4th IEEE International*, pp. 1,4, 20–23 May 2012.

[20] Park, Y.; Lee, J. Device considerations of planar NAND flash memory for extending towards sub-20 nm regime, *Memory Workshop (IMW), 2013 5th IEEE International*, pp. 1,4, 26–29 May 2013.

[21] Seo, J.; Han, K.; Youn, T.; Heo H.-E.; Jang, S.; Kim, J.; Yoo, H.; Hwang, J.; Yang, C.; Lee, H.; Kim, B.; Choi, E.; Noh, K.; Lee, B.; Lee, B.; Chang, H.; Park, S.; Ahn, K.; Lee, S.; Kim, J.; Lee, S. Highly reliable M1X MLC NAND flash memory cell with novel active air-gap and *p*+ poly process integration technologies, *Electron Devices Meeting (IEDM), 2013 IEEE International*, pp. 3.6.1,3.6.4, 9–11 Dec. 2013.

[22] Aritome, S.; Kirisawa, R.; Endoh, T.; Nakayama, R.; Shirota, R.; Sakui, K.; Ohuchi, K.; Masuoka, F. Extended data retention characteristics after more than 10^4 write and erase cycles in EEPROMs, *International Reliability Physics Symposium, 1990. 28th Annual Proceedings*, 1990, pp. 259–264, 1990.

[23] Kirisawa, R.; Aritome, S.; Nakayama, R.; Endoh, T.; Shirota, R.; Masuoka, F.; A NAND structured cell with a new programming technology for highly reliable 5 V-only flash EEPROM, *1990 Symposium on VLSI Technology, 1990. Digest of Technical Papers*, 1990, pp. 129–130, 1990.

[24] Aritome, S.; Shirota, R.; Kirisawa, R.; Endoh, T.; Nakayama, R.; Sakui, K.; Masuoka, F.; A reliable bi-polarity write/erase technology in flash EEPROMs, *International Electron Devices Meeting, 1990. IEDM '90. Technical Digest, 1990*, pp. 111–114, 1990.

[25] Aritome, S.; Shirota, R.; Hemink, G.; Endoh, T.; Masuoka, F.; Reliability issues of flash memory cells, *Proceedings of the IEEE*, vol. 81, no. 5, pp. 776–788, 1993.

[26] Tanaka, T.; Tanaka, Y.; Nakamura, H.; Oodaira, H.; Aritome, S.; Shirota, R.; Masuoka, F. A quick intelligent program architecture for 3 V-only NAND-EEPROMs, *VLSI Circuits, 1992. Digest of Technical Papers, 1992 Symposium on*, pp. 20–21, 4–6 June 1992.

[27] Choi, J.-D.; Lee, J.-H.; Lee, W.-H.; Shin, K.-S.; Yim, Y.-S.; Lee, J.-D.; Shin, Y.-C.; Chang, S.-N.; Park, K.-C.; Park, J.-W.; Hwang, C.-G. A 0.15 μm NAND flash technology with 0.11 μm² cell size for 1 Gbit flash memory," *Electron Devices Meeting, 2000. IEDM '00. Technical Digest. International*, pp. 767,770, 10–13 Dec. 2000.

[28] Arai, F.; Arai, N.; Satoh, S.; Yaegashi, T.; Kamiya, E.; Matsunaga, Y.; Takeuchi, Y.; Kamata, H.; Shimizu, A.; Ohtami, N.; Kai, N.; Takahashi, S.; Moriyama, W.; Kugimiya, K.; Miyazaki, S.; Hirose, T.; Meguro, H.; Hatakeyama, K.; Shimizu, K.; Shirota, R. High-density (4.4F²) NAND flash technology using super-shallow channel profile (SSCP) engineering, *Electron Devices Meeting, 2000. IEDM '00. Technical Digest International*, pp. 775,778, 10–13 Dec. 2000.

[29] Choi, J.-D.; Cho, S.-S.; Yim, Y.-S.; Lee, J.-D.; Kim, H.-S.; Joo, K.-J.; Hur, S.-H.; Im, H.-S.; Kim, J.; Lee, J.-W.; Seo, K.-I.; Kang, M.-S.; Kim, K.-H.; Nam, J.-L.; Park, K.-C.; Lee, M.-Y. Highly manufacturable 1 Gb NAND flash using 0.12 μm process technology, *Electron Devices Meeting, 2001. IEDM '01. Technical Digest International*, pp. 2.1.1,2.1.4, 2–5 Dec. 2001.

[30] Kim, D.-C.; Shin, W.-C.; Lee, J.-D.; Shin, J.-H.; Lee, J.-H.; Hur, S.-H.; Baik, I.-G.; Shin, Y.-C.; Lee, C.-H.; Yoon, J.-S.; Lee, H.-G.; Jo, K.-S.; Choi, S.-W.; You, B.-K.; Choi, J.-H.; Park, D.; Kim, K. A 2 Gb NAND flash memory with 0.044 μm² cell size using 90 nm flash technology, *Electron Devices Meeting, 2002. IEDM '02. International*, pp. 919,922, 8–11 Dec. 2002.

[31] Ichige, M.; Takeuchi, Y.; Sugimae, K.; Sato, A.; Matsui, M.; Kamigaichi, T.; Kutsukake, H.; Ishibashi, Y.; Saito, M.; Mori, S.; Meguro, H.; Miyazaki, S.; Miwa, T.; Takahashi, S.; Iguchi, T.; Kawai, N.; Tamon, S.; Arai, N.; Kamata, H.; Minami, T.; Iizuka, H.; Higashitani, M.; Pham, T.; Hemink, G.; Momodomi, M.; Shirota, R. A novel self-aligned shallow trench isolation cell for 90 nm 4 Gbit NAND flash EEP-ROMs, *VLSI Technology, 2003. Digest of Technical Papers. 2003 Symposium on*, pp. 89,90, 10–12 June 2003.

[32] Noguchi, M.; Yaegashi, T.; Koyama, H.; Morikado, M.; Ishibashi, Y.; Ishibashi, S.; Ino, K.; Sawamura, K.; Aoi, T.; Maruyama, T.; Kajita, A.; Ito, E.; Kishida, M.; Kanda, K.; Hosono, K.; Miyamoto, S.; Ito, F.; Hemink, G.; Higashitani, M.; Mak, A.; Chan, J.; Koyanagi, M.; Ohshima, S.; Shibata, H.; Tsunoda, H.; Tanaka, S. A high-performance multi-level NAND flash memory with 43 nm-node floating-gate technology, *Electron Devices Meeting, 2007. IEDM 2007. IEEE International*, pp. 445, 448, 10–12 Dec. 2007.

[33] Kamigaichi, T.; Arai, F.; Nitsuta, H.; Endo, M.; Nishihara, K.; Murata, T.; Takekida, H.; Izumi, T.; Uchida, K.; Maruyama, T.; Kawabata, I.; Suyama, Y.; Sato, A.; Ueno, K.; Takeshita, H.; Joko, Y.; Watanabe, S.; Liu, Y.; Meguro, H.; Kajita, A.; Ozawa, Y.; Watanabe, T.; Sato, S.; Tomiie, H.; Kanamaru, Y.; Shoji, R.; Lai, C.H.; Nakamichi, M.; Oowada, K.; Ishigaki, T.; Hemink, G.; Dutta, D.; Dong, Y.; Chen, C.; Liang, G.; Higashitani, M.;

Lutze, J. Floating Gate super multi level NAND Flash Memory Technology for 30 nm and beyond, *Electron Devices Meeting, 2008. IEDM 2008. IEEE International*, pp. 1,4, 15–17 Dec. 2008.

[34] Lee, C.-H.; Sung, S.-K.; Jang, D.; Lee, S.; Choi, S.; Kim, J.; Park, S.; Song, M.; Baek, H.-C.; Ahn, E.; Shin, J.; Shin, K.; Min, K.; Cho, S.-S.; Kang, C.-J.; Choi, J.; Kim, K.; Choi, J.-H.; Suh, K.-D.; Jung, T.-S. A highly manufacturable integration technology for 27 nm 2 and 3bit/cell NAND flash memory, *Electron Devices Meeting (IEDM), 2010 IEEE International*, pp. 5.1.1,5.1.4, 6–8 Dec. 2010.

[35] Hwang, J.; Seo, J.; Lee, Y.; Park, S.; Leem, J.; Kim, J.; Hong, T.; Jeong, S.; Lee, K.; Heo, H.; Lee, H.; Jang, P.; Park, K.; Lee, M.; Baik, S.; Kim, J.; Kkang, H.; Jang, M.; Lee, J.; Cho, G.; Lee, J.; Lee, B.; Jang, H.; Park, S.; Kim, J.; Lee, S.; Aritome, S.; Hong, S. and Park, S. A middle-1X nm NAND flash memory cell (M1X-NAND) with highly manufacturable integration technologies, *Electron Devices Meeting (IEDM), 2011 IEEE International*, pp. 199–202, Dec. 2011.

[36] Govoreanu, B.; Brunco, D. P.; Van Houdt, J. Scaling down the interpoly dielectric for next generation flash memory; Challenges and opportunities, *Solid-State Electronics*, vol. 49, pp. 1841–1848, Nov. 2005.

[37] Lee, J. D.; Lee, C. K.; Lee, M. W; Kim, H. S.; Park, K. C.; Lee, W. S. A new programming disturbance phenomenon in NAND flash memory by source/drain hot-electrons generated by GIDL current, *NVSMW*, pp. 31–33, 2006.

[38] Spessot, A.; Monzio Compagnoni, C.; Farina, F.; Calderoni, A.; Spinelli, A. S.; Fantini P. Effect of floating-gate polysilicon depletion on the erase efficiency of nand flash memories, *Electron Device Letters, IEEE*, vol. 31, no. 7, pp. 647, 649, July 2010.

[39] Takeuchi, K.; Kameda, Y.; Fujimura, S.; Otake, H.; Hosono, K.; Shiga, H.; Watanabe, Y.; Futatsuyama, T.; Shindo, Y.; Kojima, M.; Iwai, M.; Shirakawa, M.; Ichige, M.; Hatakeyama, K.; Tanaka, S.; Kamei, T.; Fu, J.Y.; Cernea, A.; Li, Y.; Higashitani, M.; Hemink, G.; Sato, S.; Oowada, K.; Lee S.-C.; Hayashida, N.; Wan, J.; Lutze, J.; Tsao, S.; Mofidi, M.; Sakurai, K.; Tokiwa, N.; Waki, H.; Nozawa, Y.; Kanazawa, K.; Ohshima, S. A 56 nm CMOS 99 mm^2 8Gb Multi-level NAND Flash Memory with 10MB/s Program Throughput, *Solid-State Circuits Conference, 2006. ISSCC 2006. Digest of Technical Papers. IEEE International*, pp. 507–516, 6–9 Feb. 2006.

4

ADVANCED OPERATION FOR MULTILEVEL CELL

4.1 INTRODUCTION

In order to reduce the cost per bit of flash memory, the multilevel memory cell technologies had been intensively developed [1–7], along with reduced memory cell size [8] (see Chapter 3). The multilevel cell technology was initially developed for MLC (2 bits/cell), but it was extended to TLC (3 bits/cell) and QLC (4 bits/cell). The chip size can be reduced to about 60% by using an MLC (2 bits/cell) scheme, compared with a single-level cell SLC (1 bit/cell) scheme. However, in a multilevel memory cell, a narrow threshold voltage (V_t) distribution width is necessary to have a enough margin between V_t distributions. Due to this narrow V_t distribution width, the programming time of the multilevel cell becomes longer than that of a conventional SLC. Also, reliability of the multilevel cell is worse than that of SLC due to less V_t margin (read V_t window margin). To avoid these problems, it is very important that the V_t distribution width be controlled to be as narrow as possible.

The memory cell structure and fabrication process of the multilevel cells are basically the same as that of SLC. Therefore, multilevel cell technology has been developed to focus on the operations of making narrow V_t distribution width. A lot of sophisticated techniques have been proposed and implemented to a NAND flash memory product [9]. Section 4.2 describes these techniques, such as the incremental step pulse program (ISPP), bit-by-bit verify operations, a two-step verify scheme, and a pseudo-pass scheme.

Nand Flash Memory Technologies, First Edition. Seiichi Aritome.
© 2016 The Institute of Electrical and Electronics Engineers, Inc. Published 2016 by John Wiley & Sons, Inc.

Even if narrow V_t distribution width is made during page programming, the V_t distribution is disturbed and is getting wider after programming neighbor cells due to the floating-gate capacitive coupling interference (cell-to-cell interference), and so on. Section 4.3 describes several page program sequences to reduce the effect of floating-gate capacitive coupling.

TLC (3 bits/cell) and QLC (4 bits/cell) technologies are described in Section 4.4 and Section 4.5, respectively. The three-level cell technology is introduced in Section 4.6 to compromise the performance and reliability of SLC and MLC.

Finally, in Section 4.7, the moving read algorithm is presented to compensate a V_t shift for minimizing a bit failure rate.

4.2 PROGRAM OPERATION FOR TIGHT V_t DISTRIBUTION WIDTH

4.2.1 Cell V_t Setting

Figure 4.1 shows the image of threshold voltage (V_t) setting for one program state. In order to avoid failure, V_t distribution width has to be tight enough, and a tail of distribution has to have enough margins from read voltage. However, by scaling memory cell size, V_t distribution width becomes much wider by several physical mechanisms, such as floating-gate capacitive coupling (FGC) interference, random telegraph signal noise (RTN), program electron injection spread (EIS), back pattern dependence, and so on, as shown in Fig. 4.1 (see Chapter 5 for details). The operation margins have been decreased as scaling memory cell, because each physical phenomenon becomes worse as scaling.

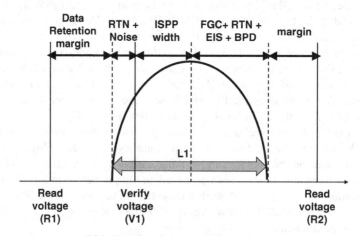

FGC: Floating-Gate capacitive Coupling
RTN: Random Telegraph sognal Noise
EIS: Electron Injection Spread
BPD: Back Pattern Dependence

FIGURE 4.1 The image of threshold voltage (V_t) setting for one program state.

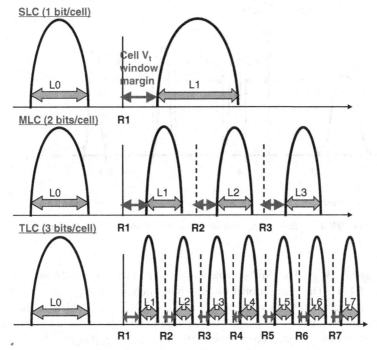

FIGURE 4.2 V_t distribution image of SLC (1 bit/cell), MLC (2 bits/cell), and TLC (3 bits/cell).

Figure 4.2 shows a V_t distribution image of SLC (1 bit/cell), MLC (2 bits/cell), and TLC (3 bits/cell). SLC has a wider cell V_t window margin, then SLC has a better reliability performance and also a better program and read performance than MLC and TLC. MLC and TLC have a very narrow margin to manage good enough reliability. To obtain a wider margin, it is important to make a tight V_t distribution width. Figure 4.3 shows one example of MLC threshold voltage (V_{th}) distributions of the four cell states [5]. Erase "11" cells are sufficiently "deep," and erase V_{th} distribution width is not needed to be controlled as tightly as the three program states. Each program state has a 0.4-V V_{th} distribution width and a 0.8-V margin separating them. The measured V_{th} distribution in a 0.4-μm cell is shown in Fig. 4.4. [5], which demonstrates that the V_{th} optimization results in a relatively tight 0.4 V V_{th} distribution width per state at normal operating condition.

4.2.2 Incremental Step Pulse Program (ISPP)

In order to make the tight programmed V_t distribution width, an incremental step pulse program (ISPP) scheme had been proposed [10, 11].

An incremental step pulse program (ISPP) scheme [10] (step-up programming scheme [11]) is shown in Fig. 2.13 (Chapter 2). The program pulses (V_{pgm}) are

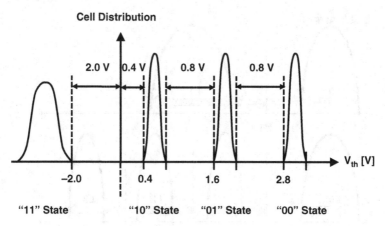

FIGURE 4.3   Target threshold voltage distribution of four states for MLC.

stepped up by ΔV_{pgm}. The ISPP scheme was compared with other schemes, as shown in Fig. 4.5 [11]. The conventional programming pulse (Fig. 4.5a) is the repeating pulses of the same program voltage V_{pp} ($= V_{pgm}$). There is a problem of increasing program time because many pulses are required to complete a page programming. On the other hand, in step-up program pulses (Fig. 4.5c), program speed çan be drastically improved because the slow cells in page can be programmed by higher V_{pp}, and then page programming can be completed by the small number of program pulses.

In the ISPP scheme, tight programmed V_t distribution width can be obtained by using narrower step ΔV_{pp} ($= \Delta V_{pgm}$) without increasing the number of program pulses, as shown in Fig. 4.6 [11]. Furthermore, the ISPP scheme has another important advantage. During programming pulse, the electric field in tunnel oxide can be reduced by the lower starting V_{pp} in comparison with the conventional

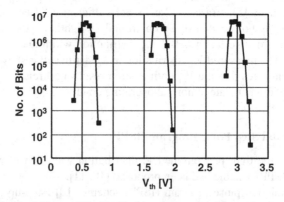

FIGURE 4.4   Measured V_{th} distribution of three programmed states for MLC.

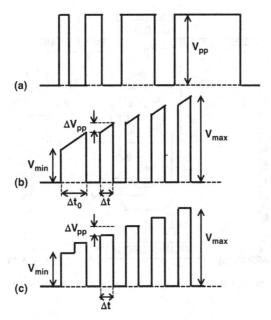

FIGURE 4.5 Program pulse waveforms: (a) conventional, (b) trapezoidal, and (c) staircase (incremental step pulse programming (ISPP)). A verify step is carried out after each pulse.

programming pulse. The degradation of tunnel oxide can be suppressed, and then the reliabilities of program/erase cycling, data retention, and read disturb can be greatly improved [12].

The ISPP scheme has been used in NAND flash memory products for a long time, more than 20 years, due to fast programming speed, tight V_t distribution, and excellent reliability.

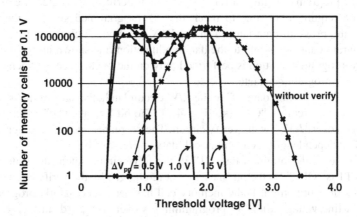

FIGURE 4.6 V_t distribution after programming in a 16-Mbit memory array, with/without verify, using staircase pulses (ISPP) with length of 20 μs and V_{pgm} step of 0.5, 1.0, 1.5 V.

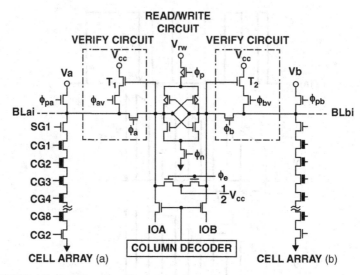

FIGURE 4.7   Intelligent verify circuit schematic for the bit-by-bit verify operation.

4.2.3   Bit-by-Bit Verify Operations

Another important and basic operation technique to make tight V_t distribution is the bit-by-bit verify operation.

An intelligent quick bit-by-bit verify circuit was proposed [13, 14] to realize fast page programming speed as well as tight programmed V_t distribution width. The new verify circuit is composed of adding only two transistors (T1, T2) to a conventional circuit, as shown in Fig. 4.7 [13, 14]. The program/verify operation could be much simplified in comparison with the conventional chip external verify operation. Detail operations are described in the following.

After the program operation, a verify operation is performed to detect the memory cells which require more time to reach the "1" programmed state. In the verify operation, the program data latched in the R/W (read/write) circuit is modified to the re-program data, according to the data modification rule shown in Fig. 4.8. As a result, a re-program operation is performed only on the memory cells which did not reach the "1"-programmed state.

In case of the program data "0" in the R/W circuit latch, the state of the transistor T1 in the verify circuit is "ON" (see Fig. 4.7). The bit line after "0"-programming is re-charged over $^1/_2V_{cc}$ by the verify circuit. Therefore, the latched re-programmed data is "0" independent of the memory cell data in Fig. 4.8a,b.

In the case of the program data "1" in the R/W circuit latch, the state of the transistor T1 is "OFF." So the bit lines are not re-charged by the verify circuits even if the clock ϕ_{av} turns high. If the memory cell has been successfully programmed "1," the bit-line voltage after "1"-programming is over $^1/_2V_{cc}$ ((d) in Fig. 4.8). On the other hand, if the memory cell does not reach the "1"-programmed state, the bit-line voltage decreases below $^1/_2V_{cc}$ (in Fig. 4.8c). The latched re-program data is

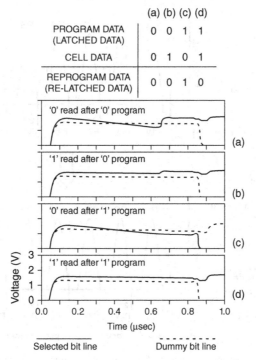

	(a) (b) (c) (d)
PROGRAM DATA (LATCHED DATA)	0 0 1 1
CELL DATA	0 1 0 1
REPROGRAM DATA (RE-LATCHED DATA)	0 0 1 0

FIGURE 4.8   Data modification rule and simulated waveform for the bit-by-bit verify operation.

"0" for the memory cell which is in the "1"-programmed state (in Fig. 4.8d). The re-programmed data is "1" for the memory cell which did not reach the "1"-programmed state yet (in Fig. 4.8c).

By using the verify circuit, the program data is automatically and simultaneously modified to the re-program data according to Fig. 4.8.

The programmed V_t distribution could be tight with quick verify operation, and programming speed became fast due to chip internal verify operation, which replaced conventional chip external verify operation.

4.2.4   Two-Step Verify Scheme

To achieve a tight programmed V_{th} distribution width, it is important to control the cell V_{th} movement during ISPP program operation. The two-step verify scheme in program verify read operation is widely used [15] for a multilevel cell (MLC, TLC, QLC) to control V_t movement, as shown in Fig. 4.9a–c [16] and Fig. 4.10 [15]. In the two-step verify scheme, two times verify read operations are performed for each program level of P1–P7 in TLC (3 bits/cell), as shown in Fig. 4.9c. For example, for the program level of P1, two times verify read of first P1V and second P1V are performed (in Fig. 4.10, First step write verify voltage and Second step write verify

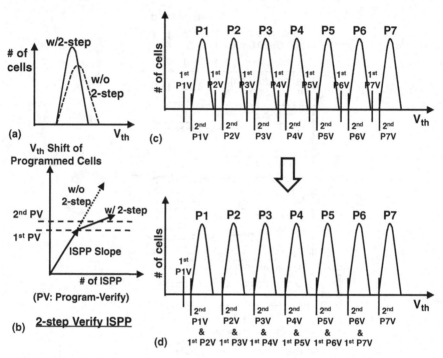

FIGURE 4.9 (a) V_t distribution image of a "with two-step verify" scheme and a "without two-step verify" scheme. (b) V_t shift of program cell in two-step verify scheme. (c) Two-step verify scheme. (d) Verify-skip two-step verify scheme.

voltage). The second P1V is the target verify voltage, and the first P1V is slightly lower than the target verify level. For cells that have V_{th} < first P1V, 0 V is applied to the bit line (BL) during the next program pulse to program (make V_{th} shift) normally. For cells that have V_{th} > second P1V, V_{cc} is applied to the bit line (BL) during the next program pulse to be the inhibit condition. For cells that have first P1V < V_{th} < second P1V, a predetermined low voltage of V_{fbl} (= 0.4 V, for example, in Fig. 4.10) is applied to the bit line (BL) during the next program pulse to make the smaller V_{th} shift than ISPP step voltage, as shown in Fig. 4.9b and Fig. 4.10. Due to the smaller V_t shift for the cells of just below target verify voltage (first P1V < V_{th} < second P1V), the programmed V_{th} distribution width of the two-step verify scheme can be tighter than that of the conventional verify scheme.

The two-step verify scheme, however, requires two times more verify operations for each target V_{th} state, causing an increase in program time. This is especially exaggerated for 3 bits/cell (TLC) NAND, where over 66% of total program time is spent in the verify operation. In order to reduce the extra verify overhead time, a verify-skip two-step tunneling ISPP scheme was proposed [16], as shown in Fig. 4.9d and Fig. 4.11. The verify-skip two-step tunneling ISPP scheme uses the second verify level of the previous target state as the first step verify for next target state. To obtain

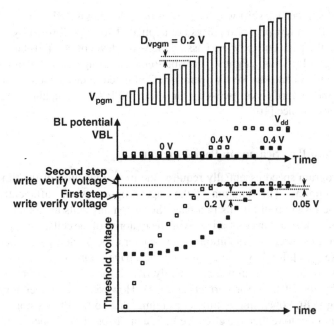

FIGURE 4.10   The two-step verify scheme of program waveform V_{pgm}, bit-line voltage V_{bl}, and V_{th} movement.

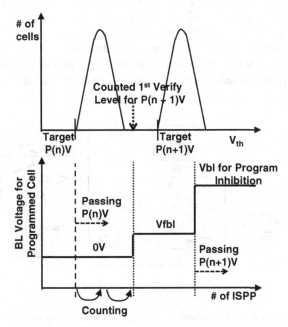

FIGURE 4.11   Forcing BL voltage by latch counting.

the effect of smaller V_{th} shift with V_{fbl}, the time of forcing the BL voltage should be delayed to after a few program pulses are applied. This is performed by a counting data latch in the page buffer, when passing the verify level of previous target state, as shown in Fig. 4.11. Thus, a tight V_{th} distribution width using a two-step tunneling rate is realized without an extra verify operation. Compared to the conventional two-step verify scheme, a verify-skip two-step tunneling ISPP scheme achieves 13% better program performance [16].

4.2.5   Pseudo-Pass Scheme in Page Program

The fast program speed essentially requires the more reduction of the time for one page programming. The duration of the page program is set sufficiently long to complete the program of all bits in a page. So, when any cells have unusually slow program characteristics in comparison with the majority of the cells, the page program speed becomes slower. As a solution to this problem, the pseudo-pass scheme (PPS) was proposed [17]. It allows the completion of a page programming operation even if a few bits are not programmed sufficiently. The error bits are corrected in a read operation by the ECC (error correction code). However, the conventional failure bit counting (FBC) operation is time-consuming, and so the PPS is not sufficiently effective. In order to realize the effective PPS, a high-speed FBC operation had been also proposed [17].

Figure 4.12a shows the flowchart of the conventional page program sequence. At first, memory cells in page are programmed according to loaded data. Then, they are verified consecutively. If all the cells, which should be programmed, are programmed, the program operation finishes and becomes a status pass. However, if they are not completed to program, the memory cells are programmed again. The judgment of "all cells programmed or not" is done by using the data that are stored in page buffer, as shown in Fig. 4.13. When the data in the buffer is "1," the cell that corresponds to the buffer is not programmed, however, when the data is "0," the cell is programmed

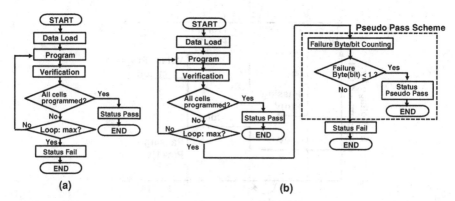

FIGURE 4.12   (a) Flowchart of a conventional page program sequence. (b) Flowchart of a page program sequence with the pseudo-pass scheme (PPS).

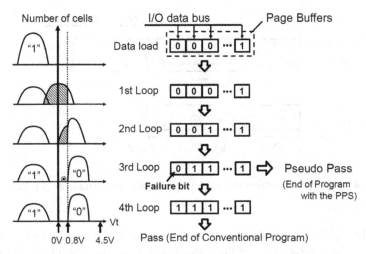

FIGURE 4.13 Change of V_t of memory cells and data in page buffers during a program sequence.

repeatedly. The data of the page buffers are revised in each verify operation. When the threshold voltage (V_t) of the programmed cell shifts up from the negative voltage of the erased status to more than the target value of 0.8 V, the data of the buffer is changed from "0" to "1" after the verify.

Figure 4.12b presents the flowchart of the pseudo-pass scheme (PPS). The PPS can be implemented just after the conventional program sequence. If the program operation doesn't complete after the predetermined iteration number of the program loops, the failure bit counting (FBC) circuit counts up the number of the page buffers whose data are "0." If the detected number of failure bits is less or equal to the allowed value, the status of "the pseudo pass" is output, and then the program sequence terminates. In Fig. 4.13, the predetermined iteration number of the program loops, which is enough programming for the majority of the cells, is assumed to be three. In this case, the program operation is finished with operation of the pseudo pass, as a result of the FBC after the third program loop without retrying an additional program loop, even though some insufficiently programmed cells are remained. Therefore, the iteration number of the program loops can be reduced by one or more in comparison with the conventional verify method, which doesn't permit any insufficient program bits.

Figure 4.14 compares the SLC program performance between the conventional program and the PPS program operation. The horizontal axis shows the worst program time. When the typical program time (tProg_typical) of the majority of cells is assumed to be 200 μs, there is a possibility that the worst program time of the conventional program becomes 250 μs or more because one or more program/verify sequence is required. However, the worst program time of the PPS operation using the new high-speed failure bit counter circuit [17] is limited to 200.8 μs, which is the sum of the typical program time of 200 μs and the counting-up time of 0.8 μs

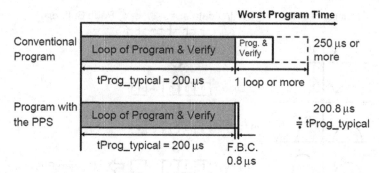

FIGURE 4.14 Program performance of a conventional page program and a page program with PPS.

in the FBC operation. In this case, the improvement of the worst program time is at least 20% in comparison with the conventional worst program time, by reducing the number of program loops for a few slowly programmed cells. The additional time of 0.8 μs for the PPS operation is negligibly small in comparison with the total program time.

The pseudo-pass scheme (PPS) has been implemented to the NAND flash product of SLC/MLC/TLC over 10 years due to fast page programming speed and less program disturb failure by avoiding excess program stress.

4.3 PAGE PROGRAM SEQUENCE

4.3.1 Original Page Program Scheme

In order to realize multi-bit cells (MLC) in scaled NAND flash memory cells, precise V_{th} distribution control is the key factor. The V_t distribution in a program state can be very tight by an ISPP and a bit-by-bit verify scheme. However, the distribution is eventually disturbed by well-known major parasitic effects, which are the background pattern dependency (BPD), source line noise (noise), and floating gate capacitive coupling interference (cell-to-cell interference), as shown in Fig. 4.15 [18, 19]. The background pattern dependency (BPD) can be minimized by various techniques such as fixed page program order and applied proper read voltage for unselected cells in a selected NAND string. And the source line noise can be also minimized by low resistance of mesh common source lines, as well as by low resistance of p-well structures of the memory array, such as retrograded doping profiled p-well. However, the cell-to-cell interference is mainly caused by floating gate capacitive coupling due to parasitic capacitances between cells, thus it is greatly affected by cell scaling (see Chapter 5). Figure 4.15b shows typical contributions of the three mentioned parasitic effects measured at a device at the 60-nm technology node [18–19]. Actually, the detailed portions of each effect can be different depending on the used NAND device structure and its operation condition, however, floating-gate

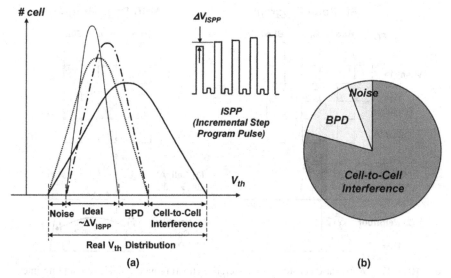

FIGURE 4.15   (a) Parasitic effects of a V_{th} distribution in NAND flash memory. (b) Contributed portion of each parasitic effect measured at a 60-nm technology node.

capacitive coupling interference is the most dominant effect, and will increase dramatically as scaling down NAND flash memory cells.

Figure 4.16a shows the memory cell array core architecture and page assignment of a conventional NAND flash memory device (original MLC product) [6, 7][18, 19]. Two BLs (bit lines) of even BL (BLe) and odd BL (BLo) are connected to sense amplifier (not shown in figure) through a switch. Either an even or odd BL cell is alternately selected and programmed sequentially in the order as described in Figure 4.16a. This BL scheme is called even/odd shield bit-line architecture [14, 21]. This scheme is effective to reduce BL noise shielding in read and program-verify

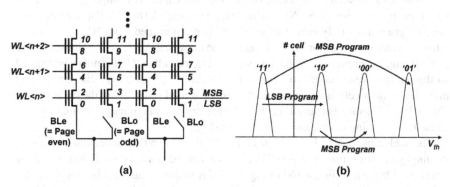

FIGURE 4.16   (a) Conventional core architecture and page assignment. (b) Conventional MLC program scheme. MSB; Most Significant Bit, and LSB; Least Significant Bit.

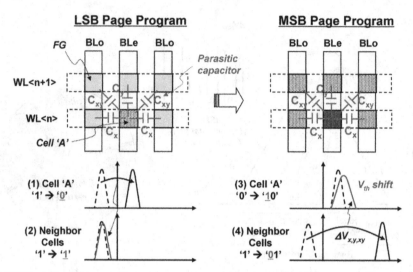

FIGURE 4.17   Worst-case cell-to-cell interference (floating-gate capacitive coupling interference) of conventional NAND architecture.

operations. A conventional MLC program scheme used in an original MLC NAND flash is shown in Fig. 4.16b. During the LSB program, V_{th} states of selected cells which have the erased V_{th} as the initial state move to the lowest programmed state '10'. Next, during the MSB program stage, two states, "00" and "01", are formed sequentially, depending on previous LSB data. After finishing the programming of four pages corresponding to a word line (WL$<n>$), the four pages corresponding to the next upper word line (WL$<n + 1>$) are programmed consecutively. It is noted that logical even and odd pages on the same word line are matched to physical even and odd BLs in the conventional architecture. The original MLC NAND architecture and page programming scheme shown in Fig. 4.16 was used in the first MLC NAND product of 0.16-μm 512-Mbit NAND flash memory in 2000.

Figure 4.17 shows the worst case of floating-gate capacitive coupling interference which occurs in original NAND architecture (see Fig. 4.16) [18, 19]. During LSB page programming, only selected the cell 'A' is programmed from '1' to '0', but all other surrounding neighbor cells are kept in the erase state ('1' → '1'). Subsequently at MSB page programming, if the data for the selected cell is '1', it is not programmed so that its state remains at '10'. Next, if the data for all neighbor cells are '0' and then all neighbor cells are programmed from the erased state '11' to the highest state '01', a large V_{th} shift is caused for the selected cell 'A' due to parasitic floating gate capacitive coupling interference, as shown in Fig. 4.17.

The widening of the distribution of the original NAND architecture caused by floating-gate capacitive coupling interference can be approximately expressed by equation "Original in Fig. 4.16" in Fig. 4.18. From the equation, it is found that not only reducing the parasitic capacitances, but also reducing the number of neighbor cells that are programmed after the programming of a selected cell and the amount

	Equation of floating-gate capacitive coupling
Original in Fig. 4.16	$\Delta V_x * (2C_x/C_{tot})$ $+ \Delta V_y * (C_y/C_{tot})$ $+ \Delta V_{xy} * (2C_{xy}/C_{tot})$
New Scheme (1) in Fig. 4.19	$(\Delta V_x/2) * (2C_x/C_{tot}) + (\Delta V_y/2) * (C_y/C_{tot})$ $+ (\Delta V_{xy}/2) * (2C_{xy}/C_{tot})$
New Scheme (2) in Fig. 4.23	$(\Delta V_y/2) * (C_y/C_{tot})$ $+ (\Delta V_{xy}/2) * (2C_{xy}/C_{tot})$

FIGURE 4.18 Approximated equations of floating-gate capacitive coupling interference in three page program schemes.

of shift at the MSB programming stage, is important to minimize the floating-gate capacitive coupling interference in a NAND flash memory cell.

4.3.2 New Page Program Scheme (1)

Figure 4.19 shows a new page program scheme (1) of new memory cell array core architecture and page assignment [22, 18, 19]. This scheme has been widely used in massproduction due to reducing V_{th} distribution width by decreasing an effect of the floating gate capacitive coupling interference. The floating-gate capacitive coupling interference by BL–BL direction (x-direction) can be reduced by performing the LSB program to the temporary state 'x0'. And the floating-gate capacitive coupling interference by WL–WL (y-direction) and diagonal neighbor cells can be reduced by performing MSB programming for a selected WL after LSB programming of its neighbor WL cell, as shown in Fig. 4.19a. The V_{th} shift by WL–WL and diagonal

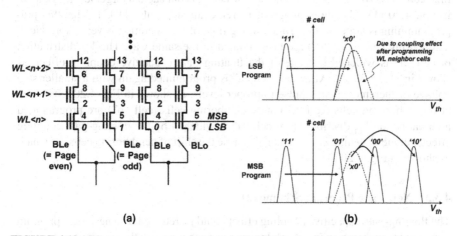

(a)   (b)

FIGURE 4.19 The new page program scheme (1). MLC program is performed after temporary LSB data storing. (a) New core architecture and page assignment. (b) New MLC program scheme.

interference can be almost reduced to half compared to the original program scheme of Fig. 4.16. With the new page program scheme (1), the achieved floating-gate capacitive coupling interference can be expressed as in "New Scheme (1) in Fig. 4.19" in Fig. 4.18. The new page program scheme (1) shown in Fig. 4.19 was first applied to a 70-nm 8-Gbit MLC NAND flash memory product in 2005 [23].

By using this new page program scheme (1), the worst case of the floating-gate capacitive coupling interference can be improved, compared with the original MLC program scheme. A new program scheme with temporary LSB data storing is used, as shown in Fig. 4.19b. At the LSB programming stage, the memory cell is programmed from "11" to "x0" as a temporary state just like SLC programming. After the WL neighbor cells are also LSB programmed, the V_{th} distribution is possibly widened as shown in Fig. 4.19b, uppergraph. Then, at the MSB programming stage, the 'x0' state is programmed to either '00' and '01' as the final state corresponding to the input data or either the '11' state is programmed to the final "01" state. All memory cells except '11' cells are programmed to their final states at the MSB programming stage from the temporary programmed state for LSB data. The V_{th} shift of neighbor cells is greatly reduced to around half in comparison with conventional page programming scheme shown in Fig. 4.16, so that the floating-gate capacitive coupling interference of neighbor cells can be greatly reduced. During MSB programming in this new page program scheme (1), a flag cell that is used for representing MSB programming and placed for each page is also programmed in order to distinguish LSB and MSB for read.

Other reports [24–26] also introduced the new scheme that reduced WL–WL interference by using programming to a temporary state. This programming scheme is that neighboring cells are roughly programmed before final programmed levels are programmed properly. Figure 4.20 shows the transient of a V_{th} distribution of cell "a" and neighboring cells, that is, cell "b". Figure 4.21 shows the programming order. At first, cell "a" is roughly programmed to lower levels than actual target level, as shown in Fig. 4.20 (1). The step voltage of incremental step pulse [10, 11] for this pre-programming is large, so the programming time of the operation is very short. Next, neighboring cells (cell "b") are programmed in the same way. The V_{th} distribution of cell "a" is widened because of the floating gate capacitive coupling effect, as shown in Fig. 4.20 (2). After this, cell "a" is programmed again with a smaller step voltage of incremental step pulse to proper levels, as shown in Fig. 4.20 (3). When next-neighboring cells (cell "c") and neighboring cells (cell "b") are programmed afterwards, the V_{th} distribution of cell "a" is widened by the floating-gate coupling effect, but the widening is very small, because the shift of neighboring cells are small, as shown in Fig. 4.20 (4), (5).

4.3.3   New Page Program Scheme (2)

The floating-gate capacitive coupling effect could be reduced by a new page program scheme (1), as shown in Fig. 4.19. In order to further reduce the floating-gate capacitive coupling interference between BLs (x-direction), the way to program adjacent cells (in both even page and odd page) at the same program pulse (sequence) is

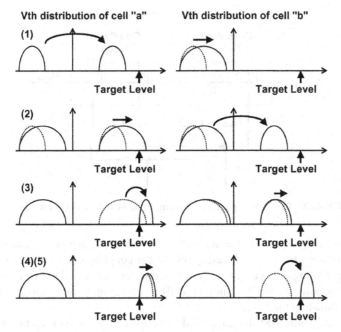

FIGURE 4.20 V_t distribution transition of cell "a" and cell "b."

effective [18, 19, 27, 28]. The concept of the new architecture is simply to reduce the number of neighbor cells as well as their amount of V_t shift at the MSB program stage. Figure 4.22 [18, 19] shows the concept of page assignment of the new page program scheme (2) to reduce the number of neighbor cells between BLs. In this new scheme (2), the logical even and odd pages on the same word line are each assigned to a physical group of memory BLs. By adopting this architecture, the same page address is assigned to adjacent memory cells on the same word line of a selected group of memory cells, which means that memory cells including adjacent cells in the BL direction can be programmed simultaneously. Accordingly, while logical odd

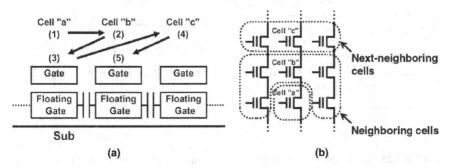

FIGURE 4.21   Programming order over word lines.

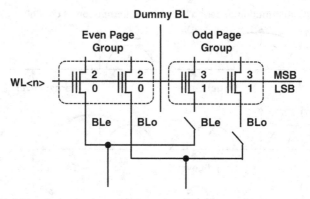

FIGURE 4.22   Concept of page assignment of the new page program scheme (2).

page data are programmed in the memory cells of the odd page group, memory cells in the even page group are not susceptible to the coupling effect at all. In order to remove the floating-gate capacitive coupling interference for edge memory cells of the page group, a dummy BL (dummy cell) can be simply used between the page groups, as shown in the figure.

Figure 4.23 shows (a) the simplified memory cell array core architecture and (b) page address ordering of the new page program scheme (2). Two BL selectors coupling to each even and odd page group are configured to transfer even and odd page data from the page buffer. The boundary between even and odd page groups is simply formed using dummy BL in order to remove the floating-gate capacitive coupling interference for edge memory cells of the page group. It should be noted that no additional area penalty arises in the proposed memory array. This is because the dummy BL which already exists for contacting the CSL (common source line) or wells in the conventional memory array can be used as the dummy BL for a page

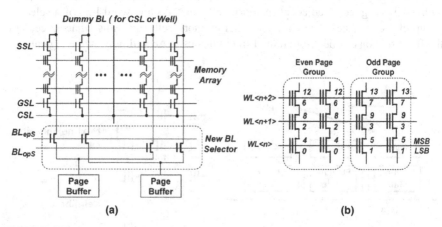

FIGURE 4.23   (a) Simplified core architecture and (b) page address ordering of the new page program scheme (2).

group boundary. With the new architecture, the floating-gate capacitive coupling interference can be much reduced as approximately expressed in equation "New Scheme (2) in Fig. 4.23" in Fig. 4.18.

4.3.4   All-Bit-Line (ABL) Architecture

The all-bit-line (ABL) architecture was firstly proposed in ISSCC 2008 [27, 28]. In ABL architecture, all cells along a selected word line are programmed simultaneously, not separated to even or odd page groups. Then, the ABL architecture could reduce the floating-gate capacitive coupling interference due to reducing BL–BL interference, and also the ABL architecture could realize high-speed page programming with double page size.

Before ABL architecture was proposed, the conventional even/odd shield bit-line scheme was used for a NAND flash product. In the NAND cell array structure, the BL–BL coupling capacitance (not floating-gate coupling capacitance) is around 90% of the total bit-line capacitance [23]. For this reason, most NAND flash products utilize a conventional shielded bit-line scheme to do sensing [13, 21], where only half of the BLs are sensed and the other half of the BLs are at 0 V. Since only half of the cells on the same WL can be sensed at the same time, the data latches are designed to be shared by even/odd pairs, to save on die size. This architecture requires the separate programming and verification of even and odd BLs. The program speed was limited by small page size of even and odd BLs. Moreover, as memory cell size scales down, the program disturb issue was more aggravated due to a longer programming time on the same WL, and then reliability was degraded.

Figure 4.24 shows a schematic diagram of memory core circuits of ABL architecture [27, 28]. The even and odd bit lines (namely, all bit lines) have their own sense

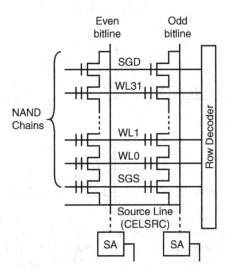

FIGURE 4.24   Simplified memory core architecture of all-bit-line (ABL) architecture.

amplifier (SA) attached. The sensing scheme was changed from voltage sensing in even/odd shield bit-line scheme to current sensing in the ABL scheme, to solve the large BL–BL coupling capacitance issues. The SA performs sensing operations in read, program verify, and erase verify operations. In ABL architecture, all bits on the same word line (WL) can be programmed and read at the same time. The total number of cells that can be programmed are double that of conventional even/odd bit-line architecture.

Therefore, the ABL architecture could lead to high performance of programming with double page size, accompanied by high reliability with shorter program disturb stress time. Also, in ABL architecture, the floating-gate capacitive coupling interference can be reduced, compared with a conventional even/odd shield bit-line scheme. Figure 4.25a,b shows the floating-gate capacitive coupling interference in a conventional even/odd shield bit-line scheme. The cells in an even bit line (even page) causes a V_t shift by programming neighbor cells in odd bit line (odd page) due to the floating-gate capacitive coupling interference, as shown in Fig. 4.25a,b. On the other hand, in an ABL scheme, all cells in even page and odd page are programmed at the same time. The V_t shift of the floating-gate capacitive coupling interference can be much reduced [27]. While the erase distribution can still encounter the full

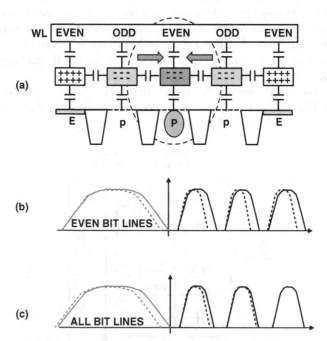

FIGURE 4.25 (a) Floating-gate capacitive coupling interference for an even page in a conventional even/odd programming scheme. (b) V_t distribution image of an even page in an even/odd programming scheme. (c) V_t distribution image in an ABL programming scheme. Floating-gate capacitive coupling interference in an all-bit-line programming scheme can be reduced, because both even page and odd page are programmed simultaneously.

the floating-gate capacitive coupling interference, the interference effect on the first programming state is reduced. The highest state has almost no V_t shift of interference, as shown in Fig. 4.25c [27].

4.4   TLC (3 BITS/CELL)

In order to decrease a bit cost of NAND flash memory, TLC (3 bits/cell) technologies had been developed [29–36, 16]. The first paper for massproduction of TLC was presented in 2008 ISSCC (International Solid-State Circuits Conference) by using 56-nm technology [30]. The key issue of TLC technologies is the page program sequence and method to achieve a very tight V_t distribution width for producing a margin between each V_t state.

The page address is assigned in a way that enables each page to be treated as an independent page for users. The same user commands can be used for all pages programmed in the array. The conventional page program sequence is shown in Fig. 4.26 [24, 25, 35, 29–31]. This sequence is implemented as applying the concept of a new program scheme (2) for MLC (Section 4.3.3) to TLC, in order to minimize an effect of the floating-gate capacitive coupling interference. The three pages on the same WL are called lower page (first), middle page (second), and upper page (third), respectively. The lower page (first) is programmed like a normal SLC program operation, where the erase cell "E" is programmed to state "A1". After the lower page programming, the middle page (second) program data can be brought in for programming. For the middle page programming, 2 bits (both lower and middle pages) are needed to program 3 program states. The lower page data can be read from a memory array into the data latch. The middle page is programmed similarly to that of 2-bit-per-cell programming that is from "E" to "B1" and from "A1" to "B2" or "B2". After the middle page programming, the upper page (Third) program data is brought in from outside. The upper page program requires the lower and middle page

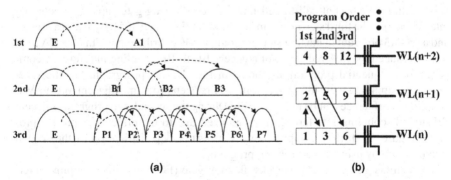

(a)   (b)

FIGURE 4.26   (a) The conventional page program sequence of TLC (3 bits/cell). Three pages of first (lower page), second (middle page), and third (upper page) are programmed on each word line. (b) Program order between word lines.

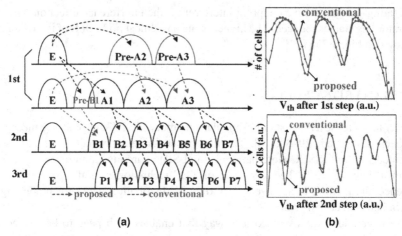

FIGURE 4.27   (a) The new page program sequence with pre-program scheme. (b) Measured V_t distribution after first (lower page) and second (middle page) program.

data to program seven program states. The lower and middle page information can be read from an array. The seven program states must fit into a V_t window similar to a MLC program V_t window, so the upper page program should have a very small V_{PGM} step size to achieve a well-controlled, narrow V_t distribution for all seven program states. Therefore, the upper page program is the slowest programming speed of the three pages. Figure 4.26b shows a page program sequence between world lines. This sequence is also implemented as applying the concept of new program scheme for MLC (Section 4.3.3) to TLC, in order to minimize an effect of the floating-gate capacitive coupling interference.

A new page program sequence with a pre-program scheme had been proposed for a 21-nm node cell [35], as shown in Fig. 4.27. In a new scheme, 5 states and 8 states are implemented in the first and second step program, respectively, so that it is minimized adjacent cell-to-cell interference (floating-gate capacitive coupling interference) at the third step program, as shown in Fig. 4.27a [35]. By using a pre-A2 and a pre-A3 program in the first step program, the V_t distribution width of A1/A2/A3 can be reduced by 15% reduction of adjacent BL-to-BL coupling interference compared to a sequential program, as shown in Fig. 4.27b, uppergraph. For the same reason, by applying a pre-B1 program in the first step program, adjacent WL-to-WL coupling interference can be reduced 10%. During the second step program, adjacent WL-to-WL coupling interference is minimized due to the effect of a pre-B1 program. Figure 4.27b (lowergraph) shows a measured V_{th} distribution of the second step program by using a pre-program program.

As memory cell geometry shrinks, floating-gate (FG) capacitive coupling interference is becoming worse. And smaller memory cells are also vulnerable to more cell-to-cell variations. These factors combine to negatively impact program performance. The FG capacitive coupling interference can be reduced by the air gap (AG)

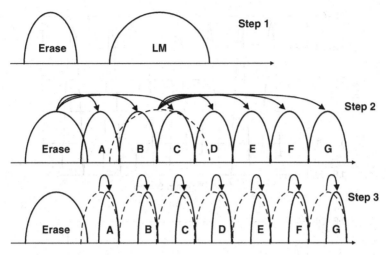

FIGURE 4.28   Three-step program algorithm for a 19-nm node cell.

process (see Section 5.3.4) between two adjacent WLs. The 19-nm AG technology has an FG-to-FG coupling ratio equivalent to that of 2X nm without AG. Also, for enhancing program performance in the 19-nm technology node, a new enhanced Three-Step Program (TSP) was applied to TLC 128-Gb NAND flash memory, as shown in Fig. 4.28 [36]. In a new enhanced TSP, cells are programmed from two states of Erase/LM to eight states of Erase/A-G, and then compaction program is performed to states of A–G (Step 3 in Fig. 4.28). Due to the skip of second page programming in the conventional program scheme in Fig. 4.26, program speed is enhanced. Therefore, a combination of a new enhanced Three-Step Program (TSP) and air gap allows fast program speed of 18 MB/s on TLC of the 19-nm technology node.

Figure 4.29 shows the measured V_t distributions of TLC in several generations of NAND flash memory cells, which are (a) 56-nm cell [29, 30], (b) 32-nm cell [31], (c) 20-nm-node cell (27-nm) [16], (d) 21-nm cell [35], and (e) 19-nm cell [36]. We can see that the read window margin is gradually degraded as scaling of memory cells, even if program operation is newly developed for each generation.

4.5   QLC (4 BITS/CELL)

QLC technology was presented in a 70-nm cell in the 2007 Symposium on VLSI Circuits [24] and 43-nm cell on 2009 ISSCC [37]. The papers were focused on the intelligent operation of achieving a tight V_t distribution width by reducing the floating-gate capacitive coupling interference. Figure 4.30 shows the V_{th} distribution transition of 16LC (16-level cell) case [24, 25]. At first, the cells are roughly programmed to lower levels (lower verify voltages) than the target levels (target verify voltages), as

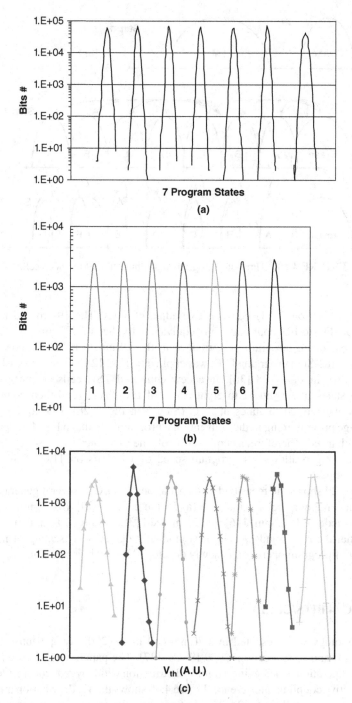

FIGURE 4.29 The V_t distribution of seven program states (TLC) in cell generations of (a) 56 nm, (b) 32 nm, (c) 20-nm node (27 nm), (d) 21 nm, and (e) 19 nm.

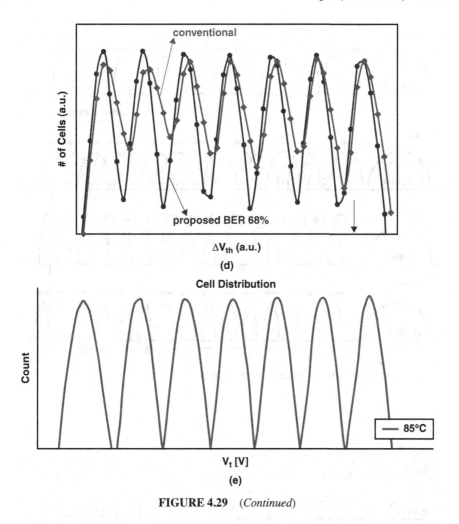

FIGURE 4.29   (*Continued*)

shown in Fig. 4.30 (1). Then, when neighboring cells are programmed, the distribution width is widened mainly due to the floating-gate capacitive coupling interference, as shown in Fig. 4.30 (2). After that, the cells are programmed again to 16 levels with target levels, as shown in Fig. 4.30 (3). And when neighboring cells are programmed again, the distribution of these cells is widened, but it is very small, because the shift of neighboring cell is small enough, as shown in Fig. 4.30 (4)(5). By using this method, very tight V_t distribution width for 16LC can be obtained.

QLC technology was developed in a 43-nm memory cell [37]. The page program sequence of 43-nm QLC is nearly the same as that of a 70-nm QLC cell [24, 25]. Figure 4.31 shows the V_T-distribution transition of cell "a" in the string and the programming order. Each cell goes through three steps of programming. First, cell "a" is programmed to three levels (Step 1), similar to a MLC device (see Fig. 4.31 (1)).

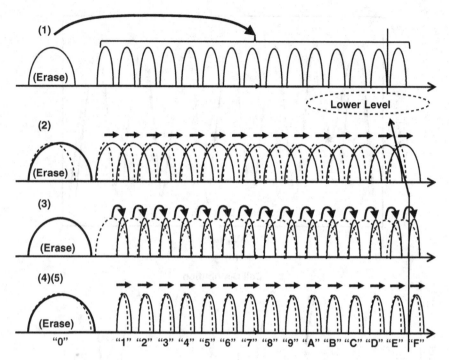

FIGURE 4.30 V_{th} distribution transition of 16LC (QLC, 4 bits/cell).

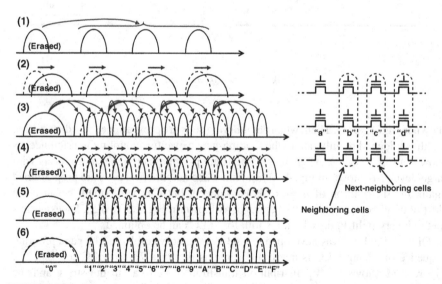

FIGURE 4.31   Three-step programming scheme of page programming order and V_t transition for QLC (4 bits/cell).

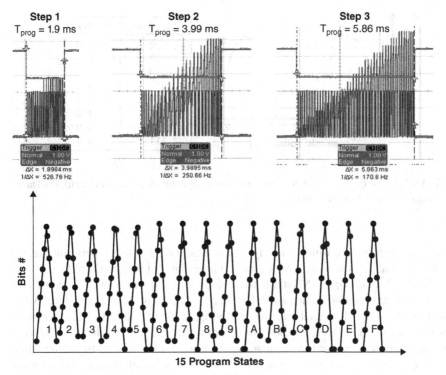

FIGURE 4.32   The measurement data of programming time for each page program step and V_t distribution for QLC.

Next, neighboring cells ("b") are programmed the same way. The V_T distribution of cell "a" is widened due to the FG coupling effect (see Fig. 4.31 (2)). Cell "a" is roughly programmed (Step 2) to 15 levels, lower than the targets (see Fig. 4.31 (3)). Next, neighboring cells ("c" and "b") are programmed to 3 and 15 rough levels, respectively, causing the V_T distribution of cell "a" to widen again (see Fig. 4.31 (4)). Finally, cell "a" is programmed (Step 3) to the 15 target levels (see Fig. 4.31 (5)). Neighboring cells ("b", "c", and "d") are then programmed again the same way (see Fig. 4.31 (6)). This page program sequence minimizes the FG coupling effect, even with large values due to technology scaling. Figure 4.32 shows the measured data of 15 program states, along with the programming time for each step. Total programming time of 11.75 ms translates to 5.6 MB/s, when two pages are programmed together in two cell arrays (two-page mode).

4.6   THREE-LEVEL (1.5 BITS/CELL) NAND FLASH

The 1.5-bit/cell technology was proposed to realize both a high performance and a low bit cost in the same product [38]. Targets of 1.5-bit/cell technology are (1) a high

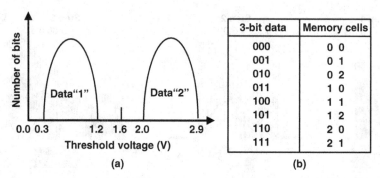

3-bit data	Memory cells
000	0　0
001	0　1
010	0　2
011	1　0
100	1　1
101	1　2
110	2　0
111	2　1

(a)   (b)

FIGURE 4.33   (a) Threshold distribution for a 3-level memory cell (1.5 bits/cell). (b) Three-bit data and three-1evel data in memory cells.

program performance which is nearly the same as SLC (2 bits/cell) performance, (2) a better reliability rather than MLC, and (3) lower bit cost than SLC product.

MLC (2 bits/cell) NAND flash technology based on four-level V_t states is a most popular solution for demands. However, it is difficult for the four-level MLC to provide as good a reliability and performance as SLC due to MLC's narrow read window margin (RWM) and slow program speed. These problems would limit its application to be used in the market. To overcome the market limitation while achieving both cost and performance, three-level memory cell technology, which has at least 2 times wider read window margin than four-level MLC, is promising.

The three-level memory cell has data "0", "1", and "2" as shown in Fig. 4.33 [38] and Fig. 4.34 [39]. A "0"-state (erase state) corresponds to a threshold voltage of less than −1 V. A pair of memory cells stores 3-bit data as shown in Fig. 4.33b. Here, 528-byte page data including parity-check bits and several flag data are simultaneously transmitted from or to 2816 memory cells through 2816 compact intelligent three-level column latches.

Four intermediate code (i.e., 6-bit data), are loaded within a 25-ns cycle time. During the data load, a charge pump generates a high voltage for the first program pulse. The setup time of the high voltage is 5 μs. The first pulse duration is 20 μs and each of the subsequent pulses has 10 μs duration. A program recovery time and the program verify time are 1 μs and 16 μs, respectively. Total program time for 512-byte data is 704 × 25 ns + 20 μs + 1 μs + 16 μs + (5 μs + 10 μs + 1 μs +16 μs) × 3 = 150.6 μs. Then, the typical program throughput is 3.4 Mbyte/s and 68% of the two-level NAND flash, as shown in Fig. 4.35. In another paper [39], the page program speed is 45% of a 1-bit/cell (two-level) SLC NAND cell. Figure 4.35 shows the estimated program speed comparison between the three-level and conventional methods.

The die size is also estimated on the assumption that, in the case of the two-level flash memory, the memory cells and the column latches occupies 66% of the die size. A number of the memory cells and an area of the column latches are increased to 133.3% when a memory capacity is doubled. The die size of the three-level flash

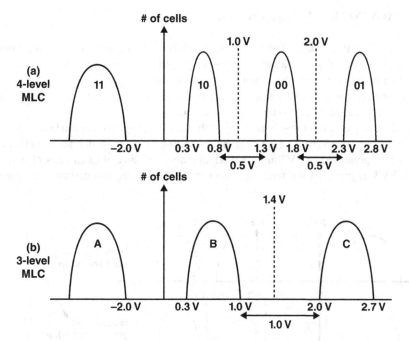

FIGURE 4.34 V_{th} distribution in NAND flash: (a) Four-level MLC. (b) Three-level cell (1.5 bits/cell).

memory chip is increased to 122%. Therefore, the die size per bit is reduced to 61%, as also shown in Fig. 4.35. Another paper shows that the estimated die size is 103 mm², which is reduced by 20% compared to SLC die, as shown in Fig. 4.35.

As shown above, a three-level cell has an advantage of 3–6 times improved program speed compared with MLC, along with 20–39% die size reduction compared with SLC. The three-level cell would have a possibility to use a certain application—for example, high-end enterprise server, and so on.

	Program Speed		Die Size/bit	
	Ref. 38	Ref. 39	Ref. 38	Ref. 39
1 bit/cell (SLC)	100%	100%	100%	100%
1.5 bit/cell	68%	45%	61%	80%
2 bit/cell (MLC)	11%	15%	50%	62%

FIGURE 4.35 Comparison of program speed and die size per bit.

4.7   MOVING READ ALGORITHM

As memory cell size is scaled down, the V_t shift of data retention becomes worse and worse, especially after a large amount of program/erase cycling. To compensate this data retention issues, the moving read algorithm was proposed [26]. Operation of moving read is to adjust a read voltage on the selected control gate according to cell V_t shift caused by data retention, and so on.

One example of the moving read algorithm for the program and read sequence is shown in Fig. 4.36 [26]. During page buffer setting in program operation, cells that will be programmed to PV3 are counted and stored in special extra cells of named FLG_PV3 in page. At the read operation, cells at PV3 are counted and compared

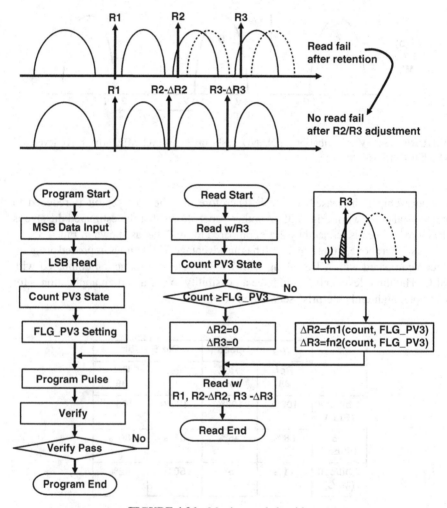

FIGURE 4.36   Moving read algorithm.

to ideal FLG_PV3. If count value meets FLG_PV3, the present read level can be applied. However, if count value does not meet FLG_PV3 (most cases are large V_t shift of data retention) to affect read results, the voltage of read level is calculated to be shifted down based on the difference between count value and FLG_PV3. The reason why PV3 cells are monitored is because data retention V_{th} shift is worst in the PV3 state. More than 30% error due to data retention V_{th} shift was improved.

Moving read operation in Fig. 4.36 is one example. There would be many alternate algorithms of the moving read to compensate cell V_t shifts which are caused not only by data retention but also by floating-gate capacitive coupling interference, program injection spread, RTN, program disturb, read disturb, and so on, as described in Chapters 5 and 6. The moving read operation has to be optimized because it is very effective to improve reliability of NAND flash memory product.

REFERENCES

[1] Bauer, M.; Alexis, R.; Atwood, G.; Baltar, B.; Fazio, A.; Frary, K.; Hensel, Ishac, M.; Javanifard, J.; Landgraf, M.; Leak, D.; Loe, K.; Mills, D.; Ruby, P.; Rozman, R.; Sweha, S.; Talreja, S., Wojciechowski, K. A multilevel-cell 32 Mb flash memory, *IEEE ISSCC*, pp. 132–133, 1995.

[2] Takeuchi, K.; Tanaka, T.; Nakamura, H. A double-level-V_{th} select gate array architecture for multi-level NAND flash memories, in *1995 Symposium on VLSI Circuits*, Technical Paper, pp. 69–70, 1995.

[3] Hemink, G. J.; Tanaka, T.; Endoh, T.; Aritome, S., Shirota, R. Fast and accurate programming method for multi-level NAND EEPROM's, in *1995 Symposium on VLSI Technology*, Technical Paper, pp. 129–130, 1995.

[4] Jung, T. S.; Choi, Y. J.; Suh, K. D.; Suh, B. H.; Kim, J. K.; Lim, Y. H.; Koh, Y. N.; Park, J. W.; Lee, K. J.; Park, J. H.; Park, K. T.; Kim, J. R.; Lee, J. H.; Lim, H. K. A 3.3 V 128 Mb multi-level NAND flash memory for mass storage applications, *IEEE ISSCC*, pp. 32–33, 1996.

[5] Jung, T.-S.; Choi, Y.-J.; Suh, K.-D.; Suh, B.-H.; Kim, J.-K.; Lim, Y.-H.; Koh, Y.-N.; Park, J.-W.; Lee, K.-J.; Park, J.-H.; Park, K.-T.; Kim, J.-R.; Yi, J.-H.; Lim, H.-K. A 117-mm² 3.3-V only 128-Mb multilevel NAND flash memory for mass storage applications, *Solid-State Circuits, IEEE Journal of*, vol. 31, no. 11, pp. 1575–1583, Nov. 1996.

[6] Takeuchi, K.; Tanaka, T.; Tanzawa, T. A Multi-page cell architecture for high-speed programming multi-level NAND flash Memories, *VLSI Circuits, 1997. Digest of Technical Papers, 1997 Symposium on*, pp. 67–68, 12–14 June 1997.

[7] Takeuchi, K.; Tanaka, T.; Tanzawa, T. A multipage cell architecture for high-speed programming multilevel NAND flash memories, *Solid-State Circuits, IEEE Journal of*, vol. 33, no. 8, pp. 1228–1238, Aug. 1998.

[8] Aritome, S.; Satoh, S.; Maruyama, T.; Watanabe, H.; Shuto, S.; Hemink, G. J.; Shirota, R.; Watanabe, S.; Masuoka, F. A 0.67 μm² self-aligned shallow trench isolation cell (SA-STI cell) for 3 V-only 256 Mbit NAND EEPROMs, *Electron Devices Meeting, 1994. IEDM '94. Technical Digest., International*, pp. 61–64, 11–14 Dec. 1994.

[9] Aritome, S. NAND Flash Innovations, *Solid-State Circuits Magazine, IEEE*, vol. 5, no. 4, pp. 21, 29, Fall 2013.

[10] Suh, K.-D.; Suh, B.-H.; Lim, Y.-H.; Kim, J.-K.; Choi, Y.-J.; Koh, Y.-N.; Lee, S.-S.; Kwon, S.-C.; Choi, B.-S.; Yum; J.-S. Choi, J.-H.; Kim, J.-R.; Lim, H.-K. A 3.3 V 32 Mb NAND flash memory with incremental step pulse programming scheme, *Solid-State Circuits, IEEE Journal of*, vol. 30, no. 11, pp. 1149–1156, Nov. 1995.

[11] Hemink, G. J.; Tanaka, T.; Endoh, T.; Aritome, S.; Shirota, R. Fast and accurate programming method for multi-level NAND EEPROMs, *VLSI Technology, 1995. Digest of Technical Papers. 1995 Symposium on*, pp. 129–130, 6–8 June 1995.

[12] Hemink, G. J.; Shimizu, K.; Aritome, S.; Shirota, R. Trapped hole enhanced stress induced leakage currents in NAND EEPROM tunnel oxides, *Reliability Physics Symposium, 1996. 34th Annual Proceedings., IEEE International*, pp. 117–121, April 30 1996–May 2 1996.

[13] Tanaka, T.; Tanaka, Y.; Nakamura, H.; Oodaira, H.; Aritome, S.; Shirota, R.; Masuoka, F. A quick intelligent program architecture for 3 V-only NAND-EEPROMs, *VLSI Circuits, 1992. Digest of Technical Papers, 1992 Symposium on*, pp. 20–21, 4–6 June 1992.

[14] Tanaka, T.; Tanaka, Y.; Nakamura, H.; Sakui, K.; Oodaira, H.; Shirota, R.; Ohuchi, K.; Masuoka, F.; Hara, H. A quick intelligent page-programming architecture and a shielded bitline sensing method for 3 V-only NAND flash memory, *Solid-State Circuits, IEEE Journal of*, vol. 29, no. 11, pp. 1366–1373, Nov. 1994.

[15] Tanaka, T.; Chen, J. US Patent, US6643188, 2003.

[16] Park, K.-T.; Kwon, O.; Yoon, S.; Choi, M.-H.; Kim, I.-M.; Kim, B.-G.; Kim, M.-S.; Choi, Y.-H.; Shin, S.-H.; Song, Y.; Park, J.-Y.; Lee, J.-e.; Eun, C.-G.; Lee, H.-C.; Kim, H.-J.; Lee, J.-H.; Kim, J.-Y.; Kweon, T.-M.; Yoon, H.-J.; Kim, T.; Shim, D.-K.; Sel, J.; Shin, J.-Y.; Kwak, P.; Han, J.-M.; Kim, K.-S.; Lee, S.; Lim, Y.-H.; Jung, T.-S. A 7 MB/s 64 Gb 3-bit/cell DDR NAND flash memory in 20 nm node technology, *Solid-State Circuits Conference Digest of Technical Papers (ISSCC), 2011 IEEE International*, pp. 212, 213, 20–24 Feb. 2011.

[17] Hosono, K.; Tanaka, T.; Imamiya, K.; Sakui, K. A high speed failure bit counter for the pseudo pass scheme (PPS) in program operation for Giga bit NAND flash, *Non-Volatile Semiconductor Memory Workshop, 2003. IEEE NVSMW 2003*. pp. 23–26, 16–20 Feb. 2003.

[18] Park, K.-T. A zeroing cell-to-cell interference page architecture with temporary LSB storing program scheme for sub-40 nm MLC NAND flash memories and beyond, *VLSI Circuits, 2007 IEEE Symposium on*, pp. 188–189, 14–16 June 2007.

[19] Park, K.-T.; Kang, M.; Kim, D.; Hwang, S.-W.; Choi, B. Y.; Lee, Y.-T.; Kim, C.; Kim, K. A zeroing cell-to-cell interference page architecture with temporary LSB storing and parallel MSB program scheme for MLC NAND flash memories, *Solid-State Circuits, IEEE Journal of*, vol. 43, no. 4, pp. 919–928, April 2008.

[20] Takeuchi, Y.; Shimizu, K.; Narita, K.; Kamiya, E.; Yaegashi, T.; Amemiya, K.; Aritome, S. A self-aligned STI process integration for low cost and highly reliable 1 Gbit flash memories, *VLSI Technology, 1998. Digest of Technical Papers. 1998 Symposium on*, pp. 102–103, 9–11 June 1998.

[21] Sakui, K.; Tanaka, T.; Nakamura, H.; Momodomi, M.; Endoh, T.; Shirota, R.; Watanabe, S.; Ohuchi, K.; Masuoka, F. A shielded bitline sensing technology for a high-density and

low-voltage NAND EEPROM design, in *International Workshop on Advanced LSI's*, pp. 226–232, July 1995.

[22] Shibata, N.; Tanaka, T. US Patent 7,245,528. 7,370,009. 7,738,302.

[23] Hara, T.; Fukuda, K.; Kanazawa, K.; Shibata, N.; Hosono, K.; Maejima, H.; Nakagawa, M.; Abe, T.; Kojima, M.; Fujiu, M.; Takeuchi, Y.; Amemiya, K.; Morooka, M.; Kamei, T.; Nasu, H.; Chi-Ming, Wang; Sakurai, K.; Tokiwa, N.; Waki, H.; Maruyama, T.; Yoshikawa, S.; Higashitani, M.; Pham, T. D.; Fong, Y.; Watanabe, T. A 146-mm^2 8-gb multi-level NAND flash memory with 70-nm CMOS technology," *Solid-State Circuits, IEEE Journal of,* vol. 41, no. 1, pp. 161, 169, Jan. 2006.

[24] Shibata, N.; Maejima, H.; Isobe, K.; Iwasa, K.; Nakagawa, M.; Fujiu, M.; Shimizu, T.; Honma, M.; Hoshi, S.; Kawaai, T.; Kanebako, K.; Yoshikawa, S.; Tabata, H.; Inoue, A.; Takahashi, T.; Shano, T.; Komatsu, Y.; Nagaba, K.; Kosakai, M.; Motohashi, N.; Kanazawa, K.; Imamiya, K.; Nakai, H. A 70 nm 16 Gb 16-level-cell NAND Flash Memory, *VLSI Circuits, 2007 IEEE Symposium on*, pp. 190–191, 14–16 June 2007.

[25] Shibata, N.; Maejima, H.; Isobe, K.; Iwasa, K.; Nakagawa, M.; Fujiu, M.; Shimizu, T.; Honma, M.; Hoshi, S.; Kawaai, T.; Kanebako, K.; Yoshikawa, S.; Tabata, H.; Inoue, A.; Takahashi, T.; Shano, T.; Komatsu, Y.; Nagaba, K.; Kosakai, M.; Motohashi, N.; Kanazawa, K.; Imamiya, K.; Nakai, H.; Lasser, M.; Murin, M.; Meir, A.; Eyal, A.; Shlick, M. A 70 nm 16 Gb 16-level-cell NAND flash memory, *Solid-State Circuits, IEEE Journal of*, vol. 43, no. 4, pp. 929–937, April 2008.

[26] Lee, C.; Lee, S.-K.; Ahn, S.; Lee, J.; Park, W.; Cho, Y.; Jang, C.; Yang, C.; Chung, S.; Yun, I.-S.; Joo, B.; Jeong, B.; Kim, J.; Kwon, J.; Jin, H.; Noh, Y.; Ha, J.; Sung, M.; Choi, D.; Kim, S.; Choi, J.; Jeon, T.; Yang, J.-S.; Koh, Y.-H. A 32 Gb MLC NAND-flash memory with V_{th}-endurance-enhancing schemes in 32 nm CMOS, *Solid-State Circuits Conference Digest of Technical Papers (ISSCC), 2010 IEEE International*, pp. 446–447, 7–11 Feb. 2010.

[27] Cernea, R.-A.; Pham, L.; Moogat, F.; Chan, S.; Le, B.; Li, Y.; Tsao, S.; Tseng, T.-Y.; Nguyen, K.; Li, J.; Hu, J.; Yuh, J. H.; Hsu, C.; Zhang, F.; Kamei, T.; Nasu, H.; Kliza, P.; Htoo, K.; Lutze, J.; Dong, Y.; Higashitani, M.; Junnhui, Yang; Hung-Szu, Lin; Sakhamuri, V.; Li, A.; Pan, F.; Yadala, S.; Taigor, S.; Pradhan, K.; Lan, J.; Chan, J.; Abe, T.; Fukuda, Y.; Mukai, H.; Kawakami, K.; Liang, C.; Ip, T.; Chang, S.-F.; Lakshmipathi, J.; Huynh, S.; Pantelakis, D.; Mofidi, M.; Quader, K. A 34 MB/s MLC write throughput 16 Gb NAND with all bit line architecture on 56 nm technology *Solid-State Circuits, IEEE Journal of*, vol. 44, no. 1, pp. 186–194, Jan. 2009.

[28] Cernea, R.; Pham, L.; Moogat, F.; Chan, S.; Le, B.; Li, Y.; Tsao, S.; Tseng, T.-Y.; Nguyen, K.; Li, J.; Hu, J.; Park, J.; Hsu, C.; Zhang, F.; Kamei, T.; Nasu, H.; Kliza, P.; Htoo, K.; Lutze, J.; Dong, Y.; Higashitani, M.; Yang, J.; Lin, H.-S.; Sakhamuri, V.; Li, A.; Pan, F.; Yadala, S.; Taigor, S.; Pradhan, K.; Lan, J.; Chan, J.; Abe, T.; Fukuda, Y.; Mukai, H.; Kawakamr, K.; Liang, C.; Ip, T.; Chang, S.-F.; Lakshmipathi, J.; Huynh, S.; Pantelakis, D.; Mofidi, M.; Quader, K. A 34 MB/s-program-throughput 16 Gb MLC NAND with all-bitline architecture in 56 nm, *Solid-State Circuits Conference, 2008. ISSCC 2008. Digest of Technical Papers. IEEE International*, pp. 420–624, 3–7 Feb. 2008.

[29] Li, Y.; Lee, S.; Fong, Y.; Pan, F.; Kuo, T.-C.; Park, J.; Samaddar, T.; Nguyen, H. T.; Mui, M. L.; Htoo, K.; Kamei, T.; Higashitani, M.; Yero, E.; Kwon, G.; Kliza, P.; Wan, J.; Kaneko, T.; Maejima, H.; Shiga, H.; Hamada, M.; Fujita, N.; Kanebako, K.; Tam, E.; Koh, A.; Lu, I.; Kuo, C. C.-H.; Pham, T.; Huynh, J.; Nguyen, Q.; Chibvongodze, H.; Watanabe, M.; Oowada, K.; Shah, G.; Byungki, Woo; Gao, R.; Chan, J.; Lan, J.;

Hong, P.; Peng, L.; Das, D.; Ghosh, D.; Kalluru, V.; Kulkarni, S.; Cernea, R.-A.; Huynh, S.; Pantelakis, D.; Wang, C.-M.; Quader, K. A 16 Gb 3-bit per cell (X3) NAND flash memory on 56 nm technology with 8 MB/s write rate, *Solid-State Circuits, IEEE Journal of*, vol. 44, no. 1, pp. 195, 207, Jan. 2009.

[30] Li, Y.; Lee, S.; Fong, Y.; Pan, F.; Kuo, T.-C.; Park, J.; Samaddar, T.; Nguyen, H.; Mui, M.; Htoo, K.; Kamei, T.; Higashitani, M.; Yero, E.; Gyuwan, Kwon; Kliza, P.; Jun, Wan; Kaneko, T.; Maejima, H.; Shiga, H.; Hamada, M.; Fujita, N.; Kanebako, K.; Tarn, E.; Koh, A.; Lu, I.; Kuo, C.; Pham, T.; Huynh, J.; Nguyen, Q.; Chibvongodze, H.; Watanabe, M.; Oowada, K.; Shah, G.; Woo, B.; Gao, R.; Chan, J.; Lan, J.; Hong, P.; Peng, L.; Das, D.; Ghosh, D.; Kalluru, V.; Kulkarni, S.; Cernea, R.; Huynh, S.; Pantelakis, D.; Wang, C.-M.; Quader, K. A 16 Gb 3 b/cell NAND flash memory in 56 nm with 8MB/s write rate, *Solid-State Circuits Conference, 2008. ISSCC 2008. Digest of Technical Papers. IEEE International*, pp. 506–632, 3–7 Feb. 2008.

[31] Futatsuyama, T.; Fujita, N.; Tokiwa, N.; Shindo, Y.; Edahiro, T.; Kamei, T.; Nasu, H.; Iwai, M.; Kato, K.; Fukuda, Y.; Kanagawa, N.; Abiko, N.; Matsumoto, M.; Himeno, T.; Hashimoto, T.; Liu, Y.-C.; Chibvongodze, H.; Hori, T.; Sakai, M.; Ding, H.; Takeuchi, Y.; Shiga, H.; Kajimura, N.; Kajitani, Y.; Sakurai, K.; Yanagidaira, K.; Suzuki, T.; Namiki, Y.; Fujimura, T.; Mui, M.; Nguyen, H.; Lee, S.; Mak, A.; Lutze, J.; Maruyama, T.; Watanabe, T.; Hara, T.; Ohshima, S. A 113 mm^2 32 Gb 3b/cell NAND flash memory, *Solid-State Circuits Conference—Digest of Technical Papers, 2009. ISSCC 2009. IEEE International*, pp. 242–243, 8–12 Feb. 2009.

[32] Nobukata, H.; Takagi, S.; Hiraga, K.; Ohgishi, T.; Miyashita, M.; Kamimura, K.; Hiramatsu, S.; Sakai, K.; Ishida, T.; Arakawa, H.; Itoh, M.; Naiki, I.; Noda, M. A 144 Mb 8-level NAND flash memory with optimized pulse width programming, *VLSI Circuits, 1999. Digest of Technical Papers. 1999 Symposium on*, pp. 39–40, 1999.

[33] Nobukata, H.; Takagi, S.; Hiraga, K.; Ohgishi, T.; Miyashita, M.; Kamimura, K.; Hiramatsu, S.; Sakai, K.; Ishida, T.; Arakawa, H.; Itoh, M.; Naiki, I.; Noda, M. A 144-Mb, eight-level NAND flash memory with optimized pulsewidth programming, *Solid-State Circuits, IEEE Journal of*, vol. 35, no. 5, pp. 682–690, May 2000.

[34] Yang, J.; Park, M.; Jung, S.; Park, S.; Cho, S.; An, J.; Lee, J.; Cho, S.; Lee, H.; Cho, M. K.; Ahn, K. O.; Jin, K.; Koh, Y. The operation scheme and process optimization in TLC (triple level cell) NAND flash characteristics, SSDM 2009.

[35] Shin, S.-H.; Shim, D.-K.; Jeong, J.-Y.; Kwon, O.-S.; Yoon, S.-Y.; Choi, M.-H.; Kim, T.-Y.; Park, H.-W.; Yoon, H.-J.; Song, Y.-S.; Choi, Y.-H.; Shim, S.-W.; Ahn, Y.-L.; Park, K.-T.; Han, J.-M.; Kyung, K.-H.; Jun, Y.-H. A new 3-bit programming algorithm using SLC-to-TLC migration for 8 MB/s high performance TLC NAND flash memory," *VLSI Circuits (VLSIC), 2012 Symposium on*, pp. 132, 133, 13–15 June 2012.

[36] Li, Y.; Lee, S.; Oowada, K.; Nguyen, H.; Nguyen, Q.; Mokhlesi, N.; Hsu, C.; Li, J.; Ramachandra, V.; Kamei, T.; Higashitani, M.; Pham, T.; Honma, M.; Watanabe, Y.; Ino, K.; Binh, Le; Woo, B.; Htoo, K.; Tseng, T.-Y.; Pham, L.; Tsai, F.; Kim, K.-h.; Chen, Y.-C.; She, M.; Yuh, J.; Chu, A.; Chen, C.; Puri, R.; Lin, H.-S.; Chen, Y.-F.; Mak, W.; Huynh, J.; Chan, J.; Watanabe, M.; Yang, D.; Shah, G.; Souriraj, P.; Tadepalli, D.; Tenugu, S.; Gao, R.; Popuri, V.; Azarbayjani, B.; Madpur, R.; Lan, J.; Yero, E.; Pan, F.; Hong, P.; Jang, Yong Kang; Moogat, F.; Fong, Y.; Cernea, R.; Huynh, S.; Trinh, C.; Mofidi, M.; Shrivastava, R.; Quader, K. 128 Gb 3b/cell NAND flash memory in 19 nm technology with 18 MB/s write rate and 400 Mb/s toggle mode, *Solid-State Circuits Conference*

Digest of Technical Papers (ISSCC), 2012 IEEE International, pp. 436, 437, 19–23 Feb. 2012.

[37] Trinh, C.; Shibata, N.; Nakano, T.; Ogawa, M.; Sato, J.; Takeyama, Y.; Isobe, K.; Le, B.; Moogat, F.; Mokhlesi, N.; Kozakai, K.; Hong, P.; Kamei, T.; Iwasa, K.; Nakai, J.; Shimizu, T.; Honma, M.; Sakai, S.; Kawaai, T.; Hoshi, S.; Yuh, J.; Hsu, C.; Tseng, T.; Li, J.; Hu, J.; Liu, M.; Khalid, S.; Chen, J.; Watanabe, M.; Lin, H.; Yang, J.; McKay, K.; Nguyen, K.; Pham, T.; Matsuda, Y.; Nakamura, K.; Kanebako, K.; Yoshikawa, S.; Igarashi, W.; Inoue, A.; Takahashi, T.; Komatsu, Y.; Suzuki, C.; Kanazawa, K.; Higashitani, M.; Lee, S.; Murai, T.; Lan, J.; Huynh, S.; Murin, M.; Shlick, M.; Lasser, M.; Cernea, R.; Mofidi, M.; Schuegraf, K.; Quader, K. A 5.6MB/s 64Gb 4b/Cell NAND Flash memory in 43 nm CMOS, *Solid-State Circuits Conference—Digest of Technical Papers, 2009. ISSCC 2009. IEEE International*, pp. 246–247, 247a, 8–12 Feb. 2009.

[38] Tanaka, T.; Tanzawa, T.; Takeuchi, K. A 3.4-Mbyte/sec programming 3-level NAND flash memory saving 40% die size per bit, *VLSI Circuits, 1997. Digest of Technical Papers., 1997 Symposium on*, pp. 65–66, 12–14 June 1997.

[39] Park, K.-T.; Choi, J.; Cho, S.; Choi, Y.; Kim, K. A high cost-performance and reliable 3-level MLC NAND flash memory using virtual page cell architecture, *Non-Volatile Semiconductor Memory Workshop, 2006. IEEE NVSMW 2006, 21st*, pp. 34–35, 12–16 Feb. 2006.

5

SCALING CHALLENGE OF NAND FLASH MEMORY CELLS

5.1 INTRODUCTION

Low-cost and highly reliable NAND flash memory technologies have been intensively developed [1–9] over 25 years, as described in Chapter 3. As a suitable memory cell structure for NAND flash, the self-aligned STI cell (SA-STI cell) had been developed [4–7] and implemented to NAND flash products [8]. This cell could reduce memory cell size to ideal $4*F^2$ [4], and had also demonstrated an excellent reliability, because the floating gate does not overlap the STI corner. Thus, the SA-STI cell structure and process have been used for more than 15 years and 10 generations of NAND flash product. The most advanced memory cell had presented as mid-1X-nm (15 to 16-nm) SA-STI memory cells [10], as shown in a cross-sectional TEM micrograph in Fig. 5.1. The effective cell size can be also reduced by multilevel cell technology, as described in Chapter 4. Therefore, the small physical cell size of $4*F^2$ combined with a multilevel cell can drastically reduce the bit cost of NAND flash memory.

However, by scaling memory cell size beyond the 20-nm generation, it is becoming very difficult to realize high-performance and highly reliable NAND flash memory, because many physical phenomena have a serious impact on the operation margin of NAND flash [11].

In Chapter 5, the scaling challenges of the NAND flash memory cell with a multilevel cell are discussed beyond 20-nm feature sizes. One important physical phenomenon is the floating-gate capacitive coupling interference [12] that causes a V_t shift by programming neighbor cells. An increase in V_t distribution width (Section 5.3)

Nand Flash Memory Technologies, First Edition. Seiichi Aritome.
© 2016 The Institute of Electrical and Electronics Engineers, Inc. Published 2016 by John Wiley & Sons, Inc.

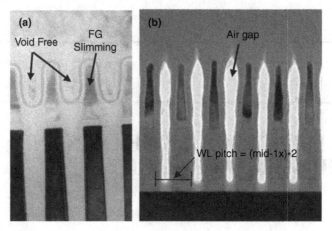

FIGURE 5.1   TEM photograph of mid-1X-nm SA-STI NAND flash cells.

will result in the degradation of read window margin (RWM). The other major physical phenomena to have an impact on RWM are electron injection spread (EIS) [13–15] (Chapter 5.4) and random telegraph noise (RTN) [16] (Section 5.5). Except for the RWM degradation, there are several other problems, such as CG formations between FGs [17] (Chapter 5.6), the WL high-field problem [11, 18] (Section 5.7), reducing the number of stored electrons [19] (Section 5.8), and so on.

The scaling capability of NAND flash memory has been discussed in several conferences and papers [20–33]. They pointed out major scaling limitations, such as floating-gate capacitive coupling interference [21–23, 25–28, 31, 32], reduced number of electrons [21–23], lithograph/patterning [22–24], RTN and RDF (random dopant fluctuation) [23], structure limitation [25, 28, 30], air gap [34, 35], V_t window margin [11, 28, 30], and so on.

In Chapter 5, several scaling problems and limitations have been widely discussed over 2X to 0X-nm generations [11]. As a result, there is a possibility that the NAND flash memory cell can be scaled down to 1Z-nm (10-nm) generation with an accurate control of FG/CG formation process and air-gap process to manage floating-gate capacitive coupling interference and the WL high-field problem.

5.2   READ WINDOW MARGIN (RWM)

The read window margin (RWM) of a self-aligned STI cell (SA-STI cell) is discussed for NAND flash memories over 2X to 0X-nm generations in Section 5.2 [11]. The RWM is investigated by extrapolating the physical phenomena of FG–FG coupling interference (floating-gate capacitive coupling interference), electron injection spread (EIS), back pattern dependence (BPD), and random telegraph noise (RTN). The RWM is degraded not only by increasing programmed V_t distribution width, but also by increasing the V_t of the erase state mainly due to the large FG–FG coupling

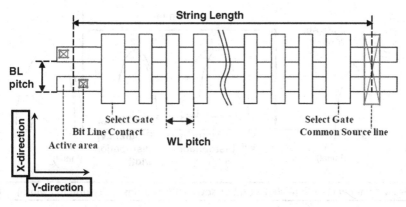

FIGURE 5.2   Top view of a NAND cell string. 64 cells are connected in series with two select gates. BL pitch and WL pitch are nearly equal to 2F (F: feature size). Then unit cell size is close to ideal $4*F^2$.

interference. However, RWM is still positive in the 1Z-nm (10-nm) generation with 60% reduction of FG–FG coupling interference by the air-gap process. Therefore, the SA-STI cell is expected to be able to scale down to the 1Z-nm (10-nm) generation, with the air gap of 60% reduced FG–FG coupling interference.

5.2.1   Assumption for Read Window Margin (RWM)

Figure 5.2 shows a top view of conventional NAND cell strings. In order to investigate the scaling of the NAND flash cell, cell dimensions beyond the 2X-nm (26-nm) generation are assumed, as shown in Table 5.1. Dimensions of 2X nm are given, 27 nm for the bit line (BL) half-pitch and 26 nm for the word-line (WL) half-pitch. Dimensions beyond 2X nm are assumed to scale down by a fixed scaling factor of ×0.85 for BL half-pitch and ×0.8 for the WL half-pitch. And also the channel width

TABLE 5.1   Assumption of Cell Dimensions and ONO (IPD) Thickness, in Generations of 2X–0X Nanometers[a]

Generation	2X	2Y	1X	1Y	1Z	0X	Scaling factor
BL half-pitch (nm)	27	23.0	19.5	16.6	14.1	12.0	×0.85 assumption
WL half-pitch (nm), Gate length L	26	20.8	16.6	13.3	10.6	8.5	×0.8 assumption
Channel W (nm)	20	18.0	16.2	14.6	13.1	11.8	×0.9 assumption
ONO thickness (nm)	12	11.4	10.8	10.3	9.8	9.3	×0.95 assumption

[a]Dimensions of 2X-nm generation are given, as 27 nm for BL half-pitch and 26 nm for WL half-pitch. Dimensions of ~2Y nm are assumed by scaling factors of ×0.85 for BL half-pitch (x-direction) and ×0.8 for the word line (WL) half-pitch (y-direction). And also, the scaling factors of channel width (W) and ONO thickness are assumed ×0.9 and ×0.95, respectively.

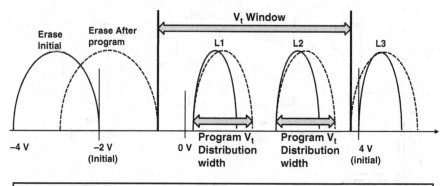

Read Window Margin (RWM) = (V$_t$ Window) – 2$*$(Program V$_t$ Distribution width)

FIGURE 5.3 Read V_t window of an MLC NAND cell. V_t distributions of erase state and programmed L1, L2, L3 states are shifted up and become wider because of electron injection spread (EIS), FG–FG coupling interference, RTN, and back pattern dependence (BPD). Read window margin (RWM) is defined as RWM = (V_t window) – 2$*$(program V_t distribution width).

W and inter-poly dielectric (IPD) thickness are assumed to scale down by the factors of ×0.9 and ×0.95, respectively.

Figure 5.3 shows an image of a read V_t window in an MLC (2 bits/cell) NAND cell [11]. The "V_t window" is defined by a right-side edge of erase distribution and a left-side edge of L3 (highest programmed state) after completing all page program operations in block (strings). Two programmed V_t distributions of L1/L2 have to be inside of the V_t window to be a reliable read operation. Read window margin (RWM) is defined by RWM = (V_t window) – 2$*$(programmed V_t distribution width), so that RWM means the separation margin of V_t distributions of each states.

The RWMs have been seriously degraded by cell scaling down from 0.7 μm to 2X-nm generation, because several physical phenomena were getting worse. Therefore, for further scaling of a NAND cell, it is very important to analyze and foresee the RWM in a future scaled NAND cell. In order to investigate RWM, the scaling trend of physical phenomena of electron injection spread (EIS) [13–15], FG–FG coupling interference [12], RTN [16], and back pattern dependence (BPD) are assumed as follows. And other assumptions of the page program sequence, parameter setting, and so on, are also shown in the following.

Assumption of RWM Calculation

(a) V_t distribution width (@ $\pm 3\sigma$) is assumed to become wider by simple summation of values of electron injection spread (EIS), FG–FG coupling interference, RTN and back pattern dependence (BPD). Each value is given for 2X-nm generation and is extrapolated for 2Y-nm to 0X-nm generations with the following formulas.

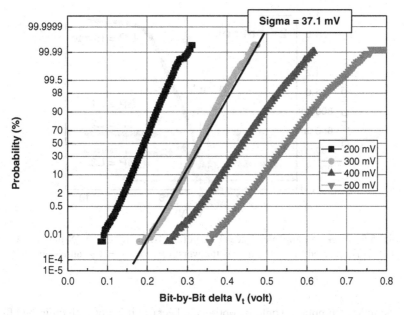

FIGURE 5.4   Electron injection spread (EIS) of a 2X-nm cell. In ISPP_step = 300 mV, the standard deviations (σ) is 37.1 mV. With the FG depletion effect, σ is assumed to be larger of 50%. Then the σ is assumed 55.6 mV for a 300-mV ISPP_Step.

(b) Program electron injection spread (EIS) [13–15] is caused during program operation due to statistical spread in a small number of injecting electrons during program pulse (see Section 5.4). The σ of EIS is linear with sqrt(q∗ISPP_Step/C_{IPD}) [13, 14], and three σ values are simply used for V_t distribution widening. Capacitance of inter-poly dielectric, C_{IPD}, is scaled down from 2X-nm generation by ×0.72 for each generation. The ISPP_step is a program voltage step of ISPP (increment step pulse program) [36,37]. The measured standard deviations (σ) is 37.1 mV for ISPP_step = 300 mV, as shown in Fig. 5.4. A value of the sigma is assumed to be 50% larger due to FG depletion effects [38], and so on. Then the σ of 2X is assumed to be 55.6 mV for ISPP_Step = 300 mV, and 78.7 mV for ISPP_Step = 600 mV.

(c) FG–FG coupling interference (floating-gate capacitive coupling interference) [12] (see Section 5.3) is scaled from 2X-nm generation to each generation by ×(1/0.9) along WL, ×(1/0.8) along BL, and ×(1/0.85) diagonal. And spread effect (additional V_t shift) is assumed 10% of the FG–FG coupling value. FG–FG coupling values of 2X nm are assumed to be 94 mV/V for two sides of x-direction (between BL–BL), 85 mV/V for one side of y-direction (between WL–WL), and 25 mV/V for two sides of xy-direction (diagonal), based on measurement results. Scaling factors of FG–FG coupling (×(1/0.9) along WL, ×(1/0.8) along BL, and ×(1/0.85) diagonal) are the simple assumption by increasing FG–FG capacitance with decreasing FG–FG distance. This

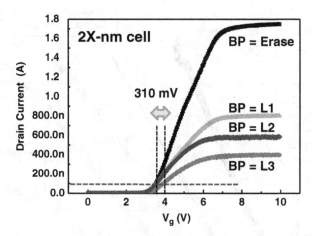

FIGURE 5.5 Back pattern dependence (BPD) of a 64-cell string in a 2X-nm cell. The V_t shift of BPD is assumed to be 310 mV as the worst case of BP = L3 (unselected cells are in L3).

simple assumption would be optimistic because it is not including the effect of decreasing total FG capacitance, which is mainly due to less scaling of IPD and tunnel-oxide thickness.

(d) RTN is linear with $1/(W*\mathrm{sqrt}(L))$ [39] (see Section 5.5). The value of 2X nm is assumed to be ± 107 mV@ 3σ, based on measurement results.

(e) Back pattern dependence (BPD) is a V_t shift, which is caused by programming series-connected cells in the same string due to increasing series resistance in string. BPD is linear with L/W. The value of 2X nm is assumed to 310 mV, as shown in Fig. 5.5.

(f) The page program sequence uses the minimized FG–FG coupling program sequence [40,41], as shown in Fig. 5.6. It means that, before the MSB program, the surrounding pages (LSB of WLn − 1, WLn, WLn + 1, and MSB of WLn − 1) have already been programmed. FG–FG coupling interference can be minimized for the programmed V_t distributions (see Section 4.3).

(g) All-bit-line scheme (ABL) [42] (see Section 4.3). The V_t shift value of the x-direction FG–FG coupling interference is assumed that it is based on neighbor cell V_t shift of $3*\sigma*\mathrm{SQRT}(2)*(1/2)$ [$\sigma = (V_t$ distribution width ($\pm 3\sigma$) of one program pulse)/6 = 3V/6 = 0.5 V] [$*(1/2)$; factor of random data pattern], because neighbor cells are programmed with target cells at the same time.

(h) Random data pattern.

(i) Erase; initial V_t distribution; −3 V ±1 V (V_t distribution width = 2 V). The right-side edge of erase initial is −2 V.

(j) L1 verify level = 0.5 V, L2 verify level = 2.25 V, L3 verify level = 4.0 V, LSB verify level = 0.8 V, except for the case of V_t setting dependence in Section 5.2.5.

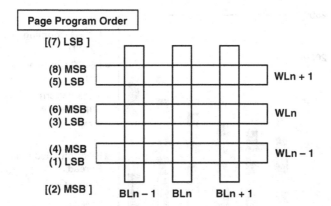

FIGURE 5.6   Page program sequence of the minimized FG–FG coupling interference. Before (6) MSB program of WLn, the surrounding pages [(1) LSB of WLn − 1, (3) LSB of WLn, (5) LSB of WLn + 1, and (4) MSB of WLn − 1] have already been programmed. Then the FG–FG coupling interference can be minimized for the programmed V_t distributions.

(k) ISPP step (increment step pulse program [36, 37]) of LSB and MSB programming are 600 mV and 300 mV, respectively.

(l) The data retention V_t shift is not included in this RWM investigation, because it is expected to be managed by the multi-times read operations (moving read algorithm), as described in Section 4.7. Also, program disturb, read disturb, and other effects are not included in this RWM investigation.

5.2.2   Programmed V_t Distribution Width

In conventional program operation, a programmed V_t distribution width can be tight by using ISPP [36, 37] (see Section 4.2.2) and a bit-by-bit verify operation [43] (see Section 4.2.3). The initial programmed V_t distribution width is determined by ISPP_step + EIS. And then it becomes wider by RTN, FG–FG coupling interference, and BPD after all pages are programmed in a block (string).

The programmed V_t distribution width after all pages have been programmed in a block have been calculated based on the assumption of Section 5.2.1, as shown in Fig. 5.7. As memory cells are scaled down from 2X nm to 0X nm, the programmed V_t distribution width is increased from 1320 mV to 2183 mV. It is clear that major reasons to increase V_t distribution width are the FG–FG coupling and RTN.

In order to obtain appropriate V_t shift values of FG–FG coupling interference, the delta V_t of the neighbor attack cell (subject to target cell) have been derived, as shown in Fig. 5.8. V_t distributions of page programming steps are also described in Fig. 5.8. For FG–FG coupling for the programmed states, delta V_t of the attack cell is described as dVt_E_L1 or dV_t_LSB_L2 + dV_t_LSB_L3, as shown at (3) after the MSB program in Fig. 5.8. A dV_t_E_L1 means delta V_t shift from erase state (@2) before MSB program) to L1 state. Larger value of dV_t_E_L1 or dV_t_LSB_L2 + dV_t_LSB_L3 is used for calculation of FG–FG coupling V_t shift.

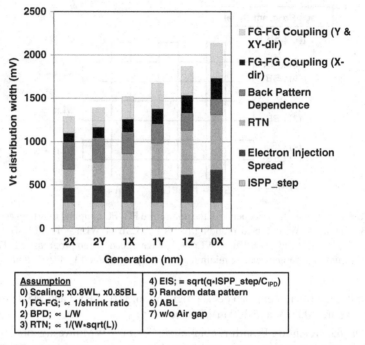

FIGURE 5.7 Calculated programmed V_t distribution width. The V_t distribution width is increased by cell dimension scaling. Major impact factors to increase V_t distributions width are the FG–FG coupling and RTN.

V_t shift value of x-direction FG–FG coupling interference is assumed that it is based on neighbor cell V_t shift of $3*\sigma*\text{SQRT}(2)*(1/2)$ [standard deviation; $\sigma = (V_t$ distribution width ($\pm 3\sigma$) of one program pulse)/6 = 3 V/6 = 0.5 V] [$*(1/2)$; factor of random data pattern], as shown in Section 5.2.1g), because neighbor cells are programmed with target cells at the same program sequence in the all-bit-line scheme. V_t shift had been assumed as follows. Cells in programmed V_t distribution (3-V width) are programmed to shift up by ISPP program. A certain cell (cell A) stops programming by passing verify at threshold voltage of $V_t\_$cell A, and neighbor cells (cell B) have not passed verify yet at threshold voltage of $V_t\_$cell B. The neighbor cells (cell B) are programmed by following ISPP steps, then it causes FG–FG coupling on cell A with a V_t difference of ($V_t\_$cell A $- V_t\_$cell B). In this assumption, the distribution of V_t difference ($V_t\_$cell A $- V_t\_$cell B) is assumed to composition of V_t distribution (3-V width), then the σ of the V_t shift is $\text{SQRT}(\sigma^2 + \sigma^2) = \sigma*\text{SQRT}(2)$. Also, we assumed that the same FG–FG coupling V_t shift occurs between L1 and L2, by assuming to use preferable program operations, such as the ABL parallel program method [44], the BC state first program algorithm [45], and the P3-pattern pre-pulse scheme [46], to reduce FG–FG coupling for both L1 and L2.

Erase V_t distribution at (2) Before MSB program in Fig. 5.8 has already shifted up as $dV_t\_E_i\_E$ from erase initial V_t distribution, by FG–FG coupling with surrounding

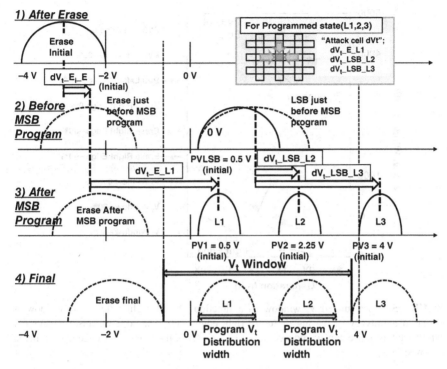

FIGURE 5.8 V_t distribution in page program steps. The attack cell delta V_t of $dV_t\_E\_L1$ or $dV_t\_LSB\_L2 + dV_t\_LSB\_L3$ are subjected to the programmed target cell of y-direction neighbor cells, resulting in wider V_t distribution width by the FG–FG coupling interference. Then distribution width of the programmed cell becomes wider from (3) After MSB program to (4) Final.

cells of LSB program (both sides of Y-direction/XY-direction/X-direction) and MSB program (one side of Y-direction/XY-direction), as shown in Fig. 5.6. By cell scaling, $dV_t\_E_i\_E$ becomes larger due to larger FG–FG coupling interference. Then $dV_t\_E\_L1$ becomes smaller in value as a result of cell scaling. Therefore, y-direction FG–FG coupling interference for programmed states is relatively smaller than expected, as shown in Fig. 5.7.

5.2.3 V_t Window

V_t window is defined from the right-side edge of erase V_t distribution to the left-side edge of L3 V_t distribution, as shown in Fig. 5.3. Figure 5.9 shows the calculation results of V_t window, the right-side edge of erase V_t distribution, and the left-side edge of L3 V_t distribution, in three cases of reducing FG–FG coupling of 0%, 30%, and 60% by air gap [34,35,47,48] or low-k dielectric. The reducing FG–FG coupling is assumed for both the x-direction (STI air-gap [35,49]) and the y-direction (WL air-gap [34,47,48]).

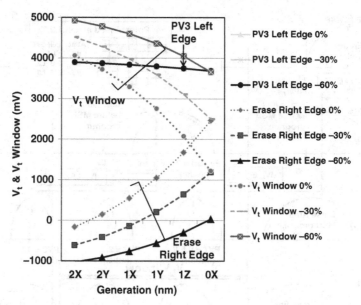

FIGURE 5.9 Calculated V_t window as a function of cell scaling down. V_t window is decreased mainly by increasing the erase right edge (right-side edge of erase). If FG–FG coupling interference can be reduced to −30% or −60%, the erase right edge can be much improved.

As shown in Fig. 5.9, the V_t window becomes seriously narrower as a result of cell scaling in the case of conventional 0% FG–FG coupling reduction (see "V_t window 0%"). This is because the right-side edge of the erase distribution is much increased as a result of scaling. However, in the case of −60% FG–FG coupling reduction, the right-side edge of erase distribution can be kept less than 0 V even in the 1Z-nm generation. Then, a V_t window can be kept more than 4000 mV in 1Z-nm generation.

In order to clarify the reason of increasing the right-side edge of erase, factors of increasing erase right-side edge are analyzed, as shown in Fig. 5.10. The erase right edge is increased mainly by FG–FG coupling, especially by Y- & XY-direction FG–FG coupling. For the erase state, the FG–FG coupling V_t shift is much larger than the FG–FG coupling V_t shift of the programmed states. Figure 5.11 shows the reasons of large FG–FG coupling for erase states. There are two reasons. One is the large delta-V_t of an attack cell, as shown in Fig. 5.11a. This is because a $dV_{t\_}E_i\_L1$, L2, L3 (attack cell delta V_t from an erase initial state to each programming state L1, L2, L3) is much larger, in comparison with attack cell V_t shift for a programmed state, such as $dV_{t\_}E\_L1$ or $dV_{t\_}LSB\_L2 + dV_{t\_}LSB\_L3$, as shown in Fig. 5.8. The other reason is that all of surrounding cells are subjected to cause FG–FG coupling V_t shift for the erase state (Fig. 5.11b). Conversely, for the programmed cell, only part of the surrounding cells (one side of y-direction [between WL–WL] and x-directions [between BL–BL]) have caused an FG–FG coupling V_t shift, as shown in Fig. 5.11b.

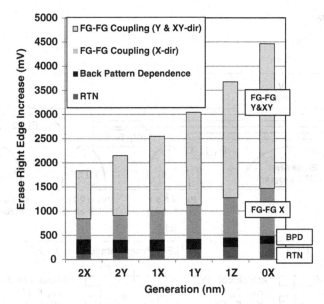

FIGURE 5.10   Increasing the right-side edge of the erase state distribution, as a function of scaling memory cells, in the case of w/o reduction of the FG–FG coupling. The erase right edge is increased mainly by the FG–FG coupling interference, especially by *Y* & *XY*–FG–FG coupling.

In order to obtain a wider V_t window for 1Y and 1Z generations, it is important to reduce FG–FG coupling, especially FG–FG coupling of the *Y*- & *XY*-directions. WL air gap (or low-*K*) [34, 47, 48] and STI air gap [35, 49] have to be implemented as small FG–FG coupling (as small as possible) for future NAND cells.

Furthermore, the optimistic scaling factors of FG–FG coupling are used in this calculation, as described in Section 5.2.1c. Even if the optimistic values are used, the dominant factor of V_t window degradation is the FG–FG coupling. Therefore it is important to reduce FG–FG coupling for future scaled cells.

5.2.4   Read Window Margin (RWM)

Figure 5.12 shows the scaling trend of RWM, which is calculated by the programmed V_t distribution width in Fig. 5.7 and V_t window in Fig. 5.9. RWMs are degraded as a cell scaling. In the case of "no air gap," 1X nm has marginal RWM, and 1Y-nm generation has negative (−719 mV) RWM. In the case of "FG–FG coupling −30% air gap," 1Y nm becomes a marginal RWM, and 1Z-nm generation has a negative RWM. Also, in the case of "FG–FG coupling −60% air gap," 1Z-nm generation has still positive RWM. This means that 30% FG–FG coupling reduction is needed to implement a 1Y-nm generation cell, and 50–60% reduction is needed for a 1Z-nm cell.

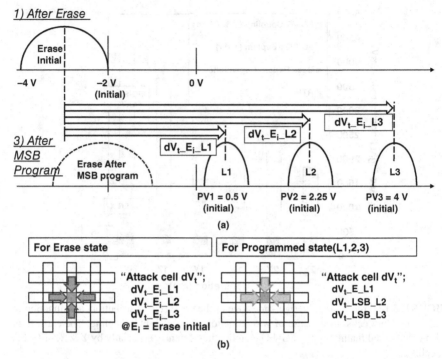

FIGURE 5.11 The FG–FG coupling for the erase state. (a) The attack cell (neighbor of target cell) V_t shift of $dV_t\_E_i\_L1$ or $dV_t\_E_i\_L2$ or $dV_t\_E_i\_L3$ are subjected to the erased target cell. The V_t shift of the erased state is larger than that of programmed state because an attack cell V_t shift for erased state is larger than that of programmed state (see Fig. 5.10), and also (b) all of the surrounding cells have been subjected to erased cells, compared to the fact that all of surrounding cells have not been subjected to a programmed state.

5.2.5   RWM V_t Setting Dependence

In order to find out other solutions for wider RWM, V_t setting dependence has been investigated. Figure 5.13 shows that RWM depends on V_t setting in the case of 1Z nm with −30% FG–FG coupling reduction. In a previous section, PV3 and the erase initial right edge are used for a fixed value of 4 V and −2 V, respectively. In this chapter, lower PV3 and lower erase initial right edge are assumed, as shown in Fig. 5.13.

RWM can be improved in the case of decreasing erase initial right edge, even if programmed V_t distribution widths are slightly increased. And RWM becomes positive in the case of erase initial right edge = −4 V. However, the decreasing erase V_t setting would degrade reliability because of subjecting higher erase voltage stress. Then, in order to obtain wider RWM in the 1Z-nm generation, the air-gap process of minimized FG–FG coupling should be combined with the decreasing erase V_t setting.

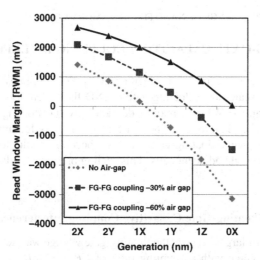

FIGURE 5.12 Calculated read window margin (RWM), in no air gap, 30% or 60% reduction of FG–FG coupling. RWM becomes less than 0 V beyond 1X-nm generation in the case of no air gap. However, by using an air gap with −60% FG–FG coupling reduction, RWM can be kept positive even if 1Z-nm generation is used.

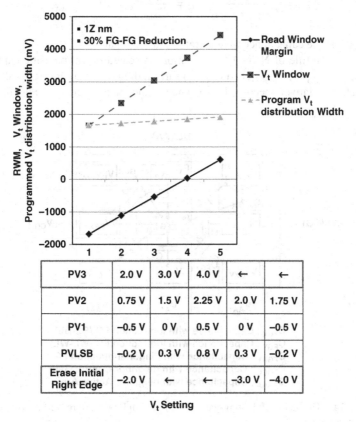

	1	2	3	4	5
PV3	2.0 V	3.0 V	4.0 V	←	←
PV2	0.75 V	1.5 V	2.25 V	2.0 V	1.75 V
PV1	−0.5 V	0 V	0.5 V	0 V	−0.5 V
PVLSB	−0.2 V	0.3 V	0.8 V	0.3 V	−0.2 V
Erase Initial Right Edge	−2.0 V	←	←	−3.0 V	−4.0 V

V_t Setting

FIGURE 5.13 RWM and V_t window in 1Z-nm generation in the case of −30% FG–FG coupling. RWM increases by decreasing erase V_t setting.

5.3   FLOATING-GATE CAPACITIVE COUPLING INTERFERENCE

Floating-gate capacitive coupling interference (FG–FG coupling) [12] is a major limitation issue to scale down floating-gate NAND flash memory cell, because the read window margin (RWM) is mainly degraded by the floating-gate capacitive coupling interference [11], as described in Section 5.2. As feature sizes (F) have scaled, the floating-gate to floating-gate space has become smaller to cause a V_t shift by V_t change of the eight adjacent cells. This scaling problem results in widening V_t distribution width.

5.3.1   Model of Floating-Gate Capacitive Coupling Interference

In the old concept of large cell size, the floating-gate voltage was determined by only the control-gate voltage with a coupling ratio of $CR = C_{IPD}/C_{total}$, where C_{IPD} is the control-gate to floating-gate capacitance and C_{total} is the total capacitance of the floating gate, as expressed in (5.1) (assuming floating-gate charge $Q_{FG} = 0$).

$$V_{FG} = \frac{C_{IPD}}{C_{TUN} + C_{IPD}} V_{CG} = \frac{C_{IPD}}{C_{total}} V_{CG} = CR * V_{CG} \tag{5.1}$$

where C_{TUN} is the capacitance of substrate to floating gate.

As the design rule of NAND flash memory is scaled down, parasitic capacitors (C_{FGX}, C_{FGY}, C_{FGXY}, C_{FGCG}, and C_{FGAA}) surrounding the floating gate, as shown in Fig. 5.14, have relatively become larger. They cannot be neglected. The floating-gate

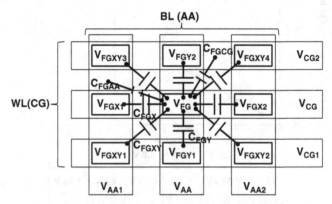

C_{FGX}; Capacitance with x-direction FG (BL-BL)
C_{FGY}; Capacitance with y-direction FG (WL-WL)
C_{FGXY}; Capacitance with diagonal-direction FG
C_{FGCG}; Capacitance with neighbor Control Gate
C_{FGAA}; Capacitance with neighbor Active Area

FIGURE 5.14   The model of floating-gate capacitive coupling interference based on parasitic capacitance coupling.

voltage is determined by not only the corresponding control-gate voltage but also the voltages of the surrounding floating gates, the control gates, and active area, as shown in (5.2) (assuming floating-gate charge $Q_{FG} = 0$),

$$V_{FG} = \frac{\begin{aligned}&C_{IPD}V_{CG} + C_{FGX}(V_{FGX1} + V_{FGX2}) + C_{FGY}(V_{FGY1} + V_{FGY2}) + \\ &C_{FGXY}(V_{FGXY1} + V_{FGXY2} + V_{FGXY3} + V_{FGXY4}) + \\ &C_{FGCG}(V_{CG1} + V_{CG2}) + C_{FGAA}(V_{AA1} + V_{AA2})\end{aligned}}{C_{TUN} + C_{IPD} + 2C_{FGX} + 2C_{FGY} + 4C_{FGXY} + 2C_{FGCG} + 2C_{FGAA}} \quad (5.2)$$

where the variables are shown in Fig. 5.14. A phenomenon called "floating-gate capacitive coupling interference," occurs, in which a cell V_t change (ΔV_t) is caused by the threshold voltage shift of the adjacent cells by floating-gate voltage shift (ΔV_{fg}). In other words, the floating-gate voltage is coupled by the floating-gate voltage changes of the adjacent cells with parasitic capacitors in the same manner as the control-gate voltage, as shown in (5.1). For example, if the floating-gate voltage of upper y-direction (V_{FGY2}) is changed by ΔV_{FGY2}, the floating-gate voltage (V_t) change of the target cell causes ΔV_{FG}, as expressed in (5.3).

$$\Delta V_{FG} = \frac{C_{FGY}}{C_{TUN} + C_{IPD} + 2C_{FGX} + 2C_{FGY} + 4C_{FGXY} + 2C_{FGCG} + 2C_{FGAA}} * \Delta V_{FGY2}$$

$$= \frac{C_{FGY}}{C_{total}} * \Delta V_{FGY2} \quad (5.3)$$

In the first report of the floating-gate capacitive coupling interferences [12], a three-dimensional (3-D) capacitance simulator was used to obtain the floating-gate capacitive coupling interference in 0.12-μm design rule cell (gate length = gate space = floating-gate height = channel width = 0.12 μm, tunnel-oxide thickness = 7.5 nm, IPD (ONO) thickness = 15.5 nm). If a neighbor cell is programmed from $V_t = -3$ V to $V_t = 2.2$ V, there are floating-gate interferences of 0.19 V in the y-direction, 0.04 V in the x-direction, and 0.01 V in the diagonal direction (xy-direction).

The floating-gate capacitive coupling interference has a linear characteristic with respect to the adjacent cell V_t change, as derived from (5.2) or (5.3). Figure 5.15 demonstrates the measurement results of the floating-gate capacitive coupling interference on a 0.12-μm design-rule cell [12]. The cell V_t shift by floating-gate interference is linearly proportional to the adjacent cell V_t change. The interference can be reduced significantly with a silicon oxide spacer as compared to a silicon nitride spacer, due to lower parasitic capacitance.

Figure 5.16 shows the simulation results of the V_{th} shift caused by floating-gate capacitive coupling interference with cell technology node scaling [50]. In 3-D TCAD simulations, a 63-nm memory cell has 8-nm tunnel-oxide thickness, 15-nm ONO thickness, and 85-nm floating-gate height. And the memory cell transistor has been scaled down from 63 to 20 nm. Along with cell size reduction, the field oxide recess is kept as +5 nm, and the doping concentration is adjusted to prevent

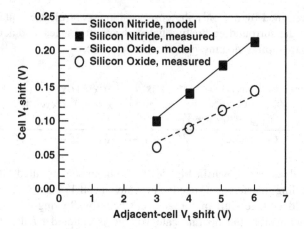

FIGURE 5.15 Floating-gate interference measurement results measured on a 0.12-μm design-rule cell array. The interference was measured on a WL8 cell as a function of WL9-cell V_t change for silicon nitride and silicon oxide spacer samples. Threshold voltage shift of the WL8 cell is monitored before and after WL9-cell programming from $V_t = -3$ V to 2.2 V. Each data consists of 15 points of a WL8 cell.

the cell transistor from punch-through. V_{th} shift by floating-gate capacitive coupling interference is drastically increased as technology node scaling.

Figure 5.17 shows the floating-gate capacitive coupling interference as a function of the technology node [23]. Floating-gate capacitive coupling interference as a percentage of the total V_t shift is remarkably increased by scaling the technology

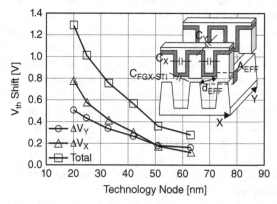

FIGURE 5.16 Simulation results of the V_{th} shift caused by cell-to-cell interference with cell size reduction. By changing the neighboring cell transistor V_{th} from −5 to 5 V, the V_{TH} shift of the reference cell is measured from 1.0 V of the initial V_{th}. Other word lines possess 6.5 V of the pass-gate voltage in a read operation. ΔV_X is the cell V_{th} shift induced by two adjacent cell transistors in the x-direction (word-line direction), while ΔV_Y is induced by a cell transistor in the y-direction (bit-line direction).

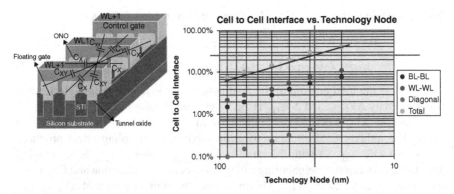

FIGURE 5.17 FG–FG coupling interference. Floating-gate interference as a percentage of the total V_t shift is shown as a function of the lithographic node. The BL–BL, WL–WL, and diagonal terms are per edge. FG–FG capacitive coupling interference exceeds 30% of FG capacitance beyond 30-nm generation. Need a solution beyond 30 nm for MLC 2 bits/cell, if assumed total interference <1 V in adjacent cell $dV_t = 4$ V → total <25%.

node. FG capacitive coupling interference exceeds 30% of FG capacitance beyond 30 nm generation. The interference approaches 50% of the total V_t shift of the cell at the 20-nm node. We need a solution to manage interference beyond 30 nm. It is estimated for MLC 2 bit/cell that the FG–FG capacitance has to be less than 20% of total FG capacitance, if assumed total interference is <1 V in an adjacent cell $\Delta V_t = 4$ V.

5.3.2 Direct Coupling with Channel

Based on the conventional theory of floating-gate capacitive coupling interference, the V_{th} shift in the y-direction (ΔV_Y) of a 63-nm technology node is more severe than that in the x-direction (ΔV_X), as shown in Fig. 5.16 [50], since the floating gates face each other directly in the y-direction, while it is shielded by a recessed control gate in the x-direction. However, it was observed that ΔV_X exceeds ΔV_Y at the node size of 50 nm, and it increases drastically as the technology node size reduces to 20 nm.

In the sub-50-nm technology nodes, the distance between the channel edge of a cell transistor and the floating gate of a neighboring cell is very close that the floating-gate voltage of the neighboring cell directly influences the channel edge, changing the electric field distribution on the channel edge. Then, V_{th} shift is caused by the direct field effect of the floating-gate potential of the neighboring cell. Since about 70% of the cell current flows on the channel edge, the V_{th} of the cell transistor is determined mostly on the condition of electric field crowding and the doping concentration of the channel edge [51]. Therefore, the memory cell suffers an intense V_{th} shift, particularly in the x-direction, where the floating gate faces the whole surface of the channel edge. This means that the observed floating-gate capacitive coupling interference is including the channel edge coupling with a neighbor cell.

The interference in the x-direction (ΔV_X) can be expressed as follows [50]:

$$\Delta V_X = \Delta V_{X-\text{Indirect}} + \Delta V_{X-\text{Direct}}$$

$$= 2*(C_{\text{FGX}}/C_{\text{Tot}})*\Delta V_{\text{FGX}} + \alpha*C_{\text{FGX-STI}}*\Delta V_{\text{FGX}} \tag{5.4}$$

where ΔV_X is the total amount of the floating-gate capacitive coupling interference effect caused by ΔV_{FGX}, and ΔV_{FGX} is the V_{th} change of the adjacent cell transistor in the x-direction. ΔV_X is decomposed into two terms caused by indirect and direct field effects. The indirect field effect or parasitic capacitance-coupling effect produces a V_{th} shift ($\Delta V_{X-\text{Indirect}}$) with the ratio of $C_{\text{FGX}}/C_{\text{Tot}}$, where C_{FGX} is the FG–FG capacitance between two neighboring cell transistors in the x-direction and C_{Tot} is the total amount of capacitance of FG. This means that the indirect V_t shift of $\Delta V_{X-\text{Indirect}}$ is conventional floating-gate capacitive coupling interference. The direct field effect causes a V_{th} shift with the amount of $\alpha*C_{\text{FGX-STI}}*\Delta V_{\text{FGX}}$, where $C_{\text{FGX-STI}}$ is the capacitance between the floating gate of a neighboring cell transistor and the channel edge. α is constant, defining the influence of direct field effect, representing the doping profile and tunnel-oxide thickness on the channel edge. In the sub-100-nm technology nodes, $C_{\text{FGX-STI}}$ has been negligibly small due to a long distance between the floating gate of a neighboring cell and the channel edge. However, as the cell size reduces to below 50 nm, $C_{\text{FGX-STI}}$ increases in a large amount and builds up a large electric field on the channel edge. Therefore, combined with boron segregation on the channel edge, a large $C_{\text{FGX-STI}}$ causes an intense V_{th} shift on the channel edge, leading that $\Delta V_{X-\text{Direct}}$ exceeds $\Delta V_{X-\text{Indirect}}$ in the sub-50-nm technology nodes.

This effect was confirmed with 3D TCAD simulations [50]. The simulated cell had a 45-nm design rule with a gate pitch and an active pitch of 90 nm, as shown in Fig. 5.18. The potential distribution of the tunnel oxide and field oxide of a selected

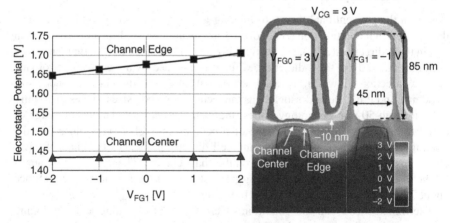

FIGURE 5.18 Simulation results depicting the potential distribution of the tunnel/field oxide of a selected cell transistor FG0 and its change with the floating-gate potential of a neighboring cell transistor FG1. The figure on the right side shows a representative electrostatic potential distribution in the case of $V_{\text{FG1}} = -1$ V.

cell transistor FG0 and its change with the floating-gate potential of a neighboring cell transistor FG1 are shown in Fig. 5.18 [50]. On the change of V_{FG1} from -2 to 2 V, it was observed that the potential on the channel edge increases from 1.65 to 1.71 V, while the potential on the channel center is kept constant on 1.43 V, showing the influence of the neighboring cell transistor potential on the channel edge. V_t is also simulated to be 0.62 V at $V_{FG1} = 2$ V, and it increases to 0.82 V in the case where $V_{FG1} = -2$ V. While the potential on the channel edge changed small from 1.65 to 1.71 V, the cell transistor V_{TH} is shifted largely from 0.62 to 0.82 V. This is because severe boron segregation occurs on the channel edge. Practically, the V_{th} shift will be twofold when considering the capacitance-coupling ratio as 0.5. This result indicates that the direct field effect of the adjacent cell transistor changes the cell V_{th} intrinsically, and it is larger than the effect of the FG–FG capacitive coupling.

Experimental data of cell-to-cell interference in a 45-nm cell are shown in Fig. 5.19 [50]. As the field oxide recess decreases, the direct field effect of the neighboring cell on the channel edge increases so that the V_{th} shift becomes larger. There are three lines in Fig. 5.19, namely, the V_{th} shift of conventional floating-gate capacitive coupling interference (ΔV_X—Indirect), V_{th} shift of direct field effect (ΔV_X—Direct), and V_{th} shift measured in experiments (ΔV_X). Both (ΔV_X—Indirect) and (ΔV_X—Direct) are calculated and classified by a 3D device simulator. While (ΔV_X—Indirect) varies a little with the field oxide recess and shows 0.28 V at -25 nm of field oxide recess, it is seen that (ΔV_X—Direct) increases drastically and reaches 0.67 V at -25 nm of field oxide recess. Moreover, the summation of two terms generates 0.95 V with an error of 0.08 V when compared with the experimental result, ΔV_X of 0.87 V. This experimental result demonstrates the strong influence of the direct field effect on floating-gate capacitive coupling interference in the x-direction. Therefore, in order

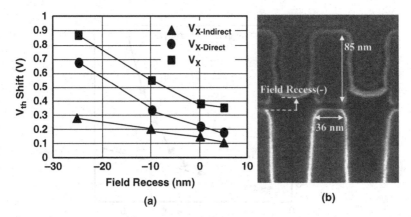

(a)   (b)

FIGURE 5.19   (a) V_{th} shift dependence on field recess. Four field recess conditions are prepared with field oxide recesses of -25, -10, 0, and $+5$ nm and the effect of cell-to-cell interference is measured by changing the neighboring cell transistor V_{th} from -5 to 5 V. (b) A representative cross-sectional SEM photograph of a NAND flash cell transistor with a 45-nm node size.

to reduce this effect in a NAND flash cell below a 50-nm cell, it is necessary to maximize the field oxide recess because keeping its balance to avoid the abnormal negative V_t shift effect in a large (deep) field recess [52], as described in Section 6.7.

5.3.3 Coupling with Source/Drain

A new cell-to-cell interference phenomenon of floating-gate induced barrier enhancement (FIBE) had been reported [53] in scaled cells below the 40-nm design rule. Unlike conventional capacitive coupling between floating gates, the threshold voltage (V_{th}) shift of the interfered cell becomes significantly large beyond some V_{th} of the interfering cell. This is due to modulation of the conduction band at the source and drain regions by capacitive coupling between source/drain and the floating gate of the interfering cell. The model was confirmed by experiment and simulation. In order to reduce the FIBE effect, the higher doping for S/D junction and the higher V_{read} scheme in neighbor WL could be effective.

Figure 5.20 shows the cell-to-cell interference between WLs (y-direction) [53]. The V_{th} of an interfered cell increases abnormally in region B (higher Interfering cell V_{th}), while the V_{th} of an interfered cell has a linear dependency on the lower V_{th} of the interfering cell at region A, which shows the conventional floating-gate capacitive coupling. This abnormal V_t increase at region B is observed only in higher V_t of an interfering cell for scaled dimension of NAND flash memory cell beyond 40 nm.

A new model of floating-gate induced barrier enhancement (FIBE) was proposed [53] for this phenomenon. The programming of an interfering cell causes the higher potential of the channel conduction band of an interfered cell at the drain-side region due to direct capacitive coupling between the source/drain region and the floating gate of an interfering cell, resulting in the increase of the cell V_{th}, as shown in Fig. 5.21a.

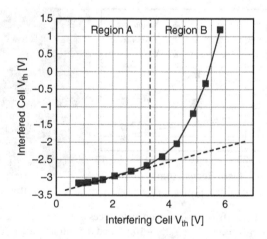

FIGURE 5.20   The V_{th} shift of interfered cell dependence on the interfering cell V_{th}. The initial V_{th} of an interfered cell is set to -3 V and interfering cell is 0.6 V. Linearity of region A can be explained by the conventional parasitic capacitors coupling.

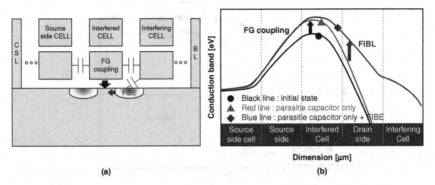

FIGURE 5.21   (a) Simplified image of parasitic coupling capacitances in the WL direction of a NAND flash cell. The source/drain junction is coupled with a floating gate of an interfering cell. (b) Conduction band profile beneath tunnel oxide. The word-line interference (y-direction interference) consists of two factors of the conventional floating-gate capacitive coupling and the FIBE (floating-gate induced barrier enhancement).

Figure 5.21b shows the simulated conduction band contour of the interfered cell in the case of a pre-programming interfering cell, with parasitic capacitance and with parasitic capacitance + FIBE effect. The FG potential change of the interfered cell with conventional floating-gate capacitive coupling appears to increase only the conduction band at the channel center as shown in Fig. 5.21b. However, the FG potential change of the interfering cell enhances the conduction band of the drain region and affect the channel conduction band of an interfered cell as shown in Fig. 5.21b.

The FIBE appears at a relatively high V_{th} region of the interfering cell where the drain region conduction band was enhanced sufficiently. Figure 5.22 shows the measured I_d–V_g curve influenced by the floating-gate coupling and FIBE in 27-nm node NAND flash cell [53]. If only the conventional floating-gate capacitance coupling interference is considered, we can see just mid-gap voltage shift in Line 2 from the original Line 1 without the slope change of the I_d–V_g curve in Fig. 5.22. However, in Line 3, the saturation region of I_d–V_g curve is distorted, and this distortion is extended even to the linear region of I_d–V_g curve when the interfering cell is programed highly to 5.5 V. This phenomenon makes V_t abnormally higher, as shown in Fig. 5.20.

5.3.4   Air Gap and Low-k Material

It had been introduced that the floating-gate capacitive coupling interference improved by using a low-k dielectric of gate spacer, such as low-k oxide and air gap [47,48,54, 55].

One example of a process flow to form an air gap between gates [47] is shown in Fig. 5.23A. After gate patterning, buffer oxide/nitride is deposited (Fig. 5.23A, part b), and then oxide is deposited to fill gate space (Fig. 5.23A, part c). After that,

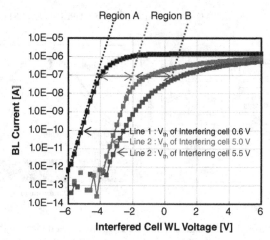

FIGURE 5.22 BL current versus WL voltage of an interfered cell. The V_{th} of an initial interfered cell is set to be -4 V, and the V_{th} of an interfering cell is set to 0.6 V. After the interfering cell is programmed up to 5.0 V, the BL current is parallel shifted up (Line 2). And after the interfering cell is programmed up to 5.5 V (V_{read}: 7 V), BL current is distorted (Line 3) by the FIBE effect.

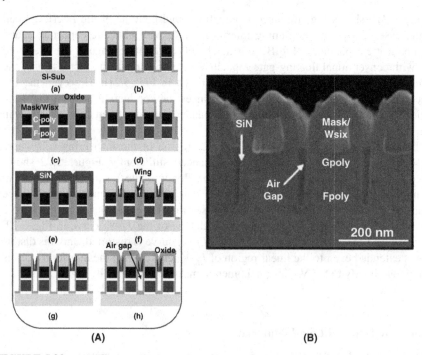

FIGURE 5.23 (A) The process flow of the air gap: (a) The gate patterning and the barrier silicon dioxide deposition (150 Å), (b) the barrier SiN deposition (200 Å), (c) thick-oxide deposition (1000 Å), (d) thick oxide is removed by dry etch, (e) SiN deposition (150 Å), (f) The wing is formed, (g) The thick oxide inside gate to gate space is removed by wet etch. (h) The air gap is formed. (B) The SEM image of air gap in a 90-nm cell (gate length = gate space = floating gate height = channel width = 90 nm, tunnel oxide thickness = 6.5 nm, ONO thickness = 16 nm).

the oxide over the gate poly is removed by dry etching (Fig. 5.23A, part d). Then, to form the gate wing inside gate-to-gate space, the SiN is deposited and etched (Fig. 5.23A, part f). After the oxide inside the spacer wings are removed by wet etching (Fig. 5.23A, part g), air gaps inside gate-to-gate space are formed by oxide deposition (Fig. 5.23A, part h). In Fig. 5.23B, the air gap can be clearly observed from SEM image of the fabricated device.

Figure 5.24a–c shows the cell V_{th} distribution of WL30 and WL31 in 90-nm cells with gate-space materials of SiN, oxide, and air gap, respectively [47]. The cell V_{th} on WL30/even bit line are shifted by programming the adjacent cells in the same word line (WL30)/odd bit line and WL31/even bit-line cell. As a cell is programmed from $V_{th} = -3$ V to $V_{th} = 1.5$ V, the threshold voltage changes for gate space material of SiN, oxide, and air gap are 0.16 V, 0.07 V, and 0.02 V, respectively. The V_{th} shift nearly corresponds to those of the dielectric constant (SiN:oxide:air = 8:4:1). The reduced V_{th} shift with air gap is due to its lower parasitic capacitance between floating gates. Figure 5.24d compares the cell V_{th} distribution of a 1Gbit cell by a single pulse program. The cell V_{th} distribution is improved with air gap due to the improved floating-gate capacitive coupling interference.

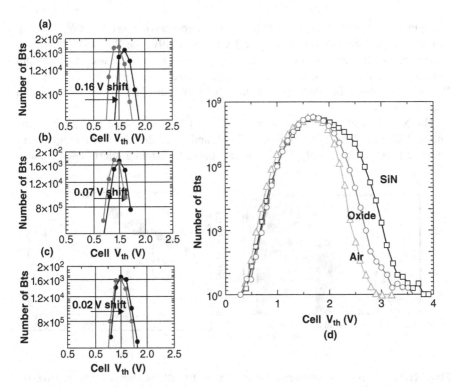

FIGURE 5.24 Threshold voltage shift by floating-gate interference was measured on WL30. (a) SiN spacer, (b) oxide, (c) air gap, and (d) the cell V_{th} distribution shifted by floating-gate interference.

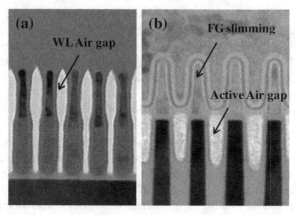

FIGURE 5.25   The cross-sectional TEM micrograph of (a) word-line air gap and (b) STI air gap in a mid-1X-nm SA-STI NAND flash cell. Air gap is key technology to improve a floating-gate capacitive coupling interference, a WL high-field problem, and a program disturb. WL air gap and STI air gap are applied to the product from 25-nm generation and 20-nm STI half-pitch, respectively.

Figure 5.25 shows the cross-sectional TEM micrograph of word-line (WL) air-gap and STI air-gap structure in the middle 1X-nm cell [49]. The WL and STI air gap were successfully fabricated. The WL air gap was started to be used from 25-nm generation product to reduce floating-gate capacitive coupling interference [34]. And WL air gap could also improve WL high-field problem [10, 11], as described in Section 5.7.

The STI air gap was started to be used from 20-nm bit-line half-pitch in the middle 1X-nm cell, as shown in Fig. 5.25 [49] and Fig. 5.26 [49]. STI air gap is very effective to improve not only floating-gate capacitive coupling interference but also

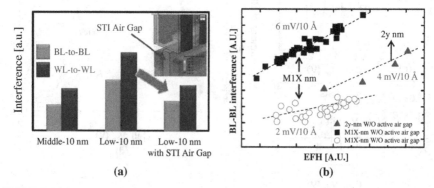

FIGURE 5.26   (a) A simulation result shows that the BL–BL interference value in a low-10-nm cell could be similar to that of a middle-10-nm cell (middle 1X-nm cell) by using STI air gap. (b) BL–BL interference of middle 1X-nm and 2y-nm design rule. EFH is "effective field height" of STI buried oxide (see FH in Fig. 6.72).

program disturb, which is related on channel–channel coupling [31,49,56] described in Section 6.5.3.

These air-gap technologies are the key to implement small cell size below 20-nm design rule, because of improvement of floating-gate capacitive coupling interference [35], WL high-field problem [10], and program disturb [49].

5.4   PROGRAM ELECTRON INJECTION SPREAD

5.4.1   Theory of Program Electron Injection Spread

By scaling memory cell size, the number of stored electrons in floating gate is reduced, as shown in Fig. 5.27 [19,57]. In 1X-nm memory cell, number of stored electron reaches close to 100, which is corresponding to 3-V V_t shift. It means that only 10 electrons are injected to floating gate in one programming pulse which has 300-mV step-up between each pulse. A small number of 10 electrons should make a large statistical variation on number of injected electrons to floating gate, resulting in wider V_t distribution width of program states.

It had been reported that the programmed V_t distribution width became wider by statistical electron injection spread during program pulse [13–15]. The electron injection process is ruled by the Poisson statistics when small number of electrons is injected during program pulse. This can be explained by the reduction of the tunnel-oxide field that follows the electrons injection to floating gate and then reduced the electron injection rate. The results are explained by means of a Monte Carlo model, which is able to correctly describe the main physics behind the program operation.

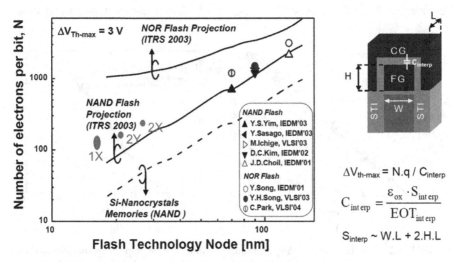

FIGURE 5.27   Number of stored electrons in FG, as a function of the flash memory technology node according to the ITRS 2003 edition. The number of electrons is decreased as scaling memory cell size.

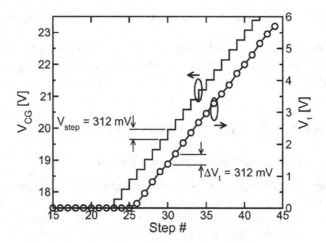

FIGURE 5.28 Example for a control-gate voltage waveform used to program a NAND cell and resulting V_t transient on a 60-nm device. Note that only positive V_t values can be sensed in a conventional NAND cell array.

The distribution is shown to broaden as a consequence of the injection statistical spread, leading some cells to displace from the verify level more than V_{step}. The injection statistical spread is considered to be larger as scaling the NAND memory cell due to reduced number of electrons in one program pulse.

An experiment of the ΔV_t spread had been studied by using the ramped programming (ISPP: incremental step pulse programming [36,37,58]), as shown in Fig. 5.28 [13, 14]. ΔV_t was defined as the V_t shift obtained after n_s programming pulse steps. For example, Fig. 5.29a shows $\Delta V_t = V_{t,29+ns} - V_{t,29}$ transients, obtained using V_t at

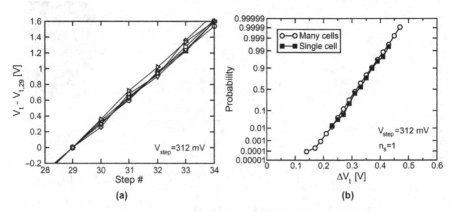

FIGURE 5.29   (a) Example of ΔV_t transients (assuming V_t at step 29 of the staircase as reference) measured on the same 60-nm technology NAND cell. (b) ΔV_t evaluated using many programming ramps on the same cell or a single programming ramp on a large number of cells, for $V_{step} = 312$ mV and $n_s = 1$.

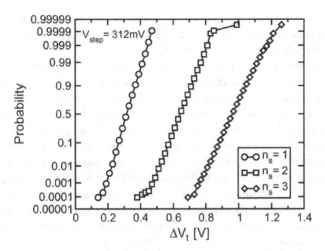

FIGURE 5.30   Experimental ΔV_t distributions for $V_{step} = 312$ mV and increasing n_s, showing the increase of $\sigma_{\Delta V_t}$ that follows the larger average ΔV_t value.

the 29th step of the control-gate ramp as a reference, on the same NAND memory cells in 60-nm technology. The statistical character of ΔV_t is clearly appeared. And the cumulative distribution of ΔV_t is shown in Fig. 5.29b for $n_s = 1$ (i.e., $V_{t,30}$ − $V_{t,29}$), using nearly 100 V_t transients on the same cell. These results of single cell are compared with the ΔV_t distribution obtained from a V_t programming transient on a page (16 kb) of the NAND memory cell array. It can be seen that there is very good agreement between the two distributions, confirming that we are observing the same statistical distribution on single- and many-cell results. Also, it can be seen that the distributions clearly show a Gaussian behavior, with a spread nearly equal to the standard deviation $\sigma_{\Delta V_t} = 41$ mV.

Figure 5.30 [14] shows the ΔV_t distribution from many cell statistics in the case of $V_{step} = 312$ mV. The distribution spread of ΔV_t is clearly increased with increasing staircase pulse, along with increasing the average of ΔV_t. The injection spread is strictly related to the ΔV_t statistics, based on the following relation:

$$\sigma_{\Delta V_t} = \frac{q}{C_{pp}} \sqrt{\sigma_n^2}$$

where q is the electron charge, n is the number of injected electrons, and C_{pp} is inter-poly capacitance. By assuming that n is ruled by Poisson statistics, its variance σ_n^2 is equal to its average value n, and the previous equation becomes

$$\sigma_{\Delta V_t} = \frac{q}{C_{pp}} \sqrt{\bar{n}} = \sqrt{\frac{q}{C_{pp}} \overline{\Delta V_t}} \qquad (5.5)$$

A square-root dependence of $\sigma_{\Delta V_t}$ on ΔV_t is expected from (5.5). However, the further consideration derives from the hypothesis of Poissonian injection. In fact, when an electron is injected to the floating gate, the floating-gate potential energy rises. This reduces the tunnel-oxide field and the electron injection rate, thus causing a sub-Poissonian electron injection process.

In order to involve the effect of the tunnel-oxide field feedback, Monte Carlo simulations of the electron injection process had been performed. For each floating-gate potential, the average electron injection rate can be calculated as JA/q, where A is the cell area and J is the tunneling current density through the tunnel oxide. To obtain a reasonable accuracy, the tunneling current–floating-gate voltage (J–V_{FG}) characteristics were experimentally extracted from constant control-gate voltage programming transients [13]. The average injection rate was used to extract the time of the next electron injection event from the substrate. Simulations are performed with considering the control-gate steps (increasing the floating-gate potential and the average injection rate) and the electron injection events (decreasing the tunnel-oxide field and the average injection rate), causing a nonstationary Poisson process. The $\sigma_{\Delta V_t}$ is then extracted from the simulation of many V_t transients.

Figure 5.31 shows the experimental and calculated $\sigma_{\Delta V_t}$ as a function of ΔV_t, extracted in the stationary part of the programming transient for different V_{step}, n_s, and step durations. As the ΔV_t spread depends only on the electron injection process, $\sigma_{\Delta V_t}$ depends only on ΔV_t and does not depend on the duration of the steps of the control-gate staircase or the number of steps used to reach the ΔV_t value [13, 14]. For low ΔV_t, both experiments and Monte Carlo calculations well match the values predicted assuming a Poisson statistics for the electron injection process. This is because the feedback to the tunnel-oxide field is negligibly small when the number of injected electrons is small. However, when ΔV_t is increased, a saturating behavior clearly appears both from experiments and from Monte Carlo simulation.

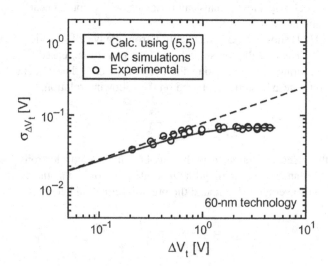

FIGURE 5.31   Experimental and calculated $\sigma_{\Delta V_t}$ as a function of the average ΔV_t in 60-nm NAND flash cell technology.

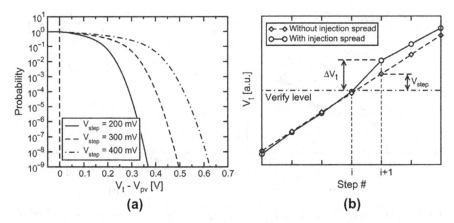

FIGURE 5.32 (a) Calculated 1-cumulative probability for $V_t - V_{pv}$ assuming constant-current NAND programming with different V_{step} values in the 60-nm NAND flash technology. (b) V_t evolution for increasing step numbers with and without considering the injection spread. In the presence of a verify level, the worst situation is obtained when, at step i, cell V_t is slightly lower than V_{pv}, thus requiring an additional program step. This shifts cell V_t to $V_{pv} + V_{step}$ when the injection spread is not considered, but to larger values when the injection spread is included.

This reveals the sub-Poissonian nature of the electron injection process, determined by the tunnel-oxide field feedback following each electron injection event. The starting point separation between the $\sigma_{\Delta V_t}$ curve from the Poissonian spread and its saturation are dependent on the shape of the J–V_{FG} characteristics, whose slope around the programming condition determines the field variation.

Figure 5.32a shows the effect of the injection spread on the V_t distribution by using bit-by-bit program verify operation [43] with program-verify voltage of V_{pv}. Results had been calculated in the case of three staircase steps, 200, 300, and 400 mV on a 60-nm NAND flash cell. If the injection spread is neglected, the ISPP programming algorithm should make all V_t's to fit between V_{pv} and $V_{pv} + V_{step}$ in principle [59] (see Section 2.2.3). Cells which have slightly lower than V_{pv}, are required to be subjected an additional program pulse to be higher than the verify level. This additional program pulse shifts V_t by V_{step}, projecting the cell to $V_{pv} + V_{step}$, as shown in Fig. 5.32b. However, when the injection spread is included, V_t values larger than $V_{pv} + V_{step}$ are caused. An example of this situation (shown in Fig. 5.32b) is due to the possibility to have single-step ΔV_t values larger than V_{step}, thus causing a cell to move further away from the verify level.

As a scaling of the NAND cell, inter-poly capacitance C_{pp} is decreased (namely number of stored electrons are decreased), then $\sigma_{\Delta V_t}$ is increased. This trend is clearly shown in Fig. 5.33a, where $\sigma_{\Delta V_t}$ is shown for different NAND cell technologies of 90-nm, 70-nm and 60-nm nodes. As the tunnel-oxide conduction characteristics are kept the same with technology scaling, the same saturating behavior is observed for all the curves. The spread corresponding to ΔV_t equal to 200, 300, and 400 mV is

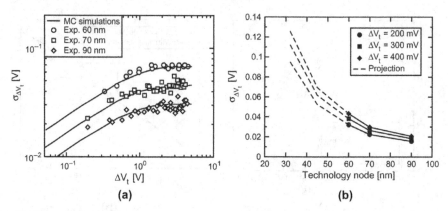

FIGURE 5.33 (a) Experimental and calculated $\sigma_{\Delta V_t}$ as a function of the average ΔV_t value for different technology nodes from 90 to 60 nm. (b) $\sigma_{\Delta V_T}$ for different ΔV_t values in the Poissonian region of the electron injection process as a function of the NAND technology nodes. Continuous curves and symbols refer to the available technologies; dashed curves are calculated projections assuming that C_{pp} scales with cell area. As number of stored electron in FG is decreased (C_{pp} decrease) as memory cell scaling, program V_t width is larger (worse) due to program variation.

shown in Fig. 5.33b for the available technologies, drawing also possible scaling projections. The $\sigma_{\Delta V_t}$ is drastically increased with scaling technology node. This result shows that electron injection spread would present a serious problem to make a tight V_t distribution width in multilevel cell for the scaled NAND flash memory. In order to keep the V_t distribution width as close as possible to the verify level without using very small V_{step} amplitudes, the scaling of C_{pp} should be carefully considered.

5.4.2 Effect of Lower Doping in FG

The electron injection spread is enhanced by lower doping concentration in floating gate (FG) [38]. Larger ΔV_t distribution has been observed in a cell which has a low floating-gate doping concentration in a 40-nm design rule cell, as shown in Fig. 5.34a, where ΔV_t means the V_t shift from j to $j + 1$ step-up programming pulses [$\Delta V_t \equiv V_t (j + 1) - V_t(j)$]. The ΔV_t distribution of the low FG doping shows a wider spread than that of the high doping, and tail bits are observed at the higher ΔV_t in the case of the low FG doping.

The reason why ΔV_t distribution in low doping is larger can be explained by the dynamics of forming an inversion layer in FG and electron–hole generation by FN (Fowler–Nordheim) tunneling electron injection, as follows. The details of the time dependence of the FG potential are considered in each programming step. At the beginning of the Nth programming pulse duration, the tunnel-oxide interface in the FG is deeply depleted due to the large electric field in the tunnel oxide. And a large band bending is caused at the tunnel-oxide interface in the FG, as schematically shown in Fig. 5.35a. The tunneling electrons become energetic at FG, and they

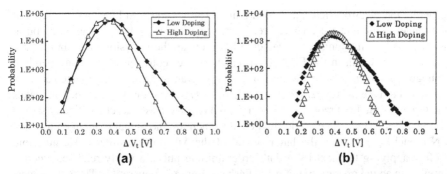

FIGURE 5.34   (a) Measurement results of bit-by-bit V_t transient (ΔV_t) distribution from i to $i + 1$ staircase programming pulses. ΔV_t distribution sampling points are accumulated by adding the data of $i = 12$–17. V_{step} is 400 mV. The NAND cell array with low FG phosphorus doping shows wider ΔV_t distribution than that with high doping. (b) Calculated ΔV_t distribution considering the effect of FN tunneling statistics in both cases with low and high FG phosphorus doping.

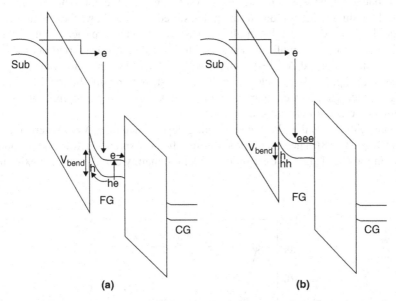

FIGURE 5.35   Schematic 1-D band diagram for the NAND cell during programming. (a) Band diagram at the initial stage of programming and the process of FN tunneling currents that generate e–h pairs. Generated holes move to the tunnel-oxide interface, and generated electrons move to the interface of ONO. (b) Band diagram after the passage of programming time. The inversion layer of holes has formed and degenerated the deep-depletion region (band bending height becomes smaller, as compared with that of initial case).

generate the electron–hole pairs (e–h pairs). These generated holes are gathered at the tunnel-oxide interface. Therefore, as the programming time (programming pulse width) has become longer, the gathered holes create the inversion layer and reduce the depletion width in the FG, which results in the reduction of the band bending voltage V_{bend}. Generally, in the case of longer programming time, the injected charges to the FG reduce the tunnel-oxide electric field. However, the V_{bend} reduction makes the tunnel-oxide field reduction slower. This V_{bend} reduction is schematically shown in Fig. 5.35b. Thus, the electric-field enhancement on the tunnel oxide increases the FN tunneling current at the latter period of the Nth programming pulse duration. Before applying the next $(N + 1)$th programming pulse, a verify read sequence is inserted in actual operation of NAND flash programming operation. Then, the generated holes during programming pulse diffuse into the entire FG area and will almost recombine with electrons during the relatively long verify read period. Therefore, the deep depletion in the FG repeatedly occurs at the beginning of the next $(N + 1)$th programming pulse duration. This electric-field enhancement effect through the tunnel oxide is exaggerated to appear in the case of lower phosphorus doping in the FG because of the larger V_{bend} at the beginning of each programming pulse. The wider ΔV_t distribution in the lower doping FG can be analyzed by this effect.

This phenomenon was simulated based on this model of combining the effect of the band bending reduction due to the holes stored in the FG and the FN tunneling statistics [13,59]. Monte Carlo simulation was carried out where each program pulse was divided into many small segments, and the calculation was carried out by each segment. The simulated ΔV_t distributions with high and low FG phosphorus doping are shown in Fig. 5.34b. It is clear that the ΔV_t distribution with the low FG doping shows a wider spread in comparison with the case of the high doping, due to the tunnel-oxide electric-field enhancement effect.

Figure 5.36 shows the phosphorus doping dependence of $\sigma(\Delta V_t)$, where V_{step} is fixed at 400 mV [38]. The ΔV_t distribution significantly spreads in lower phosphorus doping in the FG. This ΔV_t distribution widening mainly comes from the existence

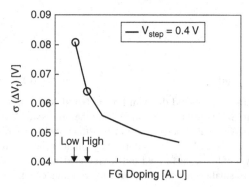

FIGURE 5.36   Calculated $\sigma(\Delta V_t)$ as a function of FG phosphorus doping. V_{step} is fixed to 400 mV.

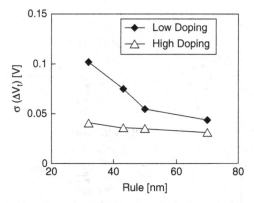

FIGURE 5.37   Calculated $\sigma(\Delta V_t)$ as a function of technology nodes. $\sigma(\Delta V_t)$ values with low and high FG phosphorus doping are compared, where V_{step} is fixed at 400 mV.

of the tail bits at the high ΔV_t side, as shown in Fig. 5.34b. Conversely, the amount of tail bits at the high ΔV_t side can be reduced as the FG doping increases. It means that, when the FG doping is higher, the σ plot of the ΔV_t distribution shows clearer linearity. This result provides a new guideline for the design of NAND cell process.

Generally, the $\sigma(\Delta V_t)$ is increased as scaling down of the NAND cell, because of the reduction of C_{pp}. Also, the tunnel-oxide electric-field enhancement effect at the lower FG impurity doping accelerates the increase in $\sigma(\Delta V_t)$ as the cell size scaling, as shown in Fig. 5.37, where the effect of FN tunneling statistics is considered. The increase in $\sigma(\Delta V_t)$ introduces a new reliability constraint to the design of read window margin (RWM) of the future NAND technologies. The upper limit of the impurity doping would come from the tunnel-oxide reliability degradation. Conversely, the lower limit would come from this ΔV_t distribution spread.

5.5   RANDOM TELEGRAPH SIGNAL NOISE (RTN)

5.5.1   RTN in Flash Memory Cells

Random telegraph noise (RTN) in a MOSFET is the drain current or threshold voltage fluctuation caused by electron capture and emission events at a charge trap site near the gate-oxide interface, as shown in Fig. 5.38 [69]. The amplitude of the threshold voltage fluctuation by each trap site ($\Delta V_{t_{trap}}$) in a flash memory cell is approximately [60–62]

$$\Delta V_{t_{trap}} = \frac{q}{L_{eff}W_{eff}\gamma C_{ox}} \tag{5.6}$$

where q is the elementary charge, L_{eff} and W_{eff} are the effective channel length and width respectively, γ is coupling ratio between the control and floating gates, and C_{ox}

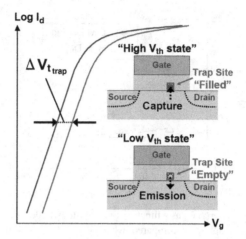

FIGURE 5.38 Threshold voltage fluctuation due to random telegraph noise (RTN) in a MOSFET is caused by electron capture and emission events at an oxide electron trap site.

is the gate capacitance. The amplitude of RTN is generally larger in a floating-gate flash memory cell than in a CMOS logic device, because of the very small C_{ox}, due to the relatively thick tunnel oxide (~10 nm thick). Also, in NAND flash memory, memory cell size has been intensively scaled down; thus dimensions of L and W, especially W, are much smaller than conventional CMOS logic device. Moreover, the amplitude of RTN can be larger than expected from Eq. (5.6) due to current-path percolation mechanism [63]. Therefore, RTN is a potential source of read failure in scaled NAND flash memory.

It had been reported for the first time that the threshold voltage (V_{th}) fluctuation due to random telegraph signal noise (RTN) had been observed in flash memory [16]. Figure 5.39 shows an example of RTN measured in 90-nm cell. Drain current shows a switching behavior as same as RTN in logic CMOS transistor.

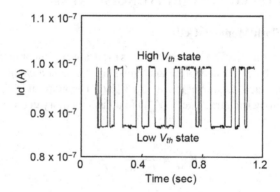

FIGURE 5.39 An example of time-series change in drain current in 90-nm-node flash memory.

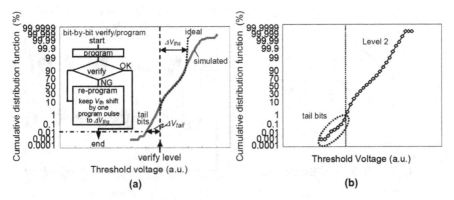

FIGURE 5.40   (a) Monte Carlo simulation results of programmed V_{th} distribution using bit-by-bit verify scheme. (b) Measured V_{th} distribution of Level 2 enlarged.

The influence of RTN for the multilevel flash memory was investigated by Monte Carlo simulation [16]. In multilevel flash memory, it is necessary to control the V_{th} precisely and make a tight V_{th} distribution. For such requirements, a bit-by-bit program/verify technique has been used [43] (Section 4.2.3). In this technique, the programming bias is applied only to the memory cells judged to "fail" in the verify operation. In addition, to keep V_{th} shifts by one programming pulse constant, the ISPP (incremental step pulse program) scheme (Section 4.2.2) is used. Figure 5.40a shows the simulated results of V_{th} distribution that is programmed by using bit-by-bit program/verify technique and ISPP scheme. In the simulated distribution with the RTN model, the tail bits appear in upper and lower of the V_{th} distribution in contrast with the ideal distribution without the RTN model. Figure 5.40b shows the measured V_{th} distribution. By comparison between Fig. 5.40a and 5.40b, we confirmed the existence of the tail bits generated by RTN in flash memory for the first time [16].

The properties of traps in the SiO_2 was investigated by means of a statistical analysis of random telegraph signal noise in flash memory arrays [65, 66]. A new physical model for the statistical superposition of the elementary Markov processes describing traps occupancy was developed. The comparison of modeling results with measured data is able to estimate the energy and space distribution of oxide defects, which are related to cell threshold voltage instability.

The random telegraph signal process [67] is schematically described in Fig. 5.41a [65]. An oxide trap has a distance x_t from the substrate/SiO_2 interface and energy E_t from the SiO_2 conduction band. An oxide trap can capture and emit single electrons from/to the substrate with average time constants τ_c and τ_e, giving rise to the typical behavior for the drain current shown in Fig. 5.41b and affecting V_t [67]. The properties of the trap responsible for the RTN can be experimentally extracted from the V_G dependences of τ_c and τ_e. Flash memory array was used to collect data, which could effectively evaluate a large number of devices. The RTN statistical distribution is then directly extracted from the V_t distribution, without any need to thoroughly characterize τ_c and τ_e for any single trap.

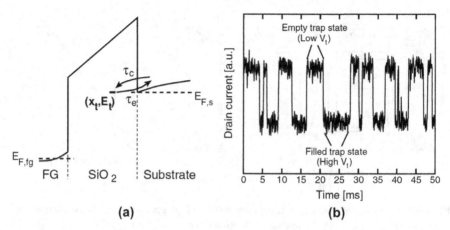

(a)   **(b)**

FIGURE 5.41 (a) Conduction band profile for a MOS structure under positive gate bias (read operation on a flash memory cell) and trap capture/emission processes. (b) Drain current in a flash memory cell as a function of time for fixed control-gate bias. Two-state RTN fluctuations can be clearly seen, associated to the empty and filled trap states.

A 512-Kbit NOR flash memory array in 65-nm technology was used to evaluate by sequentially reading the V_t for all the cells in the array up to 1000 times [66]. Figure 5.42a shows the measured V_t cumulative distribution at the first and the 100th read access on the array. No significant change in the V_t distribution between the first and the 100th read can be observed in the main distribution and in the tail. However, when ΔV_t between two read operations is evaluated for each cell, the V_t instability becomes clear, as shown in Fig. 5.42b, where the cumulative distribution (F) of ΔV_t between the first and the nth map. F is markedly different from the ideal step-like function centered in $\Delta V_t = 0$ that would be obtained for a stable V_t. And the cumulative distribution (F) also presents nearly exponential tails in its lower and upper parts. Moreover, these tails drift with time moving upward in the distribution.

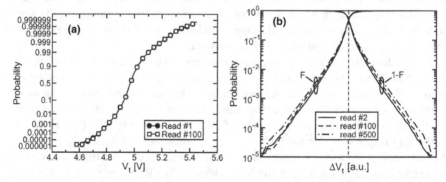

FIGURE 5.42 (a) V_t cumulative distribution at the first and the 100th read access to the array. (b) Experimental results for the cumulative probability F (and $1 - F$) of $\Delta V_t = V_t(n) - V_t(1)$, for read number $n = 2, 100, 500$. Cell V_t may shift randomly as $1/F$ noise.

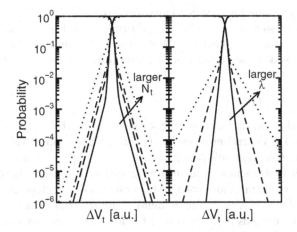

FIGURE 5.43 Cumulative probability distribution of ΔV_t calculated according to the model for different values of (left) N_t and (right) λ.

Figure 5.43 shows the cumulative distribution of ΔV_t, which is calculated by the model [66], as a function of the RTN-trap density N_t and the decay constant λ. Note that the model well reproduces the exponential behavior of the experimental ΔV_t distribution shown in Fig. 5.42b, with the tail amplitude determined by N_t and the tail slope related to λ. An increase in trap density N_t causes only an increase of the distribution tails, while the decay constant λ causes the "slope" of the exponential tails [68].

The model can be used for analysis of the traps location and energy in the oxide involved in the RTN V_t instability, as shown in Fig. 5.44. The increase in the elapsing time between the two read operations results in the activation of more traps, having

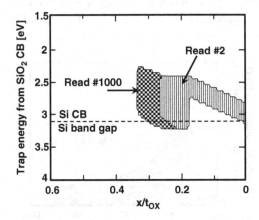

FIGURE 5.44 Tunnel-oxide region where traps that are active in the RTN V_t instability are located for read numbers 2 and 1000. An enlargement of the region inside the tunnel oxide appears for the increasing read number, corresponding to the activation of new traps in the RTN process. t_{ox} is the tunnel oxide thickness.

longer capture and emission time constants. This effect increases the randomness of cells V_t, thus increasing the width of the ΔV_t distribution. In Fig. 5.44, the RTN active traps are shown in the second and the 1000th read access, displaying an enlargement of the involved oxide region toward larger depths inside the tunnel oxide.

5.5.2   Scaling Trend of RTN

There are several reports that describe the scaling trend of RTN in flash memory cells [16, 39, 69, 70]. The dependences on gate length (L) and channel width (W) do not follow the same trend in these reports.

The RTN in flash memory becomes large as the device size is scaled down. Figure 5.45 shows the threshold voltage shifts estimated in each process node [16]. It was estimated that, if sense budget in multilevel flash memory is limited to about 1 V, total V_{th} shift exceeds the limitation of 1 V in a 45-nm process node.

Fukuda et al. [69] presented the statistical model of V_t fluctuation (ΔV_{tcell}) in 20 to 90-nm design rule floating-gate NAND flash memory cells. It considers current-path percolation, which generates a large-amplitude-noise tail, caused by dopant-induced surface potential nonuniformity.

The scaling cell size reduces the average number of trap sites in a memory cell and increases the noise contribution of each trap site, as shown in Fig. 5.46. It is interesting to note that smaller cells have larger 3-σ ΔV_{tcell} but smaller mean ΔV_{tcell} than larger cells. This results in widening of ΔV_{tcell} distribution with cell size scaling, as shown in Fig. 5.47a. The 3-σ ΔV_{tcell} extracted from Fig. 5.47a increases by 1.8× from 90 nm to 20-nm technology nodes as shown in Fig. 5.47b. This is a much smaller increase than $\propto 1/LW$ suggests (>10×) and $\propto (LW)^{-1/2}$ suggests (>3x). In other words, the prospect of scaling is less pessimistic than $\propto 1/LW$ and $\propto (LW)^{-1/2}$. It would indicate $\Delta V_{tcell} \propto (LW)^{-0.24}$, which is a much slower scaling trend than the commonly accepted $1/L_{eff}W_{eff}$ trend, as shown Eq. (5.6).

Ghetti et al. [39, 70] had also shown the scaling trend of NAND and NOR floating-gate cells. The scaling trend of RTN instabilities was investigated by using the Monte

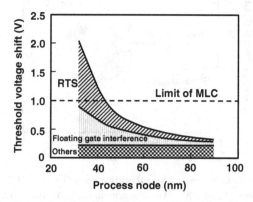

FIGURE 5.45   Estimation of threshold voltage shift as a function of process node.

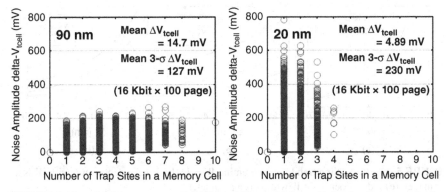

FIGURE 5.46   16-Mbit memory cell (1 Kbit × 100 page) Monte Carlo simulation results for 90 nm and 20-nm technology nodes. $N_D = 7E + 17/cm^3$ and $N_{trap} = 2E + 10/cm^2$ are assumed.

Carlo procedure with varying L, W, t_{ox}, and N_a, assuming discrete dopant atoms randomly placed according to a uniform distribution. The calculated slope of the RTN tails was divided by the control-gate to floating-gate capacitive coupling ratio α_G, to determine the real λ value of the flash cell.

Figure 5.48a shows the scaling trend for slope λ (see Fig. 5.43; unit is mV/dec) assuming $W = L$: a power-law $(W = L)^{-1.5}$ can well describe the dependence of λ on cell dimensions. This dependence is lower than the $(W = L)^{-2}$ (i.e., 1/WL) expected from pure 1D electrostatics, but stronger than the $1/\sqrt{WL}$ dependence proposed in

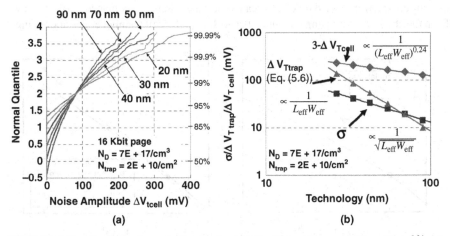

FIGURE 5.47   Random telegraph signal noise (RTN). dV_t is a linear with $(L_{eff} * W_{eff})^{-0.24}$. (a) Typical noise distributions of six technology nodes. Each trace represents the noise distribution of one 16-Kbit page. $N_D = 7E + 17/cm^3$ and $N_{trap} = 2E + 10/cm^2$ are assumed for all technology nodes. (b) Scaling trends of ΔV_{ftrap} of Eq. (5.6), σ, and 3σ of ΔV_{Tcell}. $N_D = 7E \pm 17/cm^3$ and $N_{trap} = 2E + 10/cm^2$ are assumed for all technology nodes.

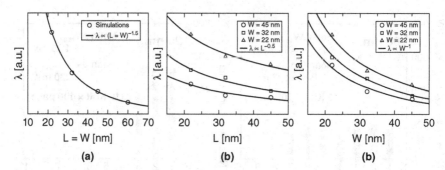

FIGURE 5.48 (a) Scaling trend for λ assuming $W = L$. (b) λ dependence on L with W as a parameter. (c) λ dependence on W with L as a parameter.

reference [69]. The separate dependences of λ on W and L were also investigated, as shown in Fig. 5.48b,c. Results indicate that the $(W = L)^{-1.5}$ power law can be decomposed in the form $W^{-1} \times L^{-0.5}$. This means that W has a stronger impact on λ due to the higher probability for a trap to effectively quench a percolation conduction path in narrower channels.

Figure 5.49a shows that the larger average substrate doping causes an increase of λ, due to the possibility to cause larger dis-uniformities of number of dopants in the channel inversion layer. This enhances the percolation effect and the current crowding at the cell channel edges which are responsible for the slope of the RTN exponentials, resulting in a square-root dependence of λ on Na.

Figure 5.49b shows that the scaling of the tunnel-oxide thickness t_{ox} reduces λ according to a slightly sublinear dependence $t_{ox}^{0.9}$. This is attributed to a less uniform current conduction over the active area as the gate electrode is placed further away from the channel, increasing the current crowding at the cell channel edges and giving

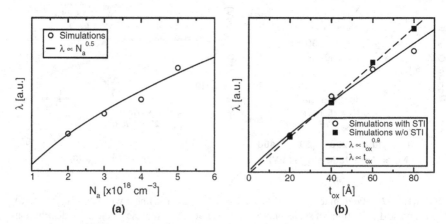

FIGURE 5.49 (a) λ dependence on Na for fixed cell geometry. (b) λ dependence on t_{ox} from 3D simulations including STI isolations (o) and for 2D-extruded structures neglecting cell active area edges (■).

a slightly weaker dependence with respect to the law of $\lambda \propto t_{ox}$. To confirm this result, Fig. 5.49b also shows simulation results obtained neglecting STI edges, that is, using 2D structures extruded in the W direction; in this case, a perfectly linear dependence of λ on t_{ox} is observed. The exponent 0.9 is therefore strictly related to the STI corner geometry, and a more general dependence t_{ox}^{α} with α slightly less than 1 can be adopted.

In summary of references [70 and 39], RTN scaling can be described by the following compact expression for λ that captures its dependence on all the main cell parameters:

$$\lambda = \frac{K}{\alpha_G} \frac{t_{ox}^{\alpha} \sqrt{N_\alpha}}{W \sqrt{L}} \tag{5.7}$$

This equation represents a powerful result to investigate the scaling trend of the RTN instabilities and to derive scaling guidelines to optimize the design of future technologies with respect to RTN.

Figure 5.50 [39] shows a comparison between simulations for λ and the experimental data for NAND and NOR technologies in different feature sizes. It can be seen that a good agreement is reached between experimental results and TCAD simulation value of λ. In addition, solid lines show the dependence predicted by (5.7) for a constant value of K, determined by fitting all the simulated cases in Figs. 5.48 and 5.49; and using the real values for cell parameters of NAND and NOR devices. A good agreement of (5.7) with the experimental trend of λ is achieved on both technologies of NAND and NOR, demonstrating its validity to make RTN extrapolations on future technology nodes.

In order to investigate the RTN phenomena in future scaled NAND flash memory cells, there are many reports regarding RTN, such as random discrete doping [71],

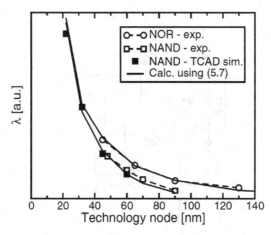

FIGURE 5.50 Experimental and simulation results for λ for different technology nodes. Results calculated by means of Eq. (5.7) are also shown and used for scaling projections.

non-equilibrium trap state [68], analytical investigation of the special and energetic position of trap [60], cycling impact on RTN [72], RTN impact on ISPP distribution [73], quick electron detrapping and random discrete dopants [74], tunnel-oxide nitridation effect [75], and inverse scaling phenomena due to the source/drain implantation condition effect in a 25-nm cell [76]. These reports are results of relatively larger device dimension (>25-nm cell); a practical data below 20-nm dimension cell will be presented in the future to clarify RTN mechanism in extremely scaled device.

5.6 CELL STRUCTURE CHALLENGE

A structural challenge of the SA-STI cell is also investigated, based on assumption in Table 5.1 in Section 5.2.1. The critical structure of the SA-STI cell is "CG formation margin," which is fabrication margin of CG between FGs [17]. Figure 5.51 shows an estimation of CG fabrication margin in the FG slimming structure beyond 2X-nm generation. FG width and CG width are assumed to be equal in this estimation. From

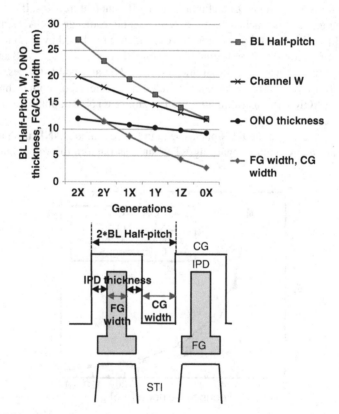

FIGURE 5.51 Estimated margin of CG fabrication between FGs. Very narrow CG and FG width of around 5 nm have to be controlled in 1Z-nm generation.

this estimation, as scaling down of the SA-STI cell, FG width and CG width are decreased to less than 10-nm width in 1X-nm cell. The FG and CG width have to be controlled around 5 nm in 1Y-nm and 1Z-nm generation, even ONO thickness is scaled down by the ratio of ×0.95 for each generation. The depletion effects in FG [38] and CG during programming and erasing have to be also suppressed. Metal or silicide material [77] will be applicable to FG and CG in future NAND cell.

In order to solve cell structure issues, so-called "Planar FG cell" has been proposed [28]. The planar FG cell has very thin FG thickness ($\sim$10 nm) with high-k inter-poly dielectric (IPD), as described in Section 3.5.

5.7 HIGH-FIELD LIMITATION

A program and erase voltage of NAND flash memory is high ($\sim$22 V) and cannot be drastically decreased because a high electric field ($\sim$10 MV/cm) in tunnel oxide is required for the Fowler–Nordheim tunneling mechanism during program and erase.

It had been reported that the new program interference phenomenon [18] occurred due to the high electric field between the program word line (WL) and the adjacent WL. This new program interference is that the V_{th}'s of the adjacent word lines are decreased during programming. This program interference became more severe as scaling memory cell, because this phenomenon is seriously aggravated as the gate space is decreased.

Figure 5.52 shows measurement conditions of new program interference phenomena. When WL(n) is programmed to the high V_t state, the high program voltage

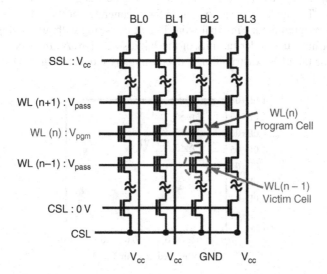

FIGURE 5.52 The schematic diagram for the test module with 4 bit lines and the basic program condition for the self-boosting scheme.

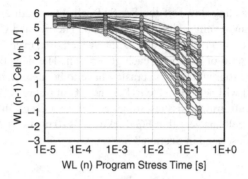

FIGURE 5.53   The V_{th} reduction of the victim cells in WL(n − 1), with the program stress times for 40 cells. The program stress voltage is 26 V and the pass voltage is 4.5 V.

(V_{pgm}) is applied to the control gate of WL(n) and the pass voltage (V_{pass}) is applied to other control gates for the program inhibit cells in the selected and unselected strings. Also normally, NAND flash memory is programmed from a lower word line which is closer to the CSL (common source line) to a higher word line which is closer to the BL (bit line). This new program interference phenomenon had been observed in the sub-30-nm memory cell under applying the relatively low pass voltage.

Figure 5.53 shows the victim cell V_t under acceleration condition of WL(n) program stress of V_{pgm} = 26 V and V_{pass} = 4.5 V. V_{th} reductions are clearly observed as increasing stress time with large variations in the measured 40 cells. These new program interference phenomena are observed in cell array as the under tail bits of the V_{th} distribution when all word lines are programmed to high state in the multilevel cell operation, as shown in Fig. 5.54. The gate design rule of the cell array is sub-30 nm. The under tail bits of Fig. 5.54 is generated at the WL(n − 1) when the WL(n) is programmed. The final word line in the string without the upper word line does not have the under tail bits of the V_{th} distribution. As the pass voltage is increasing, the tail of V_{th} distribution is decreasing, as shown in Fig. 5.54.

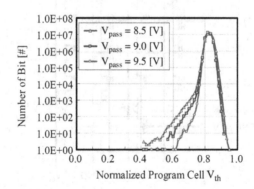

FIGURE 5.54   The distribution of program cell V_{th} after program of the cell array, where V_{pass} is 8.5 V, 9.0 V, and 9.5 V, respectively.

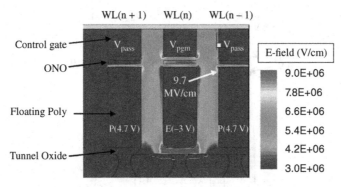

FIGURE 5.55 The simulation result of the electric field distribution under the program operation at $t = 0$, where the V_{pgm} is 24 V and the Vpass is 8.5 V.

Figure 5.55 shows a simulation result of the electric field in WLs for a 30-nm memory cell. The simulation was performed in 3D structure with practical dimension reflecting doping on Si channel, poly-Si floating gates, and control gates. The target V_{th}'s of the floating gates are adjusted from I_d–V_g curve by controlling charges of the floating gates. The maximum electric field of 9.7 MV/cm is observed in between the top edge of the floating gate and the bottom edge of the control gate, in condition of $V_{pgm} = 24$ V and $V_{pass} = 8.5$ V. It is confirmed that the edge field is large enough to generate FN tunneling current between a control gate and a neighbor floating gate.

V_{th} reductions had been evaluated in different gate design rules which are from sub-30 nm to sub-50 nm. Figure 5.56a shows the gate space dependence of this phenomenon. The space between the adjacent WL gates is very critical to this phenomenon. Although the V_{th} reductions are measured at different program voltage

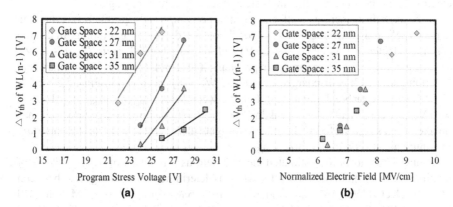

FIGURE 5.56 (a) The V_{th} reduction of WL($n - 1$) for samples with different gate space after 0.2 s of program stress; the pass voltage is 4.5 V. (b) The V_{th} reduction of WL($n - 1$) for samples with different gate space. The x-axis is the electric field between the control gate of WL(n) and the WL($n - 1$).

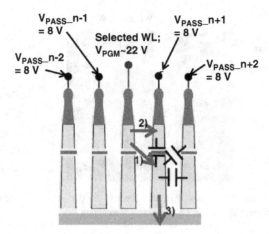

FIGURE 5.57 Word line (WL) high-field problem. (1) Charge loss from neighbor FG to selected WL (CG), (2) leakage or breakdown between selected WL and neighbor WL, and (3) program disturb in neighbor cell. Charges (electron) has injected from substrate to FG.

according to the gate design rule, all of the measured results are simply generalized with the electric field between the control gates. The normalized results with the electric field are shown in Fig. 5.56b, where the y-axis is the decrease of the victim cell V_{th} after program stress for 0.2 s. As shown in Fig. 5.55, the electric field of the WL edge is susceptible to the shape and profile. If the coupling ratio of the floating gate cell is similar, the electric field between the control gates is a simple and proper parameter for representing this phenomenon. In Fig. 5.56b, the V_{th} reduction in WL$(n - 1)$ was observed above 6.0 MV/cm regardless of the gate design rules. This means that the V_{th} reduction of WL$(n - 1)$ is the general phenomenon related to the electrical field between the floating gate and the adjacent control gate.

In NAND flash memory cell, the most serious high-field problem is caused in between selected-word line (WL) and neighbor-WL during programming. In 2X-nm cell, the selected WL is in V_{pgm} (~22 V) and the neighbor WL is in V_{pass} (7–10 V), as shown in Fig. 5.57. There are three problems: (1) charge (electron) loss; charges in FG of neighbor cell is discharged to selected WL [18], as shown in Figs. 5.52–5.56; (2) WL leakage or breakdown; high field between WLs ($V_{PGM}-V_{PASS\_n} + 1/n - 1$ >10 V) may cause leakage or breakdown; (3) program disturb; charge (electron) has injected from substrate to FG. In order to mitigate these problems, it will be important for future generation to optimize $V_{PASS\_n} + 1/n - 1$.

Figure 5.58 shows the estimated electric field as a function of V_{pgm}. Criteria of maximum electric field between WLs, which is determined by (1) charge loss between FG and selected WL [18], can be increased from 6 MV/cm [18] to 9.5 MV/cm [10] by using a WL air gap. Even if WL air gap is used, an available range of ($V_{pgm}-V_{pass}$) is reduced from 15 V of 2X-nm cell to 10 V of 1Z-nm cell ($V_{pgm} = 23$ V/$V_{pass} = 8$ V to $V_{pgm} = 18$ V/$V_{pass} = 8$ V). However, if 1Z-nm generation (word-line half-pitch; 10.6 nm, see Table 5.1) is used, $V_{PGM} = 18$ V is available to program in case of

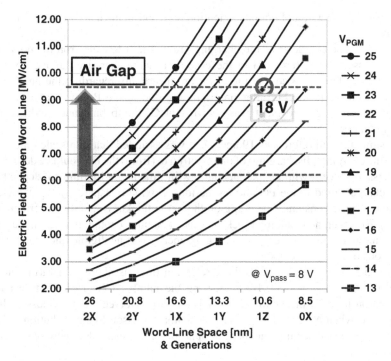

FIGURE 5.58   Estimated electric field between word lines during programming. In 1Z-nm generation (word-line space 10.6 nm), high 18 V can be applied due to using an air gap in the word-line space.

neighbor $V_{PASS\_n} + 1/n - 1 = 8$ V. It means that enough high voltage of $V_{PGM} = 22$ V/$V_{PASS\_n} + 1/n - 1 = 12$ V is available for programming with decreasing $V_{PASS\_n} + 2/n - 2$ to prevent (3) program disturb to manage high-field problems.

5.8   A FEW ELECTRON PHENOMENA

By scaling down memory cell size, the number of electrons stored on the floating gate is significantly decreased due to the decrease of inter-poly dielectric ONO capacitance. Figure 5.27 represents the number of electrons per bit (for ΔV_t of 3 V) as a function of the technology node for NAND and NOR flash memory cells [19, 57]. It is expected that around 100 electrons are stored in 1X-nm design rule cell. By scaling down memory cell size further, the number of stored electrons will be much less than 100 electrons. It will be sufficiently small enough to make few electron phenomena observable. Then the impact of these single electron phenomena on the performance of floating-gate (FG) memory cell had been studied [19, 57]. The charging and discharging of scaled FG memory cells should no longer be considered as a continuous phenomenon but as a sum of discrete stochastic events. This results in an intrinsic dispersion of both the retention time and of the memory programming window.

The stochastic character of the charging process in a few-electron memory had been addressed in reference 78 in the case of a nanometer-size storage node. It had been also demonstrated that there was an uncertainty regarding the number of charged electrons in the FG after programming due to the Poisson nature of the electron charging. Moreover, it was shown that Coulomb blockade modified the charging kinetic.

For simulation in references 19 and 57, it was assumed that in a scaled memory device, the charging (discharging) of one electron to (from) the FG could be described by a Poisson process with an exponential law in time and a lifetime τ_d, depending on the charge stored in the FG and on the tunnel-oxide transparency (τ_c, the electron capture time constant, depending on the oxide thickness and the programming voltage). However, no Coulomb blockade was taken into account. Indeed, in the case of continuous FG memory cell, the charging energy was negligible due to the large dimensions of the storage node.

Figure 5.59a represents the calculated retention-time distribution for various numbers (N) of electrons per bit. By decreasing the number (N) of electrons per bit, the retention time T_R probability density is strongly evolved from a Gaussian-like distribution (when $N \sim 250$) to a pure exponential/ Poisson-like distribution (when $N \sim 5$). We can also see that the dispersion around the mean value increases as N is decreased.

Figure 5.59b shows the cumulative probabilities of the retention time T_R for different values N of electrons per bit. For large values of N, the cumulative-probability evolution is very tight distribution. On the other hand, we can see that as N decreases, the distribution tails have much shorter retention time, which means that the number of discharged memory cells will become critical. For example, in the case of a 1-Mb memory array with memory cells having a mean retention time of 10 years and

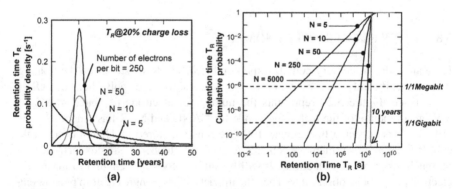

(a)   (b)

FIGURE 5.59   (a) Probability density of the retention time T_R for memories with reduced number of electrons per bit N. The mean T_R is fixed at 10 years. (b) Cumulative probability of retention time T_R (from (a)) of memories with reduced number of electrons per bit N.

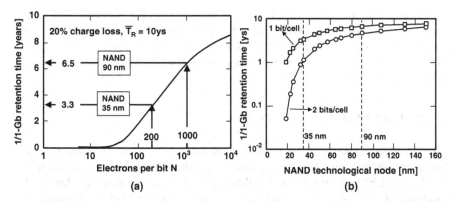

FIGURE 5.60   (a) Calculated failure time (i.e., retention time of 1 bit over 1-Gb array) due to single-electron discharging of the FG plotted as a function of the number of electrons per bit. The mean retention time is ten years for each cell array. (b) Calculated failure time (i.e., retention time of 1 bit over 1-Gb array) due to single-electron discharging of the FG plotted as a function of the technological node for 1-bit/cell and 2-bit/cell memory devices.

containing five electrons per bit, we can observe that one cell will be discharged after only a few hundreds of seconds.

These dispersion results were extrapolated to future memory generations with a smaller number of stored electrons in FG [19, 57].

Figure 5.60a shows the evolution of this failure time as a function of the number (N) of electrons per bit. The failure criterion is defined as the retention time of 1 bit over a 1-Gb array, and the retention criterion corresponds to 20% of the charge loss. We can see that the failure time reduction can become relevant when the number of electrons per bit is decreased and becomes really critical in few electron memories. If we consider the 90-nm NAND flash technology node, a threshold-voltage shift of 3 V corresponds to about 1000 electrons per bit. Thus, in this case, the retention time of one erratic bit over 1 Gb will be equal to 6.5 years. However, if the 35-nm NAND flash technology node, which corresponds to about 200 electrons per bit, is considered, the retention time of one erratic bit decreases drastically to 3.3 years, which could be very critical.

Moreover, in multilevel cells, the retention-time operation margins will be further reduced, with the number of electrons per bit being decreased by $2^{bit/cell}-1$. Figure 5.60b shows the retention time of 1 bit/1 Gb as a function of the NAND flash technology node for 1-bit/cell and 2-bit/cell memory technologies. This plot illustrates that in future technology nodes, the multilevel memories will decrease critically the failure time of high-density memory arrays. Thus, for the 35-nm memory node, the introduction of 2 bits/cell induces a reduction of the failure time from 3.3 years to one year.

As shown above, the decreasing stored electron has intrinsically degraded the data retention time of tail bits in scaled memory cells. These tail bits would not be a serious problem in an actual NAND flash usage with system solutions, such as ECC, and so on, if the number of stored electrons is more than 50. However, in a future device—for example, a 3D SONOS cell with very small channel diameter and short channel—the number of stored electrons would be much reduced. A few electron phenomena have to be considered as being one of the scaling obstacles to keeping the appropriate reliability.

5.9   PATTERNING LIMITATION

A NAND flash memory cell can be easily scaled down as scaling a minimum device dimension, without any electrical, operational, and reliability limitations due to an excellent scalability of the SA-STI cell [4] as well as an excellent gate length (L) scalability of the uniform program/erase operation scheme [79–83]. Therefore, the memory cell size could decrease straightforward as feature size decreased from 0.7 μm to current 1Y-nm generation, as shown in Chapter 3.

The cell size of the NAND flash became ideal $4*F^2$ by the SA-STI cell. The feature size (F) is normally determined by the capability of the lithography tool. At present, the most advanced lithography tool is the ArF immersion (ArFi) stepper. Minimum feature size is 38–40 nm. Then the scaling of feature size (F) had been limited by 38–40 nm. In order to accelerate to scale-down NAND flash memory cell size further, the double patterning process [84, 85] had started to be used from 3X-nm generation. The side-wall spacer was used as a patterning mask in the conventional double patterning process, as shown by the SPT process in Fig. 5.61 [10]. Thanks to double patterning, feature size could be scaled down from 38–40 nm to 19–20 nm. Furthermore, quadruple (×4) patterning has been used beyond 20 nm, as shown by the QSPT process in Fig. 5.61 [10]. Feature size (F) can be scaled down from 19–20 nm to 9.5–10 nm by using ArFi if serious physical limitations described in this chapter

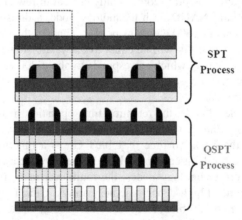

FIGURE 5.61   Schematic diagram of SPT (double patterning) and QSPT (quadruple patterning) key fabrication steps. Two times spacer patterning (QSPT) are used to make mid-1X patterning. SPT; Spacer Patterning Technology, QSPT; Quadruple Spacer Patterning Technology.

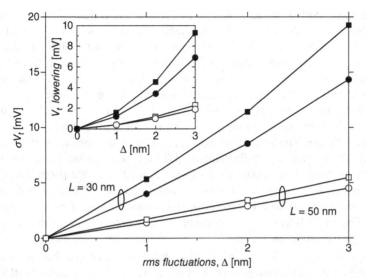

FIGURE 5.62 Standard deviation of threshold voltage σV_t for 30×50 and 50×50 nm MOSFETs as a function of RMS fluctuations Δ, at $V_D = 1.0$ V (squares) and $V_D = 0.1$ V (circles).

are not considered. For scaling beyond 9.5–10 nm, new tool or technology to make fine pattern is absolutely required. One candidate is the EUV (extreme ultraviolet) lithograph tool. However, EUV tool is not available at least in 2014.

Line edge roughness (LER) is one of the serious problems in fine patterning [86]. The LER has tended to stay relatively constant as a device scale. In the case of an aggressive scaling, the LER has become a larger fraction of the channel width or length. Figure 5.62 shows the variation in standard deviation of V_t as a function of RMS LER parameters [86]. As RMS LER increased, V_t variation is increased. This would cause a large variation of cell V_t in a <20-nm NAND flash cell.

5.10   VARIATION

One of the obstacles to scale down memory cell size is the increasing role played by variability effects [87], that strongly influence the threshold voltage (V_t) distribution of NAND flash memory cells [88, 89], affecting their performance and reliability.

To investigate the variation effect, the compact model for the NAND flash memory array was presented [90]. The model includes 3 NAND strings of 32 cells with the select transistors, and it accounts for floating-gate capacitive coupling interference effects. Only the central string is simulated, while the two neighbor NAND strings set the boundary conditions for the electrostatic couplings among the cells. The floating-gate devices are described via capacitors in series to MOS transistors, whose parameters were extracted as detailed in [90] for any technology node.

The compact SPICE model was used in a Monte Carlo framework to obtain the V_t distribution from the calculated string current in read conditions. In simulations, the device parameters were randomly changed to account for the different variability effects. In particular, both process-induced fluctuations in the cell geometry and more fundamental (intrinsic) ones were considered, for example, due to the discrete nature of the charge. The former of process-induced fluctuations in the cell geometry include W, L, tunnel, and inter-poly dielectric thickness fluctuations (indicated as WF, LF, TOXF, IPDF, respectively) as well as fluctuations in the control to floating-gate coupling coefficient. The latter of the fundamental (intrinsic) ones account for random dopant (RDF) and oxide trap fluctuations (OTF). Process-induced fluctuations are directly inserted in the compact model by changing the device parameters (W, L, etc.), according to Gaussian distributions whose spreads are extracted from process data. The implementation of the fundamental (intrinsic) contributions is carried out as follows: The RDF effect on V_t was accounted for by the analytical formula reported in reference 87, while the V_t variability due to OTF was implemented as $\sigma_{\mathrm{OTF}} = K_{ox}Q^\alpha_{ox}/\sqrt{WL}$ with $\alpha \approx 0.5$ and K_{ox} and Q_{ox} fitted on cycled distribution data.

Figure 5.63 shows the simulation results including the spread of neutral cell V_t with the experimental V_t distribution measured on a page of a 41-nm NAND flash array [88, 89]. This result shows a good agreement between measurement and simulation with support of the correctness of the variability models. The slight underestimation of the spread is probably caused by the soft erase operation from programmed V_t distribution.

Figure 5.64 shows the simulated and experimental standard deviation of the V_t distribution σ_{V_t} as a function of the NAND flash technology node from 100-nm to 25-nm rule. The results show an increase in V_t variability as scaling memory cell due to the degradation of all the fluctuations factors impacting array functionality. Figure 5.65 shows the detailed relative weight of the major variability factors as a function of the technology node, represented by RDF (random dopant fluctuation), OTF (oxide trap fluctuation), and fluctuations in W and L. The results clarify that the variability is dominated by several factors, not dominated by single killer factor.

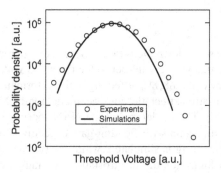

FIGURE 5.63 V_t distribution for a page of a 41-nm NAND flash array and the corresponding simulation results.

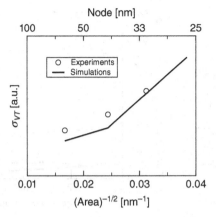

FIGURE 5.64 Comparison between modeling results and experimental data for σ_{V_t} as a function of technology node.

In a future device of less than 25-nm technology node, RDF is expected to play a more important role on V_t statistical dispersion. The consequent increase in the V_t spread should be carefully considered when the array functionality is implemented in program, erase, and read conditions [91].

The model, which was used to investigate the behavior of the neutral cell V_t, was expanded to simulate the program and erase operation. One of the most important parameters that play a fundamental role in the σ_{VT} during P/E operations is the control-gate to floating-gate coupling coefficient (α_G). This parameter is related to the structure adopted for the floating-gate definition, and its fluctuations depend on the spread in the geometrical parameters, which are schematically shown in Fig. 5.66a. The contributions of such factors to the spread in α_G are shown in Fig. 5.66b on a normalized scale. In this simulation results, fluctuations in t_{FG} and t_R play the major role for all technology nodes.

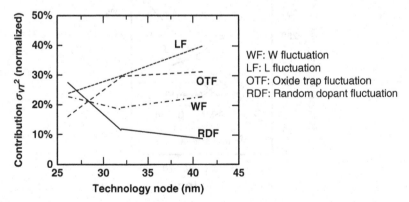

FIGURE 5.65 Most important contributions to σ_{VT}^2 of neutral cells (normalized) for different technology nodes.

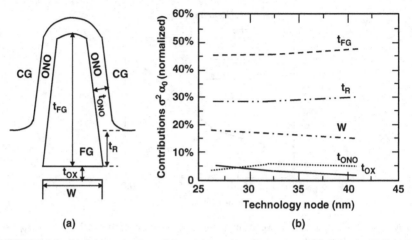

FIGURE 5.66   (a) Schematic view of the cross section of a memory cell along the W direction, showing the floating-gate geometry. (b) Individual contributions to the spread in the coupling coefficient α_G for the different technology nodes.

An RDF (random dopant fluctuation) is major contributor to neutral V_t variation in less than 25-nm memory cells, as shown in Fig. 5.65. As memory cells are scaled down, the number of dopant atoms per cell decreases, resulting in a larger standard deviation in the threshold voltage. Figure 5.67 shows the number of boron atoms per cell versus technology node and increasing variation in dose per cell at smaller nodes

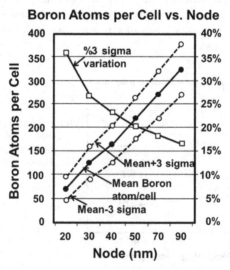

FIGURE 5.67   Number and 3σ of channel dopant atoms versus transistor size with constant V_t scaling.

[23,92]. The atomistic nature of substrate doping has been clearly shown to result in a fundamental V_t spread for MOS field effect transistors given by reference 87.

$$\sigma_{RDF} = 3.19 \times 10^{-8} \frac{t_{ox} N_A^{0.4}}{\sqrt{WL}}$$

5.11 SCALING IMPACT ON DATA RETENTION

It had been reported that data retention characteristics had the neighbor cells data pattern (back-pattern) dependence [93]. A programmed cell has a larger threshold-voltage (V_t) loss when its neighbor cells are in the erased state than when they are in the program state. The cells on the same bit line and word line have a similar impact on the acceleration of the V_t loss. This phenomenon is explained by an influence of charge in a neighbor cells, so that a stored charge in a neighbor cell has an impact on the electric field of tunnel oxide at a corner in the gate and active area of target cell.

The cell arrays in 60-nm technology were used to analyze for the electrical characterization of data retention on single cells. These arrays allow the arbitrary bias conditions of three adjacent WLs and BLs in the central part of the NAND cell matrix, as shown in Fig. 5.68a. The shaded circle in the figure shows the selected cell

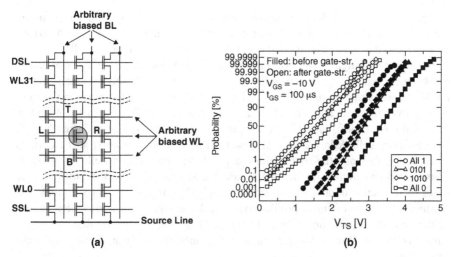

(a)   (b)

FIGURE 5.68   (a) Schematic for the cell connection in the NAND array, evidencing the nine cells whose WL and BL can be arbitrarily biased in the analytical cell array. The shaded circle highlights the selected cell whose V_{TS} was monitored during the gate-stress experiments for different V_{TA}'s of the adjacent cells at its (L) left, (T) top, (R) right, and (B) bottom. (b) Cumulative V_{TS} distributions measured on a 70-nm NAND test chip (filled symbols) before and (open symbols) after a 100-μs negative gate-stress experiment at $V_{GS} = -10$ V. Results for different background patterns are shown.

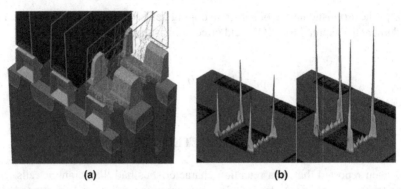

(a)   **(b)**

FIGURE 5.69   (a) Template NAND device structure for 3D TCAD (Technology Computer Aided Design) simulations. (b) Calculated tunneling current density at the edge of the selected cell active area along the WL direction, for the All 0 and the All 1 background patterns. A fixed negative V_{FG}, s is used.

for the gate-stress experiments by applying V_{GS} to its WL with all the other WLs of the NAND string at V_{pass}. Prior to the gate stress, specific array V_t data patterns were set by selective programming of the cells at the left (**L**), top (**T**), right (**R**), and bottom (**B**) of the selected one. Cell V_t is set from low-V_t erased level nearly equal to -3.5 V (referred to as state 1) to a high-V_t level nearly equal to 3 V (referred to as state 0).

Figure 5.68b shows that the negative shift of the V_{TS} distribution is larger in the All 1 (erase) than in the All 0 (program) case, with intermediate results obtained for the 0101 and the 1010 patterns [left (**L**), top (**T**), right (**R**), and bottom (**B**)].

In order to investigate neighbor cell data pattern dependence, 3D TCAD simulation had been carried out, as shown in Fig. 5.69. The template NAND device structure for 3D TCAD simulations is shown in Fig. 5.69a. The calculated tunneling current density over the selected cell active area is presented in Fig. 5.69b for neighbor cells of the All 0 (left) and the All 1 (right) patterns in 60 to 70-nm design NAND flash memory cell. A larger current flow with four peaks is caused at the corners of the active area, and the source/drain junction overlaps with the cell floating gate. These peaks are higher for the All 1 case because an electrostatic profile at the edges of the active area is strong when the neighboring floating gates are charged positive. The current density profile along the BL and WL directions at the corners of the cell active area are shown in Fig. 5.70. It was confirmed that the larger tunneling current flows over the source/drain junctions in the negative gate-stress conditions.

The above results were presented to be analyzed on 60 to 70-nm NAND flash memory cells. By scaling a memory cell, this phenomenon of tunneling current confining at the FG corner and edge has been exaggerated. The current 1X-nm memory cell would have a strong impact on this phenomenon. Data retention characteristics would be much worse in future memory cell scaling.

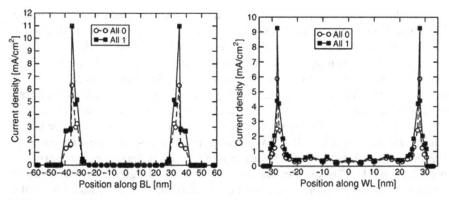

FIGURE 5.70 Calculated tunneling current density at the edge of the selected cell active area along the BL direction (left) and the WL direction (right), for the All 0 and the All 1 background patterns. A fixed negative V_{FG},s is used.

5.12 SUMMARY

The scaling limitations and challenges over 90 to 0X-nm generations was discussed for two-dimensional (2D) floating-gate NAND flash memories. The scaling challenges were categorized as (1) narrow read window margin (RWM) problem, (2) floating-gate capacitive coupling interference, (3) program electron injection spread, (4) random telegraph signal noise (RTN), (5) cell structural challenge, (6) high-field (5–10 MV/cm) limitation, (7) a few electron phenomena, (8) patterning limitation, (9) variation, and (10) scaling impact on data retention.

First, (1) the narrow RWM was discussed by extrapolating physical phenomena of FG–FG coupling interference, electron injection spread (EIS), back pattern dependence (BPD), and random telegraph noise (RTN). The RWM is degraded not only by increasing programmed V_t distribution width, but also by increasing V_t of erase state mainly due to large FG–FG coupling interference. However, RWM is still positive in 1Z-nm (10-nm) generation with 60% reduction of FG–FG coupling by the air-gap process.

Second, floating-gate capacitive coupling interference, which was a major contributor to degrade RWM, was discussed. Air gaps between word lines and in STI are the solutions to improve of floating-gate capacitive coupling interference.

Next, (3) program electron injection spread and (4) random telegraph signal noise (RTN) were described as contributors to RWM.

Then, (5) structural challenge was discussed. The control gate (CG) fabrication margin between floating gates (FGs) is becoming much more severe beyond 1X-nm generation. Very narrow 5-nm FG width/space has to be controlled. And for (6) the high-field problem, the high field between CGs (word lines: WLs) is critical during the program. By using WL air gap, the high-field problem can be mitigated, and 1Y/1Z-nm generations could be realized.

After that, several scaling problems of (7) a few electron phenomena, (8) patterning limitation, (9) variation, and (10) scaling impact on data retention were discussed. These problems are inevitable to scale down NAND flash memory cells.

To improve RWM margin and reliability margin, operational techniques and system solutions are effective to manage these margins. One example is the "randomization" [94,95]. The data pattern that is programming to memory cell is randomized by using code data. The "0" and "1" data become random data, and nearly 50% for both "0" and "1". Therefore, worst case of data pattern can be avoided. The randomization can improve RWM with mitigating floating-gate capacitive coupling (Section 5.2), back pattern dependence (Section 5.2), and scaling impact on data retention (Section 5.11). The other example is the moving read algorithm, as described in Section 4.7. The moving read operation can greatly improve the failure rate of the V_t shift of data retention as well as that of V_t distribution widening by floating-gate capacitive coupling interference, program electron injection spread, and so on.

REFERENCES

[1] Masuoka, F.; Momodomi, M.; Iwata, Y.; Shirota, R. New ultra high density EPROM and flash EEPROM with NAND structure cell, *Electron Devices Meeting, 1987 International*, vol. 33, pp. 552–555, 1987.

[2] Aritome, S. NAND Flash Innovations, *Solid-State Circuits Magazine, IEEE*, vol. 5, no. 4, pp. 21, 29, Fall 2013.

[3] Aritome, S.; Hatakeyama, I.; Endoh, T.; Yamaguchi, T.; Shuto, S.; Iizuka, H.; Maruyama, T.; Watanabe, H.; Hemink, G.; Sakui, K.; Tanaka, T.; Momodomi, M., and Shirota, R. An advanced NAND-structure cell technology for reliable 3.3 V 64 Mb electrically erasable and programmable read only memories (EEPROMs), *Japanese Journal of Applied Physics*, vol. 33, part 1, no. 1B, pp. 524–528, Jan. 1994.

[4] Aritome, S.; Satoh, S.; Maruyama, T.; Watanabe, H.; Shuto, S.; Hemink, G. J.; Shirota, R.; Watanabe, S.; Masuoka, F. A 0.67 µm² self-aligned shallow trench isolation cell (SA-STI cell) for 3 V-only 256 Mbit NAND EEPROMs, *Electron Devices Meeting, 1994. IEDM '94. Technical Digest., International*, pp. 61–64, 11–14 Dec. 1994.

[5] Shimizu, K.; Narita, K.; Watanabe, H.; Kamiya, E.; Takeuchi, Y.; Yaegashi, T.; Aritome, S.; Watanabe, T. A novel high-density 5F² NAND STI cell technology suitable for 256 Mbit and 1 Gbit flash memories, *Electron Devices Meeting, 1997. IEDM '97. Technical Digest., International*, pp. 271–274, 7–10 Dec. 1997.

[6] Takeuchi, Y.; Shimizu, K.; Narita, K.; Kamiya, E.; Yaegashi, T.; Amemiya, K.; Aritome, S. A self-aligned STI process integration for low cost and highly reliable 1 Gbit flash memories, *VLSI Technology, 1998. Digest of Technical Papers. 1998 Symposium on*, pp. 102–103, 9–11 June 1998.

[7] Aritome, S. Advanced flash memory technology and trends for file storage application, *Electron Devices Meeting, 2000. IEDM Technical Digest. International*, pp. 763–766, 2000.

[8] Imamiya, K.; Sugiura, Y.; Nakamura, H.; Himeno, T.; Takeuchi, K.; Ikehashi, T.; Kanda, K.; Hosono, K.; Shirota, R.; Aritome, S.; Shimizu, K.; Hatakeyama, K.; Sakui, K. A

130-mm^2, 256-Mbit NAND flash with shallow trench isolation technology, *Solid-State Circuits, IEEE Journal of*, vol. 34, no. 11, pp. 1536–1543, Nov. 1999.

[9] Ichige, M.; Takeuchi, Y.; Sugimae, K.; Sato, A.; Matsui, M.; Kamigaichi, T.; Kutsukake, H.; Ishibashi, Y.; Saito, M.; Mori, S.; Meguro, H.; Miyazaki, S.; Miwa, T.; Takahashi, S.; Iguchi, T.; Kawai, N.; Tamon, S.; Arai, N.; Kamata, H.; Minami, T.; Iizuka, H.; Higashitani, M.; Pham, T.; Hemink, G.; Momodomi, M.; Shirota, R. A novel self-aligned shallow trench isolation cell for 90 nm 4 Gbit NAND flash EEPROMs, *VLSI Technology, 2003. Digest of Technical Papers. 2003 Symposium on*, pp. 89,90, 10–12 June 2003.

[10] Hwang, J.; Seo, J.; Lee, Y.; Park, S.; Leem, J.; Kim, J.; Hong, T.; Jeong, S.; Lee, K.; Heo, H.; Lee, H.; Jang, P.; Park, K.; Lee, M.; Baik, S.; Kim, J.; Kkang, H.; Jang, M.; Lee, J.; Cho, G.; Lee, J.; Lee, B.; Jang, H.; Park, S.; Kim, J.; Lee, S.; Aritome, S.; Hong, S., and Park, S. A middle-1X nm NAND flash memory cell (M1X-NAND) with highly manufacturable integration technologies, *Electron Devices Meeting (IEDM), 2011 IEEE International*, pp. 199–202, Dec. 2011.

[11] Aritome, S.; Kikkawa, T. Scaling challenge of self-aligned STI cell (SA-STI cell) for NAND flash memories, *Solid-State Electronics*, vol. 82, 54–62, 2013.

[12] Lee, J.-D.; Hur, S.-H.; Choi, J.-D. Effects of floating-gate interference on NAND flash memory cell operation, *Electron Device Letters, IEEE*, vol. 23, no. 5, pp. 264–266, May 2002.

[13] Compagnoni, C. M.; Spinelli, A. S.; Gusmeroli, R.; Lacaita, A. L.; Beltrami, S.; Ghetti, A.; Visconti, A. First evidence for injection statistics accuracy limitations in NAND flash constant-current Fowler–Nordheim programming, *Electron Devices Meeting, 2007. IEDM 2007. IEEE International*, pp. 165–168, 10–12 Dec. 2007.

[14] Compagnoni, C. M.; Spinelli, A. S.; Gusmeroli, R.; Beltrami, S.; Ghetti, A., and Visconti, A. Ultimate accuracy for the NAND Flash program algorithm due to the electron injection statistics, *IEEE Trans. Electron Devices*, vol. 55, no. 10, pp. 2695–2702, Oct. 2008.

[15] Compagnoni, C. M.; Gusmeroli, R.; Spinelli, A. S.; Visconti, A. Analytical model for the electron-injection statistics during programming of nanoscale NAND flash memories, *Electron Devices, IEEE Transactions on*, vol. 55, no. 11, pp. 3192–3199, 2008.

[16] Kurata, H.; Otsuga, K.; Kotabe, A.; Kajiyama, S.; Osabe, T.; Sasago, Y.; Narumi, S.; Tokami, K.; Kamohara, S.; Tsuchiya, O. The impact of random telegraph signals on the scaling of multilevel flash memories, *VLSI Circuits, 2006. Digest of Technical Papers. 2006 Symposium on*, pp. 112–113.

[17] Govoreanu, B.; Brunco, D. P.; Van Houdt, J. Scaling down the interpoly dielectric for next generation flash memory: Challenges and opportunities, *Solid-State Electronics*, vol. 49, no. 11, pp. 1841–1848, Nov. 2005.

[18] Kim, Y. S.; Lee, D. J.; Lee, C. K.; Choi, H. K.; Kim, S. S.; Song, J. H.; Song, D. H.; Choi, J.-H.; Suh, K.-D.; Chung, C. New scaling limitation of the floating gate cell in NAND flash memory, *Reliability Physics Symposium (IRPS), 2010 IEEE International*, pp. 599–603, 2–6 May 2010.

[19] Molas, G.; Deleruyelle, D.; De Salvo, B.; Ghibaudo, G.; GelyGely, M.; Perniola, L.; Lafond, D.; Deleonibus, S. Degradation of floating-gate memory reliability by few electron phenomena, *Electron Devices, IEEE Transactions on*, vol. 53, no. 10, pp. 2610–2619, Oct. 2006.

[20] Kinam, K.; Jeong, G. Memory technologies in the nano-era: Challenges and opportunities, *Solid-State Circuits Conference, 2005. Digest of Technical Papers. ISSCC. 2005 IEEE International*, vol. 1, pp. 576, 618, 10–10 Feb. 2005.

[21] Kim, K. Technology for sub-50nm DRAM and NAND flash manufacturing, *Electron Devices Meeting, 2005. IEDM Technical Digest. IEEE International*, pp. 323, 326, 5–5 Dec. 2005.

[22] Kim, K.; Choi, J. Future outlook of NAND flash technology for 40 nm node and beyond, *Non-Volatile Semiconductor Memory Workshop, 2006. IEEE NVSMW 2006. 21st*, pp. 9, 11, 12–16 Feb. 2006.

[23] Prall, K. Scaling non-volatile memory below 30 nm, *Non-Volatile Semiconductor Memory Workshop, 2007 22nd IEEE*, pp. 5, 10, 26–30 Aug. 2007.

[24] Kim, K.; Jeong, G. Memory Technologies for sub-40nm Node, *Electron Devices Meeting, 2007. IEDM 2007. IEEE International*, pp. 27, 30, 10–12 Dec. 2007.

[25] Parat, K. Recent developments in NAND flash scaling, *VLSI Technology, Systems, and Applications, 2009. VLSI-TSA '09. International Symposium on*, pp. 101, 102, 27–29 April 2009.

[26] Kim, K. Technology challenges for deep-nano semiconductor, *Memory Workshop (IMW), 2010 IEEE International*, pp. 1, 2, 16–19 May 2010.

[27] Kim, K. From the future Si technology perspective: Challenges and opportunities, *Electron Devices Meeting (IEDM), 2010 IEEE International*, pp. 1.1.1, 1.1.9, 6–8 Dec. 2010.

[28] Goda, A.; Parat, K. Scaling directions for 2D and 3D NAND cells, *Electron Devices Meeting (IEDM), 2012 IEEE International*, pp. 2.1.1, 2.1.4, 10–13 Dec. 2012.

[29] Goda, A. Opportunities and challenges of 3D NAND scaling, *VLSI Technology, Systems, and Applications (VLSI-TSA), 2013 International Symposium on*, pp. 1, 2, 22–24 Apr. 2013.

[30] Goda, A. Recent progress and future directions in NAND Flash scaling, *Non-Volatile Memory Technology Symposium (NVMTS), 2013 13th*, pp. 1, 4, 12–14 Aug. 2013.

[31] Park, Y.; Lee, J. Device considerations of planar NAND flash memory for extending towards sub-20 nm regime, *Memory Workshop (IMW), 2013 5th IEEE International*, pp. 1, 4, 26–29 May 2013.

[32] Park, Y.; Lee, J.; Cho, S. S.; Jin, G.; Jung, E. S. Scaling and reliability of NAND flash devices, *Reliability Physics Symposium, 2014 IEEE International*, pp. 2E.1.1, 2E.1.4, 1–5 June 2014.

[33] Aritome, S. 3D flash memories, International Memory Workshop 2011 (IMW 2011), short course.

[34] Prall, K.; Parat, K. 25 nm 64 Gb MLC NAND technology and scaling challenges invited paper, *Electron Devices Meeting (IEDM), 2010 IEEE International*, pp. 5.2.1–5.2.4, 6–8 Dec. 2010.

[35] Seokkiu, L. Scaling challenges in NAND flash device toward 10 nm technology, *Memory Workshop (IMW), 2012 4th IEEE International*, pp. 1–4, 20–23 May 2012.

[36] Suh, K.-D.; Suh, B.-H.; Lim, Y.-H.; Kim, J.-K.; Choi, Y.-J.; Koh, Y.-N.; Lee, S.-S.; Kwon, S.-C.; Choi, B.-S.; Yum, J.-S.; Choi, J.-H.; Kim, J.-R.; Lim, H.-K. A 3.3 V 32 Mb NAND flash memory with incremental step pulse programming scheme, *Solid-State Circuits, IEEE Journal of*, vol. 30, no. 11, pp. 1149–1156, Nov. 1995.

[37] Hemink, G. J.; Tanaka, T.; Endoh, T.; Aritome, S.; Shirota, R. Fast and accurate programming method for multi-level NAND EEPROMs. *VLSI Technology, 1995. Digest of Technical Papers. 1995 Symposium on*, pp. 129–130, 6–8 June 1995.

[38] Shirota, R.; Sakamoto, Y.; Hsueh, H.-M.; Jaw, J.-M.; Chao, W.-C.; Chao, C.-M.; Yang, S.-F.; Arakawa, H. Analysis of the correlation between the programmed threshold-voltage distribution spread of NAND flash memory devices and floating-gate impurity concentration, *Electron Devices, IEEE Transactions on*, vol. 58, no. 11, pp. 3712–3719, Nov. 2011.

[39] Ghetti, A.; Compagnoni, C. M.; Spinelli, A. S.; Visconti, A. Comprehensive analysis of random telegraph noise instability and its scaling in deca–nanometer flash memories, *Electron Devices, IEEE Transactions on*, vol. 56, no. 8, pp. 1746–1752, Aug. 2009.

[40] Shibata, N.; Tanaka, T. US Patent 7,245,528. 7,370,009. 7,738,302.

[41] Park, K.-T.; Kang, M.; Kim, D.; Hwang, S.-W.; Choi, B. Y.; Lee, Y.-T.; Kim, C.; Kim, K. A zeroing cell-to-cell interference page architecture with temporary LSB storing and parallel MSB program scheme for MLC NAND flash memories, *Solid-State Circuits, IEEE Journal of*, vol. 43, no. 4, pp. 919–928, April 2008.

[42] Cernea, R.-A.; Pham, L.; Moogat, F.; Chan, S.; Le, B.; Li, Y.; Tsao, S.; Tseng, T.-Y.; Nguyen, K.; Li, J.; Hu, J.; Yuh, J. H.; Hsu, C.; Zang, F.; Kamei, T.; Nasu, H.; Kliza, P.; Htoo, K.; Lutze, J.; Dong, Y.; Higashitani, M.; Yang, J.; Lin, H.-S.; Sakhamuri, V.; Li, A.; Pan, F.; Yadala, S.; Taigor, S.; Pradhan, K.; Lan, J.; Chan, J.; Abe, T.; Fukuda, Y.; Mukai, H.; Kawakami, K.; Liang, C.; Ip, T.; Chang, S.-F.; Lakshmipathi, J.; Huynh, S.; Pantelakis, D.; Mofidi, M.; Quader, K. A 34 MB/s MLC write throughput 16 Gb NAND with all bit line architecture on 56 nm technology, *Solid-State Circuits, IEEE Journal of*, vol. 44, no. 1, pp. 186–194, Jan. 2009.

[43] Tanaka, T.; Tanaka, Y.; Nakamura, H.; Oodaira, H.; Aritome, S.; Shirota, R.; Masuoka, F. A quick intelligent program architecture for 3 V-only NAND-EEPROMs, *VLSI Circuits, 1992. Digest of Technical Papers, 1992 Symposium on*, pp. 20–21, 4–6 June 1992.

[44] Kim, T.-Y.; Lee, S.-D.; Park, J.-S.; Cho, H.-Y.; You, B.-S.; Baek, K.-H.; Lee, J.-H.; Yang, C.-W.; Yun, M.; Kim, M.-S.; Kim, J.-W.; Jang, E.-S.; Chung, H.; Lim, S.-O.; Han, B.-S.; Koh, Y.-H. A 32 Gb MLC NAND flash memory with V_{th} margin-expanding schemes in 26 nm CMOS, *Solid-State Circuits Conference Digest of Technical Papers (ISSCC), 2011 IEEE International*, pp. 202–204, 20–24 Feb. 2011.

[45] Kanda, K.; Shibata, N.; Hisada, T.; Isobe, K.; Sato, M.; Shimizu, Y.; Shimizu, T.; Sugimoto, T.; Kobayashi, T.; Kanagawa, N.; Kajitani, Y.; Ogawa, T.; Iwasa, K.; Kojima, M.; Suzuki, T.; Suzuki, Y.; Sakai, S.; Fujimura, T.; Utsunomiya, Y.; Hashimoto, T.; Kobayashi, N.; Matsumoto, Y.; Inoue, S.; Suzuki, Y.; Honda, Y.; Kato, Y.; Zaitsu, S.; Chibvongodze, H.; Watanabe, M.; Ding, H.; Ookuma, N.; Yamashita, R. A 19 nm 112.8 mm^2 64 Gb multi-level flash memory with 400 Mbit/sec/pin 1.8 V toggle mode interface, *Solid-State Circuits, IEEE Journal of*, vol. 48, no. 1, pp. 159–167, Jan. 2013.

[46] Lee, D.; Chang, I. J.; Yoon, S.-Y.; Jang, J.; Jang, D.-S.; Hahn, W.-G.; Park, J.-Y.; Kim, D.-G.; Yoon, C.; Lim, B.-S.; Min, B.-J.; Yun, S.-W.; Lee, J.-S.; Park, I.-H.; Kim, K.-R.; Yun, J.-Y.; Kim, Y.; Cho, Y.-S.; Kang, K.-M.; Joo, S.-H.; Chun, J.-Y.; Im, J.-N.; Kwon, S.; Ham, S.; Ansoo, P.; Yu, J.-D.; Lee, N.-H.; Lee, T.-S.; Kim, M.; Kim, H.; Song, K.-W.; Jeon, B.-G.; Choi, K.; Han, J.-M.; Kyung, K. H.; Lim, Y.-H.; Jun, Y.-H. A 64 Gb 533 Mb/s DDR interface MLC NAND flash in sub-20 nm technology, *Solid-State Circuits Conference Digest of Technical Papers (ISSCC), 2012 IEEE International*, pp. 430–432, 19–23 Feb. 2012.

[47] Kang, D.; Jang, S.; Lee, K.; Kim, J.; Kwon, H.; Lee, W.; Park, B. G.; Lee, J. D.; Shin, H. Improving the cell characteristics using low-*k* gate spacer in 1 Gb NAND flash memory, *Electron Devices Meeting, 2006. IEDM '06. International*, pp. 1–4, 11–13 Dec. 2006.

[48] Kim, S.; Cho, W.; Kim, J.; Lee, B.; Park, S. Air-gap application and simulation results for low capacitance in 60 nm NAND flash memory, *Non-Volatile Semiconductor Memory Workshop, 2007 22nd IEEE*, pp. 54–55, 26–30 Aug. 2007.

[49] Seo, J.; Han, K.; Youn, T.; Heo, H.-E.; Jang, S.; Kim, J.; Yoo, H.; Hwang, J.; Yang, C.; Lee, H.; Kim, B.; Choi, E.; Noh, K.; Lee, B.; Lee, B.; Chang, H.; Park, S.; Ahn, K.; Lee, S.; Kim, J.; Lee, S. Highly reliable M1X MLC NAND flash memory cell with novel active air-gap and p+ poly process integration technologies, *Electron Devices Meeting (IEDM), 2013 IEEE International*, pp. 3.6.1, 3.6.4, 9–11 Dec. 2013.

[50] Park, M.; Kim, K.; Park, J.-H.; Choi, J.-H. Direct field effect of neighboring cell transistor on cell-to-cell interference of NAND flash cell arrays, *Electron Device Letters, IEEE*, vol. 30, no. 2, pp. 174–177, Feb. 2009.

[51] Park, M.; Suh, K.; Kim, K.; Hur, S.; Kim, K., and Lee, W.; The effect of trapped charge distributions on data retention characteristics of NAND flash memory cells, *IEEE Electron Device Letters*, vol. 28, no. 8, pp. 750–752, Aug. 2007.

[52] Aritome, S.; Seo, S.; Kim, H.-S.; Park, S.-K.; Lee, S.-K.; Hong, S. Novel negative V_t shift phenomenon of program–inhibit cell in 2X–3X-nm self-aligned STI NAND flash memory, *Electron Devices, IEEE Transactions on*, vol. 59, no. 11, pp. 2950, 2955, Nov. 2012.

[53] Cho, B.; Lee, C. H.; Seol, K.; Hur, S.; Choi, J.; Choi, J.; Chung, C. A new cell-to-cell interference induced by conduction band distortion near S/D region in scaled NAND flash memories, *Memory Workshop (IMW), 2011 3rd IEEE International*, pp. 1, 4, 22–25 May 2011.

[54] Park, M.; Choi, J.-D.; Hur, S.-H.; Park, J.-H.; Lee, J.-H.; Park, J.-T.; Sel, J.-S.; Kim, J.-W.; Song, S.-B.; Lee, J.-Y.; Lee, J.-H.; Son, S.-J.; Kim, Y.-S.; Chai, S.-J.; Kim, K.-T.; Kim, K. Effect of low-*k* dielectric material on 63 nm MLC (multi-level cell) NAND flash cell arrays, *VLSI Technology, 2005. (VLSI-TSA-Tech). 2005 IEEE VLSI-TSA International Symposium on*, pp. 37–38, 25–27 April 2005.

[55] Kang, D.; Shin, H.; Chang, S.; An, J.; Lee, K.; Kim, J.; Jeong, E.; Kwon, H.; Lee, E.; Seo, S.; Lee, W. The air spacer technology for improving the cell distribution in 1 giga bit NAND flash memory, *Non-Volatile Semiconductor Memory Workshop, 2006. IEEE NVSMW 2006. 21st*, pp. 36–37, 12–16 Feb. 2006.

[56] Lee, C.; Hwang, J.; Fayrushin, A.; Kim, H.; Son, B.; Park, Y.; Jin, G.; Jung, E. S. Channel coupling phenomenon as scaling barrier of NAND flash memory beyond 20 nm node, *Memory Workshop (IMW), 2013 5th IEEE International*, pp. 72, 75, 26–29 May 2013.

[57] Molas, G.; Deleruyelle, D.; De Salvo, B.; Ghibaudo, G.; Gely, M.; Jacob, S.; Lafond, D.; Deleonibus, S. Impact of few electron phenomena on floating-gate memory reliability, *Electron Devices Meeting, 2004. IEDM Technical Digest. IEEE International*, pp. 877–880, 13–15 Dec. 2004.

[58] Suh, K.-D.; Suh, B.-H.; Um, Y.-H.; Kim, J.-K.; Choi, Y.-J.; Koh, Y.-N.; Lee, S.-S.; Kwon, S.-C.; Choi, B.-S.; Yum, J.-S.; Choi, J.-H.; Kim, J.-R.; Lim, H.-K. A 3.3 V 32 Mb NAND flash memory with incremental step pulse programming scheme, *Solid-State Circuits Conference, 1995. Digest of Technical Papers. 42nd ISSCC, 1995 IEEE International*, pp. 128–129, 350, 15–17 Feb. 1995.

[59] Kolodny, A.; Nieh, S. T. K.; Eitan, B.; Shappir, J. Analysis and modeling of floating-gate EEPROM cells, *Electron Devices, IEEE Transactions on*, vol. 33, no. 6, pp. 835–844, June 1986.

[60] Compagnoni, C. M.; Gusmeroli, R.; Spinelli, A. S.; Visconti, A. RTN V_T instability from the stationary trap-filling condition: An analytical spectroscopic investigation, *Electron Devices, IEEE Transactions on*, vol. 55, no. 2, pp. 655–661, 2008.

[61] Kirton, M. J., et al., *Advances in Physics*, vol. 38, no. 4, pp. 367–468, 1989.

[62] Roux dit Buisson, O.; Ghibaudo, G., and Brini, J. Model for drain current RTS amplitude in small-area MOS transistors, *Solid-State Electronics*, vol. 35, no. 9, pp. 1273–1276, Sept. 1992.

[63] Tega, N.; Miki, H.; Osabe, T.; Kotabe, A.; Otsuga, K.; Kurata, H.; Kamohara, S.; Tokami, K.; Ikeda, Y.; Yamada, R. Anomalously large threshold voltage fluctuation by complex random telegraph signal in floating gate flash memory, *Electron Devices Meeting, 2006. IEDM '06. International*, pp. 491–494, 11–13 Dec. 2006.

[64] Tanaka, T.; Momodomi, M.; Iwata, Y.; Tanaka, Y.; Oodaira, H.; Itoh, Y.; Shirota, R.; Ohuchi, K.; Masuoka, F. A 4-Mbit NAND-EEPROM with tight programmed V_t distribution, *VLSI Circuits, 1990. Digest of Technical Papers, 1990 Symposium on*, pp. 105–106, 7–9 June 1990.

[65] Gusmeroli, R.; Compagnoni, C. M.; Riva, A.; Spinelli, A. S.; Lacaita, A. L.; Bonanomi, M.; Visconti, A.; Defects spectroscopy in SiO2 by statistical random telegraph noise analysis, *Electron Devices Meeting, 2006. IEDM '06. International*, pp. 483–486, 2006.

[66] Compagnoni, M. C.; Gusmeroli, R.; Spinelli, A. S.; Lacaita, A. L.; Bonanomi, M.; Visconti, A. Statistical model for random telegraph noise in flash memories, *Electron Devices, IEEE Transactions on*, vol. 55, no. 1, pp. 388–395, Jan. 2008.

[67] Ralls, K. S.; Skocpol, W. J.; Jackel, L. D.; Howard, R. E.; Fetter, L. A.; Epworth, R. W.; Tennant, D. M. Discrete resistance switching in submicrometer silicon inversion layers: Individual interface traps and low-frequency (1f?) noise, *Physical Review Letters*, vol. 52, no. 3, pp. 228–231, 1984.

[68] Compagnoni, C. M.; Gusmeroli, R.; Spinelli, A. S.; Lacaita, A. L.; Bonanomi, M.; Visconti, A. Statistical investigation of random telegraph noise ID instabilities in flash cells at different initial trap-filling conditions, *Reliability physics symposium, 2007. proceedings. 45th annual. IEEE international, 2007*, pp. 161–166.

[69] Fukuda, K.; Shimizu, Y.; Amemiya, K.; Kamoshida, M.; Hu, C. Random telegraph noise in Flash memories—Model and technology scaling, *IEDM Technology Digest*, pp. 169–172, 2007.

[70] Ghetti, A.; Compagnoni, C. M.; Biancardi, F.; Lacaita, A. L.; Beltrami, S.; Chiavarone, L.; Spinelli, A. S.; Visconti, A. Scaling trends for random telegraph noise in deca-nanometer flash memories, *Electron Devices Meeting, 2008. IEDM 2008. IEEE International*, 2008, pp. 835–838.

[71] Ghetti, A.; Bonanomi, M.; Compagnoni, C. M.; Spinelli, A. S.; Lacaita, A. L.; Visconti, A. Physical modeling of single-trap RTS statistical distribution in flash memories, *Reliability Physics Symposium, 2008. IRPS 2008. IEEE International*, 2008, pp. 610–615.

[72] Compagnoni, C. M.; Spinelli, A. S.; Beltrami, S.; Bonanomi, M.; Visconti, A. Cycling effect on the random telegraph noise instabilities of NOR and NAND flash arrays, *Electron Device Letters, IEEE*, vol. 29, no. 8, pp. 941–943, 2008.

[73] Compagnoni, C. M.; Ghidotti, M.; Lacaita, A. L.; Spinelli, A. S.; Visconti, A. Random telegraph noise effect on the programmed threshold-voltage distribution of flash memories, *Electron Device Letters, IEEE*, vol. 30, no. 9, pp. 984–986, 2009.

[74] Kim, T.; He, D.; Porter, R.; Rivers, D.; Kessenich, J.; Goda, A. Comparative study of quick electron detrapping and random telegraph signal and their dependences on random discrete dopant in sub-40-nm NAND flash memory, *Electron Device Letters, IEEE*, vol. 31, no. 2, pp. 153–155, Feb. 2010.

[75] Kim, T.; He, D.; Morinville, K.; Sarpatwari, K.; Millemon, B.; Goda, A.; Kessenich, J. Tunnel oxide nitridation effect on the evolution of V_t instabilities (RTS/QED) and defect characterization for sub-40-nm flash memory, *Electron Device Letters, IEEE*, vol. 32, no. 8, pp. 999, 1001, Aug. 2011.

[76] Kim, T.; Franklin, N.; Srinivasan, C.; Kalavade, P.; Goda, A. Extreme Short-channel effect on RTS and inverse scaling behavior: Source–drain implantation effect in 25-nm NAND flash memory, *Electron Device Letters, IEEE*, vol. 32, no. 9, pp. 1185, 1187, Sept. 2011.

[77] Raghunathan, S.; Krishnamohan, T.; Parat, K.; Saraswat, K. Investigation of ballistic current in scaled floating-gate NAND FLASH and a solution, *Electron Devices Meeting (IEDM), 2009 IEEE International*, pp. 1–4, 7–9 Dec. 2009.

[78] Yano, K.; Ishii, T.; Sano, T.; Mine, T.; Murai, F.; Hashimoto, T.; Kobayashi, T.; Kure, T.; Seki, K. Single-electron memory for giga-to-tera bit storage, *Proceedings of the IEEE*, vol. 87, no. 4, pp. 633–651, April 1999.

[79] Aritome, S.; Kirisawa, R.; Endoh, T.; Nakayama, R.; Shirota, R.; Sakui, K.; Ohuchi, K.; Masuoka, F. Extended data retention characteristics after more than 10^4 write and erase cycles in EEPROMs, *International Reliability Physics Symposium, 1990. 28th Annual Proceedings*, 1990, pp. 259–264.

[80] Kirisawa, R.; Aritome, S.; Nakayama, R.; Endoh, T.; Shirota, R.; Masuoka, F. A NAND structured cell with a new programming technology for highly reliable 5 V-only flash EEPROM, *1990 Symposium on VLSI Technology, 1990. Digest of Technical Papers, 1990*, pp. 129–130.

[81] Aritome, S.; Shirota, R.; Kirisawa, R.; Endoh, T.; Nakayama, R.; Sakui, K.; Masuoka, F. A reliable bi-polarity write/erase technology in flash EEPROMs, *International Electron Devices Meeting, 1990. IEDM '90. Technical Digest., 1990*, pp. 111–114.

[82] Aritome, S.; Shirota, R.; Sakui, K.; Masuoka, F. Data retention characteristics of flash memory cells after write and erase cycling, *IEICE Transactions on Electronics*, vol. E77-C, no. 8, pp. 1287–1295, Aug. 1994.

[83] Aritome, S.; Shirota, R.; Hemink, G.; Endoh, T.; Masuoka, F. Reliability issues of flash memory cells, *Proceedings of the IEEE*, vol. 81, no. 5, pp. 776–788, May 1993.

[84] Shirota, R.; Nakayama, R.; Kirisawa, R.; Momodomi, M.; Sakui, K.; Itoh, Y.; Aritome, S.; Endoh, T.; Hatori, F.; Masuoka, F. A 2.3 μm^2 memory cell structure for 16 Mb NAND EEPROMs, *Electron Devices Meeting, 1990. IEDM '90. Technical Digest, International*, pp. 103–106, 9–12 Dec. 1990.

[85] Lee, C.-H.; Sung, S.-K.; Jang, D.; Lee, S.; Choi, S.; Kim, J.; Park, S.; Song, M.; Baek, H.-C.; Ahn, E.; Shin, J.; Shin, K.; Min, K.; Cho, S.-S.; Kang, C.-J.; Choi, J.; Kim, K.; Choi, J.-H.; Suh, K.-D.; Jung, T.-S. A highly manufacturable integration technology for 27 nm 2 and 3 bit/cell NAND flash memory, *Electron Devices Meeting (IEDM), 2010 IEEE International*, pp. 5.1.1, 5.1.4, 6–8 Dec. 2010.

[86] Asenov, A.; Kaya, S.; Brown, A. R. Intrinsic parameter fluctuations in decananometer MOSFETs introduced by gate line edge roughness, *Electron Devices, IEEE Transactions on*, vol. 50, no. 5, pp. 1254–1260, May 2003.

[87] Asenov, A.; Brown, A. R.; Davies, J. H.; Kaya, S.; Slavcheva, G. Simulation of intrinsic parameter fluctuations in decananometer and nanometer-scale MOSFETs, *Electron Devices, IEEE Transactions on*, vol. 50, no. 9, pp. 1837–1852, Sept. 2003.

[88] Spessot, A.; Calderoni, A.; Fantini, P.; Spinelli, A. S.; Compagnoni, C. M.; Farina, F.; Lacaita, A. L.; Marmiroli, A. Variability effects on the VT distribution of nanoscale NAND flash memories, *Reliability Physics Symposium (IRPS), 2010 IEEE International*, pp. 970–974, 2–6 May 2010.

[89] Spessot, A. M.; Compagnoni, C. M.; Farina, F.; Calderoni, A.; Spinelli, A. S.; Fantini, P. Compact modeling of variability effects in nanoscale NAND flash memories, *Electron Devices, IEEE Transactions on*, vol. 58, no. 8, pp. 2302, 2309, Aug. 2011.

[90] Larcher, L.; Padovani, A.; Pavan, P.; Fantini, P.; Calderoni, A.; Mauri, A.; Benvenuti, A. Modeling NAND Flash memories for IC design, *IEEE Electron Device Letters*, vol. 29, pp. 1152–1154, Oct. 2008.

[91] Miccoli, C.; Compagnoni, C. M.; Amoroso, S. M.; Spessot, A.; Fantini, P.; Visconti, A., and Spinelli, A. S. Impact of neutral threshold voltage spread and electron-emission statistics on data retention of nanoscale NAND flash, *IEEE Electron Device Letters*, vol. 31, no. 11, pp. 1202–1204, Nov. 2010.

[92] Mouli, C.; Prall, K.; Roberts, C. Trend in memory technlogy—reliability perspectives, challenges and opportunities, *Proceedings of 14th IPFA* 2007, pp. 130–134.

[93] Compagnoni, C. M.; Ghetti, A.; Ghidotti, M.; Spinelli, A. S.; Visconti, A. Data retention and program/erase sensitivity to the array background pattern in deca-nanometer NAND flash memories, *IEEE Transactions on Electron Devices*, vol. 57, no. 1, pp. 321–327, 2010.

[94] Park, B.; Cho, S.; Park, M.; Park, S.; Lee, Y.; Cho, M. K.; Ahn, K.-O.; Bae, G.; Park, S. Challenges and limitations of NAND flash memory devices based on floating gates, *Circuits and Systems (ISCAS), 2012 IEEE International Symposium on*, pp. 420, 423, 20–23 May 2012.

[95] Kim, C.; Ryu, J.; Lee, T.; Kim, H.; Lim, J.; Jeong, J.; Seo, S.; Jeon, H.; Kim, B.; Lee, I. Y.; Lee, D. S.; Kwak, P. S.; Cho, S.; Yim, Y.; Cho, C.; Jeong, W.; Park, K.; Han, J.-M.; Song, D.; Kyung, K.; Lim, Y.-H.; Jun, Y.-H. A 21 nm high performance 64 Gb MLC NAND flash memory with 400 MB/s asynchronous toggle DDR interface, *Solid-State Circuits, IEEE Journal of*, vol. 47, no. 4, pp. 981, 989, April 2012.

6

RELIABILITY OF NAND FLASH MEMORY

6.1 INTRODUCTION

Reliability of NAND flash memory [1] is more interesting than that of other semi-conductor devices. This is because a very high electric field (>10 MV/cm) is applied to tunnel oxide during program and erase operations, in comparison with a low field (<5 MV/cm) in other devices. Program and erase of NAND flash perform by electron injection and emission to/from floating gate (FG). There are several methods of electron injection and emission. For electron injection, there are two methods. One is channel hot electron (CHE) injection. The voltages (5–12 V) are applied to drain and control gate (CG), as drain current flows. A portion of the electrons of the drain current in the channel becomes hot, and it is injected to FG over the energy barrier of gate oxide. The CHE can operate the electron injection by relatively low applied voltage (~12 V); however, large channel electron current is needed to make a required number of hot electron injections. Then program efficiency (injected electron/drain electron current) is quite low (~10^{-6}). It is difficult to implement the parallel programming (page programming), where many cells are simultaneously programmed to achieve fast programming. The other electron injection method is the Fowler–Nordheim tunneling (FN-t) injection. A high voltage (~23 V) is applied to CG to inject electrons from channel to FG. The applied voltage is needed to be high (~23 V); however, program efficiency is high (~1). Therefore, it is possible to program many cells at the same time (parallel program or page programming).

Nand Flash Memory Technologies, First Edition. Seiichi Aritome.

TABLE 6.1 Program and Erase Schemes of NOR and NAND Flash

Flash	Electron Injection	Electron Emission
NOR flash	CHE	FN-t @ S
NOR flash [5]	CHE	Uniform FN-t
NAND flash [1]	CHE	—
NAND flash [2,3,8]	Uniform FN-t	FN-t @ D
NAND flash [4–8]	Uniform FN-t	Uniform FN-t

CHE, channel hot electron; FN-t @ S, FN-t at source; FN-t @ D, FN-t at drain; Uniform FN-t, FN-t entire channel (tunnel oxide).

For electron emission, there are mainly two methods. One is FN-t emission at a source or drain gate overlap area. High voltage ($\sim$20 V) is applied to source or drain to eject electrons from FG. During FN-t emission at a source or drain, a large source (or drain) leakage current is caused by band-to-band tunneling (BB-t) mechanism. By BB-t, electron–hole pairs are generated in substrate. A portion of the holes is accelerated by high field at a source (or drain), and it is injected to tunnel oxide. Then some holes are trapped in tunnel oxide. These hole traps have degraded Program/Erase cycling and data retention characteristics, as described in Sections 6.2 and 6.3. The other electron emission method is the FN-t emission entire channel region (entire tunnel oxide). The BB-t does not occur due to no voltage difference between the source (or drain) and the substrate.

The possible program and erase schemes of flash memory are summarized in Table 6.1. Reliability performances are compared between these program and erase schemes in Sections 6.2 [6,7] and 6.4 [5].

It had been reported that high field FN-t during program and erase causes oxide degradation in tunnel oxide. Figure 6.1 shows one of the degradation mechanisms [9] of FN-t. Hot electron injection from cathode (FG) to anode (substrate) makes

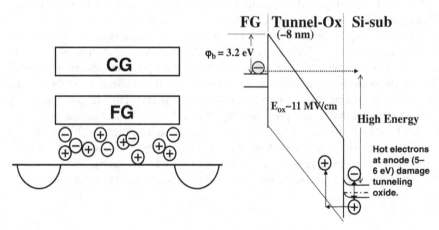

FIGURE 6.1 The degradation mechanism of Fowler–Nordheim tunneling stress on tunnel oxide.

Data Retention

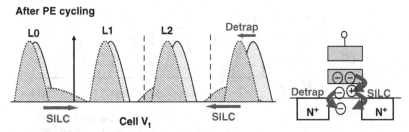

• **Data retention is degraded by P/E cycling.**
• **Two modes; Detrap and SILC**
  Detrap; trapped from bulk of tunnel-ox → V_t shift

  Interface → S-factor (subthreshold slope) recovering
  SILC; trap-assisted tunneling leakage from FG

FIGURE 6.2   Data retention phenomena.

electron–hole pairs at anode. And a portion of the hot holes has been injected to tunnel oxide and trapped inside tunnel oxide. An electron trap has also occurred in tunnel oxide during FN-t stress. These hole and electron traps have caused oxide degradation in tunnel oxide and have direct impacts on program/erase cycling endurance characteristics, such as program/erase window narrowing, as described in Sections 6.2 and 6.3.

Data retention is also degraded by electron and hole traps in tunnel oxide, as shown in Fig. 6.2. Detrapping of trapped charges in tunnel oxide is a major root cause of V_t shift during the data retention test, as described in Sections 6.2 and 6.3. And trapped holes would make a stress-induced leakage current (SILC) for tunnel oxide, because the potential barrier of tunnel oxide may locally decrease by hole traps. SILC makes tail bits in the V_t distribution, as shown in Fig. 6.2.

Figure 6.3 shows the read disturb phenomena. Read disturb failure is mainly caused in the erase state after program/erase cycling stress. The stress-induced leakage current (SILC), which is generated by program/erase cycling stress, is major root cause, as described in Section 6.4.

The program disturb issue is becoming serious as scaling memory cells. A low self-boosting voltage was the major root cause of the program disturb issue for <90-nm generation memory cells. However, below 70-nm generation, several new program disturb phenomena by the hot carrier injection mechanism were reported, as described in Section 6.5.

The erratic over-program is presented in Section 6.6. The mechanism of erratic over program is considered to be an excess electron injection at hole trap sites in tunnel oxide. The number of failure bits by erratic over program is increased by scaling memory cells. There is no good way to mitigate erratic over-program, except for slower programming operations. The strong ECC can actually save erratic over-program failure bits.

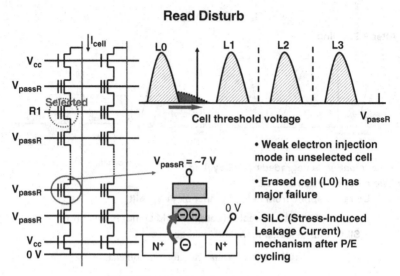

FIGURE 6.3   Read disturb phenomena.

In the NAND flash program and erase operation, a high voltage is applied to CG or substrate. This high program and erase voltage cannot be effectively decreased by the scaling feature size of the memory cell. Therefore, the high-field stress problem has been caused in memory cells [10]. One of the problems is "the negative V_t shift" during programming, as described in Section 6.7 [11].

6.2   PROGRAM/ERASE CYCLING ENDURANCE AND DATA RETENTION

6.2.1   Program and Erase Scheme

The program/erase cycling endurance and the data retention are key characteristics of floating-gate memories such as EEPROMs (electrically erasable and programmable read-only memories) and flash memories [1, 8, 12–17]. An essential requirement for the memory cell is sufficient data retention even after a large number of program/erase cycles. However, the data retention is degraded by the high-field stress in the thin tunnel oxide during the program and erase operations.

The degradation mechanisms of the flash memory cell had been studied [5–8, 18–21] to improve the reliability of the memory cell. The degradation behavior is related to the charge-trapping process in the thin tunnel oxide during both the program and erase operations [5–8, 18–21], which are performed by the Fowler–Nordheim tunneling of electrons through the thin tunnel oxide. The generation of traps in the thin tunnel oxide is strongly dependent on the program and erase conditions [5–8, 14, 20, 21]. Therefore, the program/erase cycling endurance and the data retention characteristics had been studied in several program/erase schemes [5–8], to decide a proper program/erase scheme.

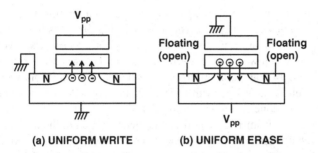

(a) UNIFORM WRITE   (b) UNIFORM ERASE

FIGURE 6.4 Operation of the uniform program (write) and uniform erase scheme. (a) Uniform program (write). A positive high voltage is applied to the control gate with the drain and substrate grounded. Electrons are injected to the floating gate over the whole channel area. (b) Uniform erase. A positive high voltage is applied to the substrate with the control gate grounded. Electrons in the floating gate are emitted uniformly over the whole channel area. Copyright © 1994 IEICE.

The endurance and data retention characteristics of the flash memory cell programmed by two different program and erase schemes are compared [6, 7, 20]. One is a uniform program (write) and uniform erase scheme (Fig. 6.4), using uniform Fowler–Nordheim tunneling over the whole channel area both during the program and erase operation. The other is a uniform program (write) and nonuniform erase scheme (Fig. 6.5), using uniform tunneling over the whole channel area during the program operation (Fig. 6.5a) and local nonuniform tunneling at the drain during the erase operation (Fig. 6.5b). A uniform program and nonuniform erase scheme could also be applied to NOR-type cells, such as the Di-NOR cell [16]. However, a uniform program and uniform erase scheme could be applied to only NAND-type cells because it was not possible to implement a selective program (program inhibit) operation in a NOR-type cell.

In experiments of program/erase cycling endurance, the programming voltages (V_{pp}) for cells of various tunnel oxide thickness were determined by the V_{pp} when the

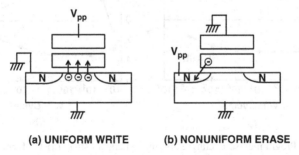

(a) UNIFORM WRITE   (b) NONUNIFORM ERASE

FIGURE 6.5 Operation of the uniform program (write) and nonuniform erase scheme. (a) Uniform program (write). A positive high voltage is applied to the control gate with drain and substrate grounded. (b) Nonuniform erase. A positive high voltage is applied to the drain with the control gate grounded. Copyright © 1994 IEICE.

threshold voltage reached 2.5 V in a program operation after 10 cycles, and −3.0 V in an erase operation after 10 cycles. The programming time was fixed at 1 ms. This method of V_{pp} determination is suitable for the comparison of these two schemes, because an initial window widening during the first 10 program/erase cycles occurs in the uniform program and nonuniform erase scheme. The threshold voltage of the flash memory cell is measured with a drain voltage of 1 V. The accelerated data retention tests were done at a temperature of 150–300°C after different numbers of program/erase cycles from 10 to 1 million had been applied to the devices.

A conventional n-channel floating gate transistor with a tunnel oxide over the entire channel region was used for experiments. The gate length is 1.0 μm. The thin gate oxide of 5.6–12.1 nm is thermally grown at 800°C. The effective oxide thickness of the oxide–nitride–oxide (ONO) inter-poly dielectric is 25 nm [22].

6.2.2   Program and Erase Cycling Endurance

The program/erase cycling endurance characteristics of the two different schemes are shown in Fig. 6.6. The uniform program and uniform erase scheme guarantees a wide cell threshold voltage (V_t) window of upto 4 V, even after 1 million program (write) and erase cycles (Fig. 6.6a). However, the threshold window obtained by the uniform program and nonuniform erase scheme begins to close rapidly after around 100 program (write) and erase cycles, and it fails after 100K program and erase cycles (Fig. 6.6b).

In the case of the uniform program and nonuniform erase scheme (Fig. 6.6b), holes are generated near the drain of the memory cell by the band-to-band tunneling mechanism [23–25] due to the presence of a high voltage over the drain junction during the erase. A part of these holes is injected into the thin gate oxide after being accelerated in the depletion region. The injection of hot holes results in hole trapping

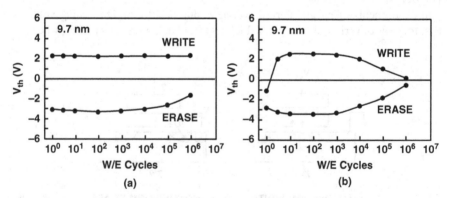

FIGURE 6.6   Program/erase cycling endurance characteristics. (a) Uniform program and uniform erase scheme, using uniform injection and uniform emission over the whole channel area. (b) Uniform program and nonuniform erase scheme, using uniform injection over the whole channel area and local nonuniform emission at the drain, respectively. Copyright © 1994 IEICE.

in gate oxide near the drain. As a result, field enhancement occurs, resulting in an initial threshold window widening, which is observed in the first 10 program/erase cycles. During the first cycle, the threshold window is small (about 1.5 V), because field enhancement has not yet occurred. After a few cycles, holes are trapped locally in the gate oxide near the drain, and the Fowler–Nordheim tunneling during both the nonuniform erase and the uniform program could be confined to a small region near the drain. Therefore, the electron trapping in the gate oxide near the drain area is enhanced, subsequently, these trapped electrons impede the tunneling of electrons between the floating gate and the substrate. As a result, window narrowing rapidly occurs.

On the other hand, in the uniform program and uniform erase scheme, the program and erase operation is performed without hot-hole generation by band-to-band tunneling mechanism; therefore, the initial window widening and rapid window narrowing do not occur.

In the uniform program and uniform erase scheme, the threshold voltage (V_{th}) of the erased cell is dependent on the number of program/erase cycles, namely V_{th} slightly shift down by 1K cycles and shift up after 1K cycles. However, the V_{th} of the programmed cell is not dependent on the number of program/erase cycles. This can be explained as follows. The oxide traps and interface traps are generated uniformly over the entire channel area because the Fowler–Nordheim tunneling of electrons is performed uniformly during uniform program and erase operations. The uniformly trapped oxide charges over the channel area affect not only the electron tunneling current through the oxide, but also the flat-band voltage. The V_{th} of the erased cell decreases slightly by 1K cycles (Fig. 6.6a) due to the hole trap generated in the oxide during the Fowler–Nordheim tunneling. The hole traps result in an increase of electron tunneling current (V_{th} decrease) as well as a decrease of the flat-band voltage (V_{th} decrease). After about 1K cycles, the V_{th} of the erased cell increases due to electron trapping. The electron traps result in a decrease of electron tunneling current (V_{th} increase) as well as an increase of the flat-band voltage (V_{th} increase).

On the other hand, the V_{th} of the programmed cell remains almost constant up to 1 million cycles in spite of the positive and negative charge trapping in the oxide. For programmed V_{th}, by 1K cycles, the hole traps result in an increase of electron tunneling current (V_{th} increase) as well as a decrease of the flat-band voltage (V_{th} decrease). After about 1K cycles, the electron traps result in a decrease of electron tunneling current (V_{th} decrease) as well as an increase of the flat-band voltage (V_{th} increase) [6, 7, 20]. Therefore V_{th} movement in programmed and erased cells in uniform program/erase scheme can be explained by the trap effect on both electron FN-t tunneling current and the flat-band voltage [6, 7, 20].

The influence of the trapped oxide charges on the flat-band voltage is confirmed by measuring the flat-band voltage shift during equivalent program and erase stress cycles of the test devices in which the floating gates are connected with the control gates, as shown in Fig. 6.7. Using the uniform program (write) and uniform erase scheme, the threshold voltage shift is negative at the beginning but becomes positive with an increasing number of stress cycles, because hole trapping mainly occurs up to the first 1K cycles while electron trapping becomes dominant after 1K cycles. The

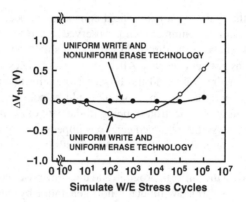

FIGURE 6.7 Threshold voltage shift of the test device in which the floating gate is connected with the control gate, during equivalent program and erase stress on tunnel-oxide. The equivqlent program/erase stress condition: In the case of uniform program and uniform erase scheme, high-voltage pulses of 12.0 V, 0.1 ms, and 13.5 V, 0.1 ms are applied to the gate and substrate, respectively. In the case of the uniform program and nonuniform erase scheme, high-voltage pulses of 10.4 V, 0.1 ms and 13.0 V, 0.1 ms are applied to the gate and drain, respectively. In the uniform program and uniform erase scheme, the trapped oxide charges are directly affecting the threshold voltage. However, in uniform program and nonuniform erase scheme, the trapped oxide charges are not directly affecting the threshold voltage. Copyright © 1994 IEICE.

influence of the trapped oxide charges on the flat-band voltage is confirmed; thus the mechanism of threshold voltage compensation with flat-band voltage and FN-t current is also confirmed [6, 7, 20].

Figure 6.8 shows the dependence of the endurance characteristics on the tunnel-oxide thickness [7]. In the case of the uniform program and uniform erase scheme, the window narrowing of the thicker tunnel oxides is larger than that of the thinner

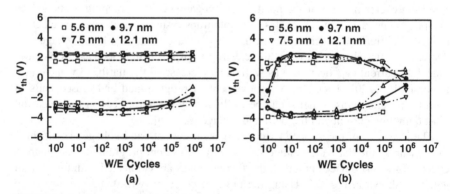

FIGURE 6.8 Dependence of the program/erase cycling endurance characteristics on the tunnel-oxide thickness. (a) Uniform program and uniform erase scheme. (b) Uniform program and nonuniform erase scheme. Copyright © 1994 IEICE.

oxides. The window widening around 1K cycles is also larger in thicker tunnel oxides, as compared with the thinner oxides. The increased hole trapping is caused by an increased hole generation in thicker oxides. Consequently, since holes are involved in the generation of traps, the electron trapping will also be enhanced for thicker oxides. In the case of the uniform program and nonuniform erase scheme, the window widening and narrowing is almost independent of the oxide thickness over the 7.5 to 12.1-nm thickness range. However, for a 5.6-nm oxide thickness, the window narrowing is strongly reduced for both program and erase operations. However, the breakdown of the 5.6-nm oxide occurs very early, before 1 million cycles, in the case of uniform program and nonuniform erase scheme.

6.2.3 Data Retention Characteristics

A. Program and Erase Scheme Dependence Figure 6.9 shows the data retention characteristics of erased cells, which are subjected to several program/erase cycles from 10 to 1 million by two program/erase schemes [6–8]. In the case of the uniform program and nonuniform erase scheme (Fig. 6.9b), the stored positive charge gradually decays as the baking time increases, so the threshold voltage window decreases. However, in the case of the uniform program and uniform erase scheme (Fig. 6.9a), the stored positive charge effectively increases up to a 100-minute baking time due to the detrapping of electrons from the gate oxide to the substrate during the retention bake, as shown in Fig. 6.10. This effective increase of the stored positive charge becomes

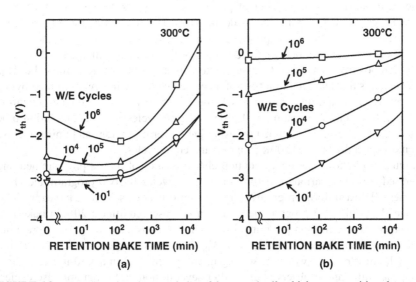

FIGURE 6.9   Data retention characteristics of the erased cell, which stores positive charges in the floating gate, as a function of retention bake time at 300°C in (a) uniform program (write) and uniform erase scheme and (b) uniform program (write) and nonuniform erase scheme, subjected to 10, 10K, 100K, 1 million program and erase cycles (P/E cycles). Copyright © 1994 IEICE.

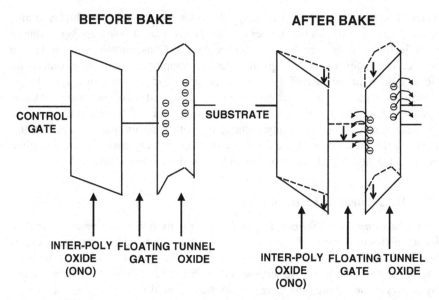

FIGURE 6.10   Band diagram before and after baking. The effect of detrapping electrons from the gate oxide to the substrate is equivalent to the effect of trapping holes in the gate oxide. Copyright © 1994 IEICE.

larger with an increasing number of program/erase cycles because the amount of trapped negative charge in the thin oxide increases. The effect of detrapping electrons is equivalent to the effect of trapping holes in the gate oxide. As a result, the detrapping of the electrons suppresses the data loss of the positively charged cell because the stored positive charge is effectively increasing at the beginning of the bake. This effect extends the data retention time of an erased cell, which is programmed by the uniform program and uniform erase scheme. Figure 6.11 shows the data retention time as a function of the number of program/erase cycles. The data retention time of an erased cell can be extended by using the uniform program and uniform erase scheme, especially beyond 100K program and erase cycles.

Figure 6.12 shows the data retention characteristics of both a programmed and an erased cell after various P/E (program/erase) cycles for two program and erase schemes [7]. In a uniform program (write) and uniform erase scheme, a large threshold voltage (V_{th}) shift in a programmed cell can be observed after a 20-min bake in a cell subjected to 1 million program/erase cycles. This is also due to the detrapping of electrons from the tunnel oxide. This V_{th} shift was observed in a 1.0-μm design rule cell, but it became worse as scaling memory cell size. In a <30-nm cell, same initial V_{th} shift can be observed even after several thousand program/erase cycles. This issue is further discussed in the next section.

B. Temperature Dependence   In order to estimate the data retention lifetime of the memory cell under the operation temperature (<85°C) in the case of the uniform

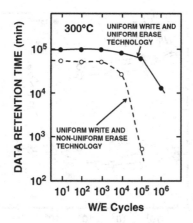

FIGURE 6.11 Data retention time of the erased memory cell after program and erase cycling. The data retention time is defined by the time that the threshold voltage reaches −0.5 V during the retention bake at 300°C. Copyright © 1994 IEICE.

program and uniform erase scheme, the data retention characteristics at different temperatures (150–300°C) had been measured, as shown in Fig. 6.13 [7]. The 1.0-μm design rule memory cells with a 9.7-nm tunnel oxide are subjected to 1 million program/erase cycles. For the programmed cell, the V_{th} monotonically shifts negative toward the neutral V_{th} (0.7 V), as baking time increases. The negative V_{th} shift after 20 min of baking increases with the number of program(write)/erase cycles, as shown in Fig. 6.14a. This is because both the charge loss from the floating gate and the electron detrapping from the tunnel oxide to the substrate are enhanced at high

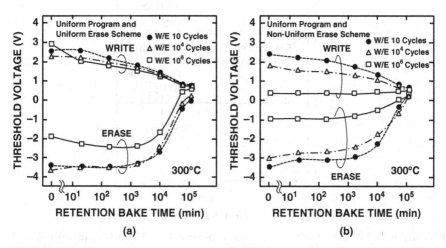

FIGURE 6.12 Data retention characteristics after different program/erase cycle. (a) Uniform program and uniform erase scheme. (b) Uniform program and nonuniform erase scheme. Copyright © 1994 IEICE.

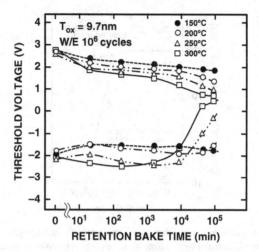

FIGURE 6.13 Dependence of the data retention characteristics on the baking temperature in the uniform program and uniform erase scheme. Copyright © 1994 IEICE.

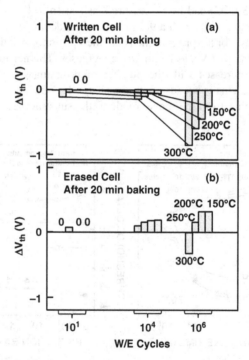

FIGURE 6.14 Threshold voltage shift after a 20-min bake at 150–300°C. (a) Programmed cell (b) Erased cell. Copyright © 1994 IEICE.

temperatures. For practical applications of the NAND flash memory, the negative V_{th} shift of the programmed cell at 100°C is estimated to be less than 0.2 V. So the V_{th} margin of the programmed cell should be determined with more than 0.2 V in this cell.

For the erased cell, the phenomenon of the effective increase of the stored positive charge can be observed at all test temperatures from 150°C to 300°C, as shown in Fig. 6.13. However, the threshold shift is strongly dependent on the temperature. For a bake of 300°C, the threshold voltage shows a negative shift until 200 min; after that, the V_{th} shows a positive shift toward the neutral V_{th} (0.7 V). However, for 150–250°C baking, the initial V_{th} shift is positive (about 0.3 V) after 20 min of baking; after that the V_{th} shift is negative for 1000 min and becomes positive again with V_{th} shifting toward the neutral V_{th}. The first positive V_{th} shift after about 20 min of baking can be explained by the charge loss from the floating gate due to the high-field-stress-induced leakage current in the tunnel oxide [5, 26, 27]. The negative shift of the V_{th} can be explained by the effect of electron detrapping from the tunnel oxide to the substrate.

Because of the lower detrapping rate of electrons at lower temperatures, the time at which the minimum V_{th} is reached during the bake is longer at lower temperature—for example, 1000 min at 250°C and 10,000 min at 200°C. Therefore, at the operation temperature (<100°C), the maximum negative V_{th} shift will occur after more than 1,000,000 min.

Figure 6.15 shows the estimation of the data retention time using the temperature dependence of the data retention of the memory cell for different numbers of program/erase cycles. Due to the program/erase cycling, the data retention time is shortened. However, 10 years of data retention time for an operation temperature of less than 100°C can be guaranteed even after 1 million cycles in the case of the

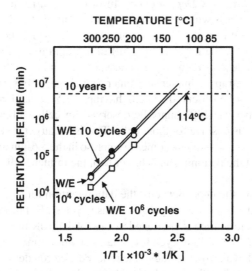

FIGURE 6.15   Estimation of the data retention time at the operation temperature. Copyright © 1994 IEICE.

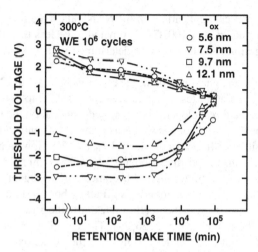

FIGURE 6.16   Dependence of the data retention characteristics on the tunnel oxide thickness in the uniform program and uniform erase scheme. Copyright © 1994 IEICE.

uniform program and uniform erase scheme. The activation energy (slope in Fig. 6.15) of data retention time is almost the same in the range of 10–10^6 program/erase cycling in the case of this experiment of a 1.0-μm rule cell. This is because data retention time is mainly determined by the mechanism of charge loss from floating gates; however, data retention time is not determined by detrapping mechanism, as shown in Fig. 6.12 and Fig. 6.13.

C. Tunnel-Oxide Thickness Dependence   In order to clarify the scaling limit of the tunnel-oxide thickness with respect to the data retention, the data retention characteristics of cells with various tunnel-oxide thicknesses had been measured, as shown in Fig. 6.16 [7]. The negative V_{th} shift of a programmed cell after a 20-min bake decreases with decreasing oxide thickness, as shown in Fig. 6.17a, because the amount of electron detrap is smaller in thinner tunnel oxide. For an erased cell, the negative V_{th} shift can be observed in cells having a 7.5- to 12.1-nm tunnel oxide; however, the negative V_{th} shift cannot be observed for cells with a tunnel oxide of 5.6 nm. The V_{th} shift is positive for those cells. This is because the stress leakage current increases as the tunnel oxide becomes thinner, so in the 5.6-nm tunnel-oxide case, the charge loss from the floating gate is larger than the influence of the detrapping of electrons.

Figure 6.18 shows the dependence of the data retention time on both the tunnel-oxide thickness and the number of program/erase cycles [7]. For the thinner tunnel oxides, the data retention time is shortened in case of 10–10K program/erase cycles; however, in the case of 1 million cycles, the data retention time is extended because of the reduced window narrowing due to the reduced electron detrapping. Therefore, the scaling of the tunnel oxide is not limited by the degradation of the data retention due to the thinning of the tunnel oxide up to a thickness of 5.6 nm.

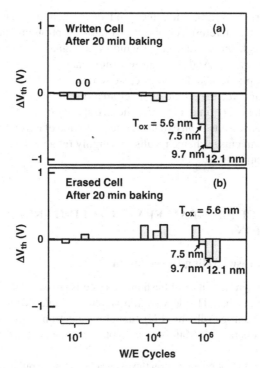

FIGURE 6.17 Threshold voltage shift after a 20-min bake at 300°C. The memory cell has a 5.6- to 12.1-nm tunnel oxide. (a) Programmed cell, (b) Erased cell. Copyright © 1994 IEICE.

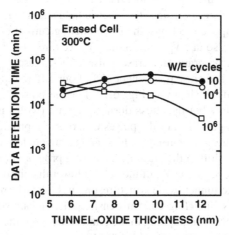

FIGURE 6.18 Data retention time at 300°C depending on the tunnel-oxide thickness and the number of program/erase cycles. Copyright © 1994 IEICE.

The program/erase cycling endurance and data retention characteristics were described in two program/erase schemes. In a uniform program and uniform erase scheme, the wide cell threshold voltage window was guaranteed even after 1 million program/erase cycles. And the data retention characteristics was improved by applying a uniform program and uniform erase scheme, which uses uniform Fowler–Nordheim tunneling over the whole channel area during both program (write) and erase. It was clarified experimentally that the detrapping of electrons from the gate oxide to the substrate results in an extended retention time of erase state. The uniform program and uniform erase scheme results in a highly reliable flash memory with an extended data retention time.

6.3   ANALYSIS OF PROGRAM/ERASE CYCLING ENDURANCE AND DATA RETENTION

6.3.1   Program/Erase Cycling Degradation

The performance and reliabilities of flash memory cells are degraded by repeating the program and erase cycling. The degradation is mainly related with the tunnel oxide degradation by Fowler–Nordheim (FN) tunneling electron injection stress during program and erase operations. Many degradation phenomena of the thin tunnel oxide had been reported.

First, the cell V_{th} shift dependence on the program and erase pulse is discussed [28]. In the program/erase scheme of the NAND cell, the V_{th} shift during program/erase cycling mainly appears on the erased V_{th}, because the erased V_{th} is very sensitive to charge trapping in the tunnel oxide, as described in Section 6.2.2. For example, if electrons are trapped in the tunnel oxide, the V_{th} of the memory cell will shift in the positive direction. Furthermore, since the electric field strength in the tunnel oxide is reduced due to the trapped electrons, the FN tunneling current during the erase will be reduced, resulting also in a V_{th} shift in the positive direction.

To investigate the program and erase pulse effect, erase V_{th} degradation was compared in program and erase pulse shape [28]. Four different program/erase pulse shapes had been used, as shown in Fig. 6.19. For the pulses A and B, the stress during erase is very low but the stress during program (write) is high for pulse A and relatively low for pulse B. For the pulses C and D, the program (write) stress is low while the erase stress for pulse C is high and relatively low for pulse D. Figure 6.19a,b shows the V_{th} shift in the erased state during program/erase cycling. During the first few hundred cycles, the V_{th} of the cells is lower than initially, this is because of hole trapping in the tunnel oxide which results in both a decrease of the V_{th} and an increase of the FN tunneling current density. After about 1000 cycles, the V_{th} increases because of electron trapping in the tunnel oxide. For the high stress pulse A and C, since more holes are generated [29], the hole trapping is about 10 times higher than for the low stress pulse B and D. Furthermore, from the differences in the slope of the V_{th} shift curves at 100K program/erase cycles, it can be concluded that the electron trap generation rate is higher for the high stress pulse A and C in comparison

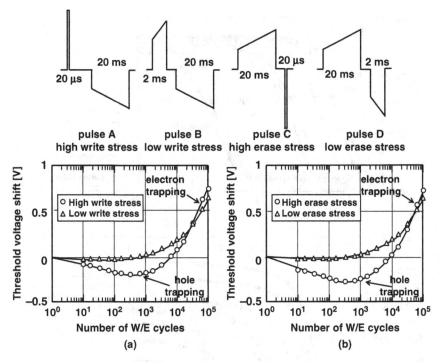

FIGURE 6.19 Program(write)/Erase pulses applied to the control gates of the memory cells. For pulses A and B the erase pulse is long (20 ms) and has a trapezoidal shape to obtain a low erase stress. For pulse A the program pulse is very short (20 μs); resulting in a high program stress. For pulse B the program (write) pulse is long (2 ms) and has a trapezoidal shape to reduce the write stress. For pulses C and D the program (write) pulse is long to obtain a low program stress. For pulse C the erase pulse is very short, resulting in a high erase stress, while for pulse D the erase stress is reduced by using a longer trapezoidal pulse. (a, b) Average erased threshold voltage shift of 96 memory cells relative to the initial erased threshold voltage during W/E cycling for (a) the high and low program (write) stress pulses A and B and (b) for the high and low erase stress pulses C and D. For the high-stress pulses A and C, hole trapping is significantly larger than for the low-stress pulses B and D. Electron trapping occurs for both the high- and low-stress pulses.

with the low stress pulse B and D. The hole trapping for high erase stress pulse C seems to be slightly larger than for the high program stress pulse A. This indicates that hole injection is increased for high erase stress pulse C; however, a more likely explanation is that for the high erase stress pulse C, the holes are trapped closer to the Si/SiO$_2$ interface (~10–20 Å), where they originated since the Si/SiO$_2$ interface corresponds with the anode during the high stress erase pulse. The hole traps near the Si/SiO$_2$ interface have much impact on the read disturb characteristics (Sections 6.4 and 6.5), the program disturb (Section 6.6), and the erratic over-program (Section 6.7) because the potential barrier for electron injection to the floating gate is reduced. Therefore the erase condition has to be carefully controlled to be better reliabilities.

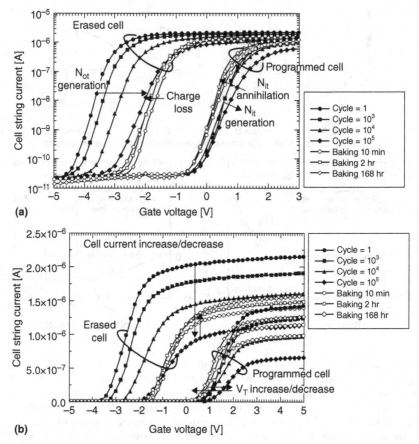

FIGURE 6.20 Drain-current–gate-voltage (I_d-V_g) curves of a cell transistor during endurance cycling and data retention test, with y-axis in (a) logarithmic scale and (b) linear scale. Programming pulses are 17 V–100 μs and erasing pulses are 17 V–1 ms in the cycling mode. The baking temperature was 250°C in the retention mode. Cell current and mobility are degraded by P/E cycling due to the interface trap generation and recovered by 250°C bake.

Next, memory cell degradation phenomena are discussed. As discussed above, the program/erase cycling has an impact on interface state and interface trap generation at the interface of tunnel oxide and substrate [30–32]. Figure 6.20 shows the cell current and mobility degradation by program/erase cycling. The origin of degradation mechanism is investigated by I_d-V_g curves of the cell transistor during the 100K program/erase cycling and 250°C 168H baking test. The oxide trap (N_{ot}) generation and charge loss can be monitored by the midgap voltage (V_{mg}) shift, and the interface trap density (N_{it}) can be monitored by the subthreshold slope of the transistor.

During program/erase cycling, oxide traps are generated in the tunnel oxide and electrons are captured at the trap sites, therefore, midgap voltage V_{mg} shifting

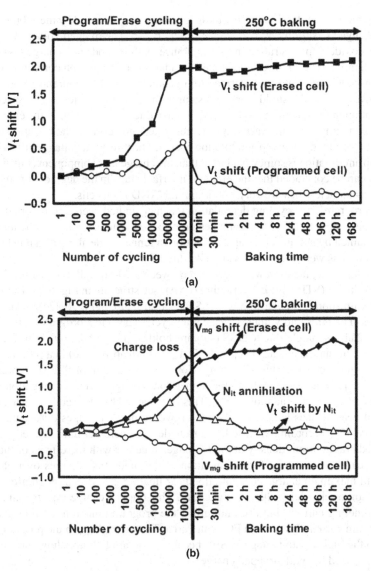

FIGURE 6.21   Analysis of threshold voltage shift in Fig. 6.20 during cycling and retention test. (a) Threshold voltage shift of the programmed cell and the erased cell. (b) Classification of threshold voltage shift by midgap voltage (V_{mg}) shift and by N_{it} generation/annihilation. V_t shift by interface trap (N_{it}) is the same both for the programmed cell and the erased cell.

toward the positive direction can be observed in the erased cells from 10^3 cycles to 10^5 cycles, as shown in Fig. 6.20. However, this phenomenon cannot be monitored in the program cell because the electron trap effects of reducing both FN current and positive V_t shift are canceled out, as described in Section 6.2.2. The threshold voltage shift in the cycling and the retention mode is shown in Fig. 6.21a. The threshold voltage shift can be classified by the midgap voltage V_{mg} shift and by the subthreshold slope change (V_t shift by N_{it}), as shown in Fig. 6.21b. The midgap voltage V_{mg} shift indicates oxide trap generation in the endurance cycling mode and charge loss in the retention mode. The subthreshold slope degradation and saturation current reduction (see Fig. 6.20b) indicate the interface trap generation in the endurance mode. And the recovery of subthreshold slope and saturation current indicates the interface trap annihilation in the retention mode of 250°C bake due to detrapping. In the endurance mode, oxide traps and interface traps are generated, however, in the retention mode, charge loss and interface trap annihilation occurs. The threshold voltage shift by interface trap annihilation is somewhat larger than the charge loss component (Fig. 6.21b). Thus, it can be concluded that the effect of interface traps in the degradation and data retention characteristics are very important in NAND flash cells.

The program/erase cycling degradation mechanism had been also reported in the scaled memory cell of 51- to 32-nm design rule [33]. The degradation phenomenon was explained by nonuniform trap distribution in tunnel oxide along a channel length (L) direction as well as channel width (W) direction.

The cycling degradation was compared between a 90-nm cell with a long shallow trench isolation (STI) edge cell structure (LSE) (see structure in Fig. 6.22a) and a 51-nm cell with short STI edge structure (SSE) (see structure in Fig. 6.22c), as shown in Fig. 6.22. The threshold voltage change after cycles (ΔV_t) in a program state of LSE (Fig. 6.22a,b) is mainly due to ΔSS (subthreshold slope degradation) resulting from generation of interface states [32]. In LSE, V_{MG} (midgap voltage) in a program state remains almost constant during cycling. However, in the case of SSE (Fig. 6.22c,d), V_{MG} in the program state shifts to the higher gate voltage during cycling along with subthreshold slope degradation (ΔSS). This is explained by the fact that a narrow FG width of SSE devices leads to high-field crowding near FG edges during the erase operation [33]. Accordingly, erase FN current density increases under FG edges. The increased erase current density at the FG edges together with the effects of the etch damage can lead to nonuniform generation of the oxide trap charges over channel area with maximum charge concentration near floating gate edges (nonuniform trap distribution of channel width (W) direction). This trapped charge affects mainly the erase tunneling current, but does not affect the program tunneling current because the program current flows at the FG center area. Therefore, V_{MG} in the program state shifts to the higher gate voltage in SEE because program FN tunneling current does not get reduced by oxide trapped charge.

This nonuniform charge distribution model was extended to the direction of gate length (L direction). It is known that oxide charges located near source/drain (S/D) junctions influence not only V_{MG} but also SS [34]. ΔSS (subthreshold slope degradation) due to the nonuniform oxide charges was simulated using negative charge clusters placed in a tunnel oxide under FG edges [33]. Simulation results reveal that the charge clusters can either increase or decrease SS depending on S/D overlap. This phenomenon can be explained by three-dimensional current flux distortion at a low current level in the presence of negative charge over the S/D region. If the S/D overlap is large, the oxide charges under FG edges are located over S/D region. Thus, the oxide charges effectively impede S/D electrons current and confine the electrons

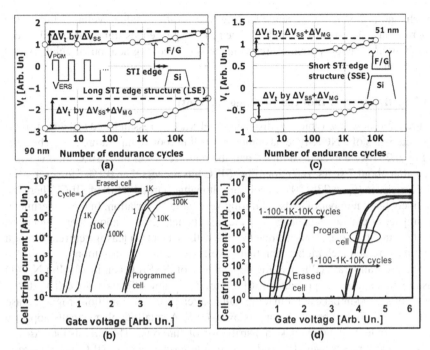

FIGURE 6.22 (a) Typical P/E cycling characteristics of LSE devices (90-nm cell). Erase state V_t degradation is caused by V_{MG} and SS change. Program state V_t is determined only by SS increase. Inset shows feature of LSE structure (bitline cross-section view). (b) I_d–V_g characteristics for various cycle amounts indicate combined V_{MG} and SS degradation in the erase state and SS degradation without V_{MG} change in the program state. (c) Typical P/E cycling characteristics of SSE structure (see inset: 51-nm cell). Erase state V_t degradation is caused by V_{MG} and SS shifts. Contrary to LSE, V_t in the program state is also determined by SS and V_{MG} change. (d) I_d–V_g characteristics for various cycle amounts indicate combined V_{MG} and SS degradation in the program and erase states.

contributing to subthreshold currents to a channel surface region. This effect leads to the improved SS. On the other hand, in the cell with a relatively small S/D overlap, the oxide charges impede both channel and S/D electrons current. Consequently, subthreshold currents arise far from the channel surface. This results in degradation of gate controllability and the deteriorated SS.

The nonuniform charge trapping model can well explain measurement results proving the dependence of ΔSS (subthreshold slope degradation) on an initial SS after cycling. For the higher initial SS value (a larger S/D overlap), SS reduces (SS improves) after cycling. However, for lower initial SS value (a smaller S/D overlap), SS increases (SS is degraded) after cycling [33].

Therefore, the model of nonuniform distribution of negative oxide charges, which are located near the floating gate edge, can explain the program/erase degradation mechanism in the scaled memory cell below the 50-nm design rule. It is shown that the nonuniformly distributed negative charges reduce erase FN current while they do not change program FN currents, leading to a positive midgap voltage shift in the program state. The localized oxide charge near the gate edges significantly influences source/drain junction potential, resulting in observed degradation of subthreshold swing value.

6.3.2   SILC (Stress-Induced Leakage Current)

Figure 6.23 shows the typical V_t distribution of SILC (stress induced leakage current) characteristics of a NOR-type flash memory cell in room temperature [35]. The cells in Fig. 6.23a were cycled 10 times, and the number of tail bits (SILC bits) is small even after about 7 years' retention time. SILC bits are less than 0.1%. However, after 100K cycles in Fig. 6.23b, 20% of cells are showing a large V_t shift of SILC. There is a relatively large number of cells exhibiting very large V_t variation. In this experiment, the impact of cycling is enhanced by the thin tunnel oxide of about 8 nm thickness. The tails in the distributions are due to cells with an electron leakage current through the tunnel oxide much higher than that of cells in the main distribution.

The program/erase (P/E) cycling dependence of the SILC cell in 16-Mbit NAND flash memory was also presented [36]. Figure 6.24 shows the memory cell V_t distribution of a 1000-hr bake at room temperature after 100K and 1 million program/erase cycles. The initial V_t before baking is over 3.9 V. A small number of cells appear a large charge loss, and they make a distribution of "tail bits." The tail bits increases as P/E cycling increases over 10^5. And the cell of large charge loss (SILC bit) has strong V_t dependence (electric field dependence). Higher V_t (higher electric field) produces a worse V_t shift of SILC. The leakage current was calculated from V_t shift of SILC bit, shown in Fig. 6.25. J/S is leakage current density. J is calculated by

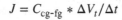

$$J = C_{\text{cg-fg}} * \Delta V_t / \Delta t$$

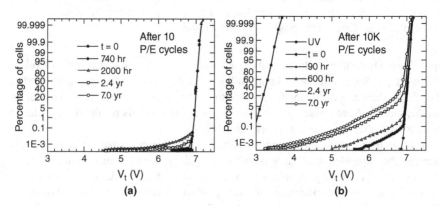

FIGURE 6.23   Data retention of SILC in the NOR flash cell. V_t distribution at different room-temperature storage times for an 8-nm-thick oxide after (a) 10 and (b) 10K cycles. Cell V_t shift-related SILC is much larger than detrapping, but a smaller percentage of cells has SILC. Strong dependence on number of cycles is observed. SILC cells disappear in high-temperature bake (250°C). Trap-assisted tunneling would be the root cause of SILC.

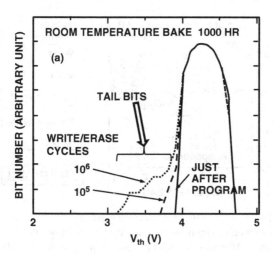

FIGURE 6.24 Data retention of SILC in NAND flash cells. V_t distribution. Solid line indicates V_t just after programming. Dotted line and dashed line are the V_t distributions after a 1000-hr bake. In programming, V_t is controlled to be more than 4.0 V.

where $C_{cg\text{-}fg}$ is the capacitance between a control gate and a floating gate. The charge loss is very small when the electric field E_{ox} is less than 1.2 MV/cm, where it corresponds to $V_t = 2.0$ V. However, it rises sharply near $E_{ox} = 1.4$ MV/cm, and it increases exponentially with increasing E_{ox}.

Repeatability of SILC bits had been also investigated [36], as shown in Fig. 6.26. Two times data retention test of first and second MEAS was performed for the same cells with recording address of bits. Tail bits behavior is categorized into two groups.

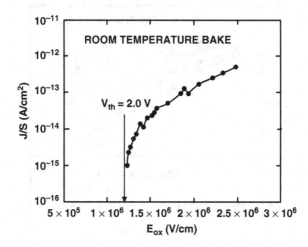

FIGURE 6.25 Charge loss rate of a typical tail bit as a function of the electric field E_{ox} in tunnel oxide.

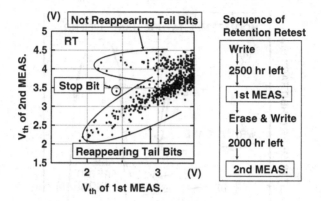

FIGURE 6.26 Reappearence of the anomalous cells when retention characteristics are measured again. After 1 million program (write)/erase cycles, memory cells were baked for 2500 hr at room temperature. After that, all cells are erased, programmed, and baked for 2000 hr again.

One group is transformed from tail bits to normal cells during reprogramming and it appears as normal cells (see "Not Reappearing Tail Bits"). The other is continuously kept as anomalous tail bit cells, which show almost the same charge loss characteristics in two measurements (see "Reappearing Tail Bits"). Many bits (~90%) appeared as the tail bits again after reprogramming and one more retention bake. After one more program/erase operation, about 10% of tail bits are transformed into normal cells. This fact indicates that tail bits are easily transformed from tail bit to normal bit. In Fig. 6.27, an exceptional cell named "stop bit" is identified. By tracking the

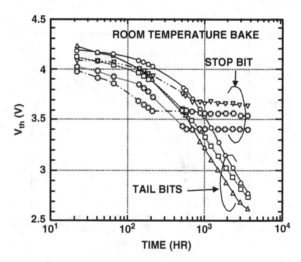

FIGURE 6.27 Charge loss characteristics of the anomalous cells of "stop bit." There are three bits whose rapid charge loss is suddenly and randomly stopped.

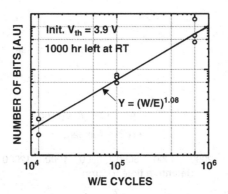

FIGURE 6.28 Program/erase cycle dependence of tail bits. The initial V_t before baking exceeds 3.9 V. Tail bits are defined as the cells whose V_t become less than 3.7 V during baking. The number of SILC bits is strongly depended on number of P/E cycles. The fail bits are almost proportional to P/E cycles. The degradation of tunnel oxide becomes worse proportional to P/E cycles.

stop bits individually, the stop bits are suddenly transformed from tail bit to normal bit during retention baking at room temperature, as shown in Fig. 6.27. Existence of stop bits strongly supports the easy transformation between the tail bit and normal bit.

This experiment result suggests that SILC is caused by electron current through trap-assisted tunneling. And trap and detrap would make reappearance and no reappearance phenomena. Also, these facts indicate that the anomalous leakage current of the tail bits flows only through one or a few spots. This leakage path can be easily transferred from inactivated to activated, or from activated to inactivated. A model of the leakage path is that electrons can easily flow from the floating gate to the substrates through the leakage path. Leakage paths are generated with a constant probability per power law of program/erase cycles, as shown in Fig. 6.28.

6.3.3 Data Retention in NAND Flash Product

Data retention performance was compared between several NAND flash products of different suppliers [37]. Figure 6.29a shows RBER (raw bit error rate) during a room-temperature bake after 10K program/erase cycles. RBER at time $= 0$ is due to program errors induced by program/erase cycling. RBER increases with retention time because of data retention errors.

Retention errors are mostly due to charge loss. A cell is losing charge and thus moves from one V_t level (e.g., L3, L2) to the one below. Two dominant mechanisms cause this retention error. The first one is a loss of FG charge via stress-induced leakage current (SILC) through the tunnel oxide [5, 27]. The second one is a detrapping of the tunnel-oxide charge that had been trapped during cycling [6–8, 38–40]. The effect on the V_t distributions is sketched in Fig. 6.29b. Detrapping causes the distributions

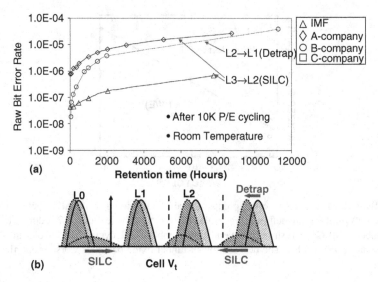

FIGURE 6.29   RBER versus room-temperature retention time following 10K P/E cycles. RBER increases as retention time.

to intrinsically broaden and shift lower. SILC causes a small number of cells to lose charge, forming tail bits in the distribution (see Section 6.3.2), as shown in Fig. 6.29b. Because SILC is more strongly field-dependent than detrapping [36, 41, 42], SILC tends to dominate the RBER of L3 cells, as shown in Fig. 6.29, which have the largest electric field in the tunnel oxide because they have the most stored electrons. Detrapping tends to dominate the RBER of L1 and L2 cells in the product of the B-company. This seems that errors are caused due to insufficient margin between L2 and below read voltage in this product. Interestingly, the same detrapping that generates retention errors also causes some program errors (from the final cycles) to recover with time because some of the tail cells that were above their intended read level drop below that read level due to charge loss. These program error bits are considered to be caused by "erratic over-program," as described in Section 6.7. Over the retention period of Fig. 6.29, about one-third of the program error recovered.

RBER of both retention mechanisms are greatly dependent on products from several suppliers. The errors were mostly L3→L2 type in two supplier's product, L2→L1 type in the third supplier's product, and both these types are available for other products. The characteristics of the retention errors can be more clearly seen by plotting only the charge-loss error, with excluding program errors, as shown in Fig. 6.30. Curve ($\triangle$) in Fig. 6.30b shows that RBER scales as a power law in cycles, which is consistent with what is known for SILC [42]. Curve (○) has a much steeper dependence on cycling count, which is what is seen for the detrapping mechanism because of its intrinsic nature. Although curve (◇) is dominated by L3→L2, the increasing cycling slope suggests that the physical mechanism may be a mix of both SILC and detrapping. From data of several suppliers, it seems that each supplier has

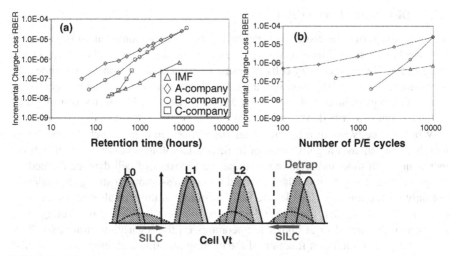

FIGURE 6.30 Charge-loss RBER as a function of (left) retention bake time after 10K P/E cycles and (right) number of P/E cycles followed by 1 year of bake. The bake was at room temperature. Data are available over only a short retention time (672 hr) for product.

a different strategy of V_t level setting. B-company of curve (o) has much less RBER in <1K cycling than other suppliers. However, at 10K cycling, L2→L1 type error is dominant as detrapping. The margin between L2 and below read voltage would not be enough. This product might have strategy to be a priority on minimizing RBER in <1K cycling, compromising L2 read margin. And other suppliers would have a priority on minimizing RBER at 10K cycling.

The relative contributions of the mechanisms also depend on the cycling and bake conditions. These devices were cycled at room temperature over several days and then baked at room temperature. If the cycling had been done at high temperature or over a longer time, then the detrapping contribution would have been reduced because some traps would have annealed in the delays between cycles [41,43]. On the other hand, if the retention bake had been done at high temperature, then the detrapping would have been larger and the SILC smaller. This is because detrapping is strongly temperature-accelerated [41,43] as shown in Section 6.3.2, whereas SILC anneals out at high temperature [41,44]. In fact, it is often thought that the detrapping mechanism is significant only at high temperature, but this discussion shows that some products under some conditions may be dominant by detrapping even at room temperature. The products dominated by detrapping charge-loss might have had substantially better retention over a more realistic time, such as year.

In order to minimize RBER in applications, it is very important to define the actual usage of NAND flash, such as temperature range, dominant temperature, number of cycles, cycling distribution, number of read, and so on. Based on this usage condition, the supplier has to optimize process and operations setting, such as V_t setting, to minimize RBER. And as applications are wide spread, product lines would be separated to satisfy criteria for each application.

6.3.4   Distributed Cycling Test

It had been reported the distributed cycling results. Compagnoni et al. [45] presented a detailed experimental investigation of the cycling-induced threshold voltage instability of NAND flash memory cells, focusing on its dependence on cycling time and temperature. This investigation was a trial to obtain a reasonable and universal test condition for guarantee quality of NAND flash products, with mechanism of SILC and detrapping described in Section 6.3.3.

When the cells are in a programmed state after cycling, the cell V_t instability mainly shows up as a negative shift of its threshold voltage cumulative distribution, increasing with time and resulting from partial recovery of cell damage created in the previous cycling period. The threshold voltage loss shows a strong dependence not only on the tunnel-oxide electric field during retention, but also on the cycling conditions. In particular, the threshold voltage transient is delayed by cycling over a longer time interval or at higher temperatures on the logarithmic time axis. The delay factor is studied as a function of the cycling duration and temperature on 60- and 41-nm technologies, extracting the parameter values required for a universal damage-recovery metric for NAND.

Figure 6.31 schematically shows the experimental procedure most commonly adopted to test V_t instabilities after cycling on multilevel NAND flash memory devices. (1) A certain number N of P/E cycles is performed in a time $t_{cyc} = N^* t_{wait}$ (t_{wait} is a constant delay time between cycles). (2) A program-and-verify (PV) algorithm is performed on the cells to a certain programmed V_t level. (3) V_t is monitored (Read) at logarithmically spaced interval times t_B since the first read operation, performed

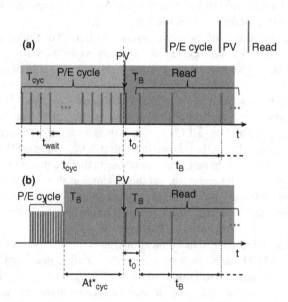

FIGURE 6.31   Schematics for the experimental procedure to investigate (a) cycling–induced V_t instabilities and (b) equivalent model for distributed cycling.

after a delay since the end of cycling. Note that the V_t monitoring (Read) phase corresponds to a data retention experiment at temperature T_B, which may be room temperature (RT) or, more generally, a selected bake temperature. In the latter case, bakes are periodically interrupted and the device cooled to room temperature for V_t reading.

The amount of cell damage present at the end of cycling in the experimental test of Fig. 6.31a is the result of damage creation by P/E cycles and damage recovery during the time elapsing in between the cycles. Assuming that damage creation by P/E cycles depends neither on t_{wait} nor on T_{cyc} and that damage recovery during cycling can be reproduced by a bake period of duration proportional to t_{cyc} at temperature T_{cyc} after damage has been created [41, 43], the testing procedure of Fig. 6.31a is equivalent to that of Fig. 6.31b; that is, the test procedure of Fig. 6.31b can be used as the shorter time evaluation procedure. In this latter experimental test, the same cell damage existing in Fig. 6.31a prior to the PV operation is obtained by a fast cycling at RT and a subsequent damage recovery period of duration At^*_{cyc}, where A is a constant to be determined from experiments. In order to deal with damage recovery at a single temperature, the time t^*_{cyc} was introduced, corresponding to the time at T_B that is required to have the same damage recovery taking place in a time t_{cyc} at T_{cyc}:

$$t^*_{cyc} = t_{cyc} \cdot \exp(E_A(1/kT_B - 1/kT_{cyc})) \qquad (6.1)$$

where an Arrhenius law of activation energy E_A was used for the time conversion. Assuming now that V_t has a logarithmic decrease due to damage recovery since the end of the damage creation period, the following formula holds for the V_t variation (ΔV_t) resulting in a time t_B since the first read operation in the experimental test of Fig. 6.31b and, in turn, of Fig. 6.31a [43]:

$$|\Delta V_t| = \alpha \ln(1 + t_B/(t_0 + At^*_{cyc})) = \alpha \ln(1 + t_B/t^*_B) \qquad (6.2)$$

where α gives the magnitude of the logarithmic decrease of V_t due to partial damage recovery and $t^*_B = t_0 + At^*_{cyc}$. From the t^*_B definition, lower V_t-loss transients should result from longer t_{cyc} and and higher T_{cyc}.

Measurement and calculation results are shown in Fig. 6.32 as a function of $1/kT_{cyc}$, referring to cell distribution probability $p = 5 \times 10^{-5}$ of lower tail of distribution. This graph is defined as the Arrhenius plot for cycling, showing a characteristic time for the data retention ΔV_t transients as a function of the reciprocal of the cycling temperature and not of the retention temperature, which is always equal to RT. Experimental data can reasonably be reproduced by the theoretical definition of t_B^* given in lines in Fig. 6.32, allowing the extraction of $E_A = 0.52$ eV, $t_0 = 0.8$ h, and $A = 0.022$ independently of the PV level and p. Note that the extracted value of t_0 well matches the experimental delay between the end of cycling and the first read operation on 60-nm NAND test-chip.

Experimental data and extracted theoretical trends in Fig. 6.32 show that for fixed t_{cyc}, t_B^* grows with T_{cyc} in the large T_{cyc} regime, where the slope of the t_B^* curve is

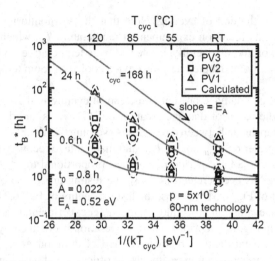

FIGURE 6.32   Arrhenius plot for cycling for the 60-nm test-chip.

given by E_A while reaching a constant value equal to t_0 for low T_{cyc}. The transition from the high to the low T_{cyc} regime depends on the t_{cyc} value, with longer cycling times allowing reaching the T_{cyc} sensitive regime at lower temperatures.

6.4   READ DISTURB

6.4.1   Program/Erase Scheme Dependence

It had been reported that the thin-oxide leakage currents, which are induced by the program and erase cycling stress, degrade the data retention and read disturb characteristics of memory cell [27]. Figure 6.33 shows the oxide current density

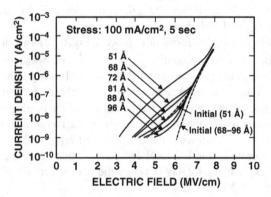

FIGURE 6.33   J–E characteristics (SILC) measured by capacitors having 51- to 96-Å oxide thickness before and after charge injection stress.

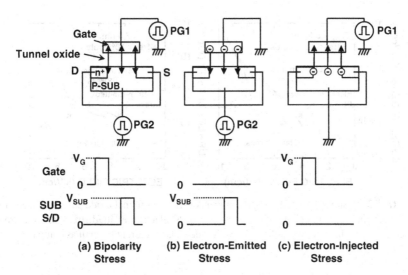

FIGURE 6.34 Setup and stressing waveform for (a) Bipolarity stress. (b) Electron-emitted stress and (c) Electron-injected stress. Stress conditions are shown in Table 6.2.

versus electric field before and after stress of electron injection from the substrate under positive gate polarity, where measurements are performed on the capacitors having 5.1- to 9.6-nm thickness oxide [27]. It can be seen that the stress induced leakage current (SILC) at low electric field is induced by the charge injection stress, and also SILC increases with decreasing the oxide thickness. The origin of the SILC was not well understood; however, it seems to be well fit by a Frenkel–Poole-type conduction. Due to SILC, it is very difficult to scale down the tunnel-oxide thickness of the memory cell [27]. An impact of SILC on NAND flash reliability was also investigated [5].

The stress-induced leakage currents (SILC) which are subjected to three types of simulated program/erase (P/E) stressing are compared. Figure 6.34 shows the stressing waveform for simulated P/E stress. Table 6.2 shows the stress conditions which correspond to program/erase conditions in NAND flash memory cells. A high voltage is applied to gate or substrate (SUB) and source/drain (S/D). The SILCs are

TABLE 6.2 Stress Conditions

T_{ox}	Gate	Sub, S/D
5.6 nm	6.79 V	8.0 V
	0.2 ms	0.2 ms
7.5 nm	7.91 V	9.15 V
	0.2 ms	0.2 ms

[a]High-voltage pulses are applied to the gate and substrate. Stress voltage V_g and V_{sub} are determined by the voltage that the opposite tunneling currents are approximately the same.

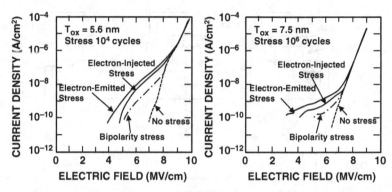

FIGURE 6.35 Stress-Induced Leakage currents (SILC) of tunnel oxide at low voltages for 5.6- and 7.5-nm oxide thickness after bipolarity stress, electron-emitted stress, and electron-injected stress. In the case of bipolarity stress, oxide leakage current is small as compared with the others.

induced by the electron injection and emission between the gate and substrate, as shown in Fig. 6.35. It was observed that the SILC induced by (a) bipolarity dynamic stressing is about one order smaller than that induced by both (b) the electron-emitted stress and (c) the electron-injected stress. This result shows that the origin of the SILC can be removed by reverse Fowler–Nordheim tunneling (FN-t) stress, and it would be the directional defect or strain or trapped holes in the tunnel oxide. This reduction of the SILC by bipolarity stress can extend the read disturb and data retention time in NAND flash memory cell.

Read disturb characteristics of the flash memory cell are compared in two program/erase (P/E) schemes. One is a bipolarity FN-t W/E technology, performed by uniform injection and uniform emission over the whole channel area of a flash memory cell (Fig. 6.36a). The other is a conventional channel-hot-electron (CHE) write and FN-t erase technology for NOR-type flash, performed by CHE injection at drain and uniform emission over the whole channel area (Fig. 6.36b). In erasing, a high voltage is applied to the substrate [6] as well as source/drain in order to prevent from causing the degradation of the thin-gate oxide due to band-to-band tunneling stress [21]. Flash memory cells which are used in this experiment have 5- to 10-nm-thick tunnel oxide, 25-nm-thick ONO inter-poly dielectric (IPD) [22], and 0.8-μm gate length.

Figure 6.37 shows the program(write)/erase endurance characteristics of two P/E schemes. The closure of the cell threshold window has not been found up to 100,000 program/erase cycles in both two schemes.

Read disturb characteristics were measured at applied various gate voltage conditions, which were the accelerated electric field test, as shown in Fig. 6.38 [5]. In the case of a CHE write and FN-t erase scheme (technology), the stored positive charges rapidly decay as stress time (retention time) increases, so the threshold window decreases. However, in the case of a bipolarity FN-t W/E scheme (technology), data loss of the stored positive charges is greatly improved. So, data retention time

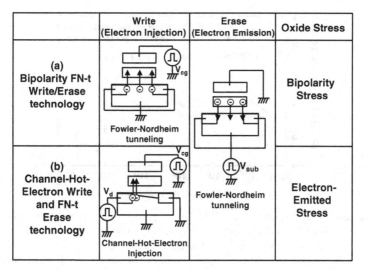

	Write (Electron Injection)	Erase (Electron Emission)	Oxide Stress
(a) **Bipolarity FN-t Write/Erase technology**	Fowler-Nordheim tunneling		**Bipolarity Stress**
(b) **Channel-Hot-Electron Write and FN-t Erase technology**	Channel-Hot-Electron Injection	Fowler-Nordheim tunneling	**Electron-Emitted Stress**

FIGURE 6.36 Comparison between (a) bipolarity FN tunneling write/erase technology, corresponding to bipolarity stress of tunnel oxide and (b) channel hot electron(CHE) write and FN tunneling erase technology, corresponding to electron-emitted stress of tunnel oxide, because of no leakage current induced by CHE injection.

of a bipolarity FN-t W/E scheme is extended about 10 times as long as that of conventional scheme. This phenomenon can be explained by the fact that the SILC is reduced by the bipolarity FN tunneling stress.

Figure 6.39 shows the data retention time under the read-disturb condition after program and erase cycling as a function of the tunnel-oxide thickness. The improvement in data retention is more effective with decreasing oxide thickness. Therefore, in the bipolarity FN-t W/E scheme, the tunnel-oxide thickness can be reduced by scaling

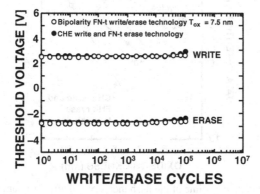

FIGURE 6.37 Endurance characteristics of flash memory cell with 7.5-nm tunnel oxide. In bipolarity FN tunneling W/E technology, write: $V_{cg} = 18$ V, 1 ms, erase: $V_{sub} = 20$ V, 1 ms. In conventional technology, write: $V_{cg} = 7$ V, $V_d = 8.5$ V, 1 ms. Erase: $V_{sub} = 20$ V, 1 ms.

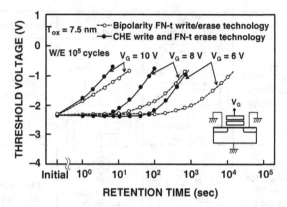

FIGURE 6.38 Read disturb characteristics at the applied various gate voltage stress. Flash memory cell having 7.5-nm-thick oxide are subjected to 100,000 program (write)/erase cycles. In bipolarity FN tunneling W/E scheme, data loss of stored positive charge is improved as compared with conventional scheme; it corresponding to results of oxide leakage currents.

down the flash memory cell. Then, it gives advantages of low-voltage program and erase operations.

Initial data loss is measured at 300°C, as shown in Fig. 6.40. It is confirmed that the initial data loss of bipolarity FN-t W/E scheme is smaller than that of the CHE write and FN-t erase scheme due to reduction of stress-induced oxide leakage current.

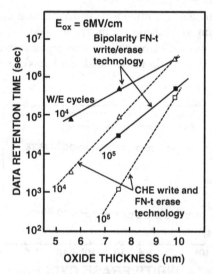

FIGURE 6.39 Data retention time of the flash memory cell after program and erase cycling as a function of tunnel-oxide thickness. Data retention time is defined by the time that V_{th} reaches -1.0 V during the applied gate voltage stress (accelerated read disturb condition). In bipolarity FN tunneling write/erase scheme, the tunnel-oxide thickness can be reduced with scaling down the flash memory cell.

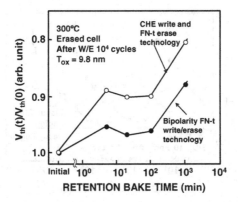

FIGURE 6.40 Initial data loss of the erased cell which stores positive charges in floating gate as a function of retention bake time at 300°C in a bipolarity FN tunneling write/erase scheme and CHE write and F–N tunneling erase scheme, subjected to 10,000 program and erase cycles.

Read disturb and data retention characteristics have been described in two different program/erase schemes on flash memories. It is clarified experimentally that flash memory cell, which is programmed and erased by bipolarity uniform Fowler–Nordheim tunneling (FN-t), has 10 times as long retention time as the conventional one, which is programmed by channel-hot-electron (CHE) injection and erased by unipolarity FN-t. This difference of data retentivity between these two W/E schemes is due to decreasing the stress-induced leakage current (SILC) in thin tunnel oxide by bipolarity FN-t stress. Also, this improvement in data retention is more remarkable, in accordance with decreases in the tunnel-oxide thickness.

6.4.2 Detrapping and SILC

The read disturb characteristics become worse after program/erase cycling due to generation of SILC (stress-induced leakage current) in tunnel oxide during program/erase cycling. The SILC mechanism had been reported by several papers [44, 46–58]. And an impact of SILC on NAND memory cell characteristics had been also reported[38–40].

Figure 6.41 shows typical read disturb characteristics with an accelerated gate voltage, after 1 million program/erase cycles at room temperature (30°C) [38, 39]. The threshold voltage V_t is increased with read disturb stress time. The stress-induced leakage current (SILC) can be directly calculated from the threshold voltage shift (ΔV_{th}) of the flash memory cell during read stress. The stress-induced leakage current can be expressed by

$$I_{leak} = C_{ono} * \Delta V_{th} / \Delta t \tag{6.3}$$

where I_{leak} is the stress-induced leakage current (SILC), C_{ono} is the capacitance of inter-poly dielectric ONO between the control gate and the floating gate, and the read disturb time (t) is read stress time.

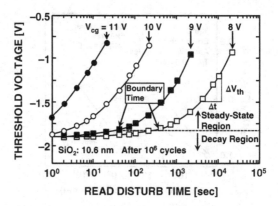

FIGURE 6.41 Read disturb characteristics with accelerated control gate voltage (V_{cg}). The threshold voltage V_t has a positive shift due to the electron injection to floating gate by SILC and the electron detrapping.

The threshold voltage of a memory cell is determined by both the floating-gate charge (Q_{fg}) and the oxide trapped charge (Q_{ot}) which consists of trapped electrons or holes in tunnel oxide. During the read disturb measurements, the Q_{fg} change is caused by the electron injection from the inversion layer and the electron trap states to the floating gate, which results in the stress-induced leakage current. And the Q_{ot} change is caused by the trapping or detrapping of carriers in the tunnel oxide. Therefore, the stress-induced leakage current calculated by the read disturb characteristics has two terms. One is the differential of the Q_{fg}, which is described as the steady-state leakage current. The other is the differential of the Q_{ot}, which is described as the decay region leakage current. As a result, the stress-induced leakage current can be written as

$$I_{leak} = -dQ_{fg}/dt + ((C_{ono} + C_{ox})/C_{ox}) * (dQ_{ot}/dt) \qquad (6.4)$$

where C_{ox} is the capacitance of the tunnel oxide. It is assumed in (6.4) that Q_{ot} is localized near the Si/SiO$_2$ interface. The electric field over the tunnel oxide (E_{ox}) is a function of the floating-gate voltage and is given by

$$E_{ox} = (V_{fg} + \phi_f - \phi_s)/T_{ox} \qquad (6.5)$$

where ϕ_f is the Fermi potential of the floating gate, ϕ_s the surface potential of the p-well, and T_{ox} the tunnel-oxide thickness. The floating-gate voltage is given by

$$V_{fg} = (C_{ono}/(C_{ono} + C_{ox})) * (V_{cg} - V_{th}) + V_{fgth} \qquad (6.6)$$

where V_{cg} is the control gate voltage during the read disturb condition, and V_{fgth} is the threshold voltage as measured on the floating gate of the memory cell. The stress-induced leakage current is calculated from (6.3), (6.5), and (6.6).

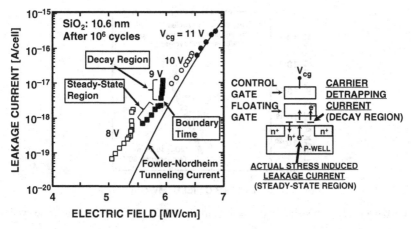

FIGURE 6.42   The calculation of the stress-induced leakage current derived from the threshold voltage shift of the flash memory cell during the read disturb condition. Quick V_t shift is caused by detrapping, and longer and large V_t shift is caused by SILC.

Figure 6.42 shows the calculated stress-induced leakage current (I_{leak}) as a function of the electric field over the tunnel oxide (E_{ox}). This leakage current is derived from the differential (dV_{th}/dt) of the threshold voltage during the read disturb condition, as shown in (6.3). It is observed that the leakage current quickly decays at the beginning of the read disturb stress (decay region). After the decay region, the leakage current reaches a certain steady value where dV_{th}/dt gradually decreases with the read disturb time (steady-state region). Two regions could be clearly confirmed by the plot of logarithmic read disturb time, as shown in Fig. 6.43. In the decay region, the decay of dV_{th}/dt is considered to be caused by both the fast decay of the stress-induced leakage current and the decay of the number of trapped carriers in the tunnel oxide, which

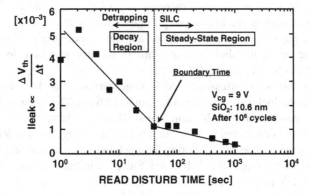

FIGURE 6.43   The differential of the threshold voltage versus read disturb time. Read disturb mechanism is explained by the detrapping and SILC mechanism. The detrapping means that a trapped carrier in tunnel oxide is detrapped in a short time. SILC (stress-induced leakage current) is dominant in a longer read disturb time.

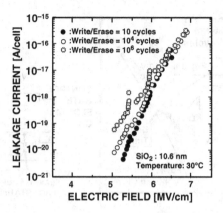

FIGURE 6.44   The stress-induced leakage current (SILC) after various write/erase cycling.

is generated during program/erase cycling. In the steady-state region, the threshold voltage shift is mainly caused by the stress-induced leakage current, which follows the same leakage current regardless of the control gate voltage of V_{cg} (the same electric field dependence in tunnel oxide), as shown in Fig. 6.42. The boundary between the decay region and the steady-state region is called "the boundary time" (see Fig. 6.43).

In this method of the leakage current derived from cell V_t shift, very low-level leakage current ($\sim 10^{-20}$ A) can be evaluated. On the other hand, in the conventional method of using capacitors as a test device, it is not possible to evaluate extremely low-level stress-induced leakage current. Therefore, using a memory cell is more practical and reliable than using a capacitor, when the stress-induced leakage current needs to be investigated.

Figure 6.44 shows the stress-induced leakage current after a number of program (write) /erase cycles (10–10^6 cycles). It was observed that the stress-induced leakage current increases with increasing the number of program/erase cycles. Also, in the decay region, the leakage current (emerging as an initial threshold voltage shift) increases with increasing the number of program/erase cycles. This result indicates that the charge traps in the tunnel oxide, which cause both stress-induced leakage current and the initial threshold voltage shift in the decay region, increase with increasing the number of program/erase cycles.

Figure 6.45 shows the stress-induced leakage current after 10^6 program/erase cycles for an oxide thickness range from 5.7 nm to 10.6 nm. The stress-induced leakage current increases greatly as the tunnel-oxide thickness decreases. The threshold voltage shift of the decay region after 10^6 program/erase cycles is independent of both the control gate voltage (V_{cg}) during the read disturb condition and the tunnel-oxide thickness for the range of 5.7–10.6 nm. Since the initial threshold voltage shift is only about 0.1 V after 10^6 program/erase cycles, the read disturb lifetime is not determined by the decay region but mainly by the steady region. Therefore, with respect to the read disturb lifetime, it is important to reduce the saturated leakage current (steady-state region) rather than the time-dependent leakage current (decay region).

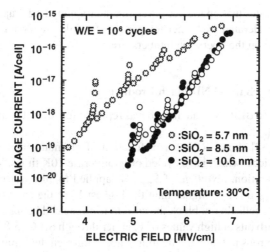

FIGURE 6.45   The stress-induced leakage current in the tunnel oxide with a thickness from 5.7 nm to 10.6 nm.

The high-temperature (125°C) operation (the temperature during program/erase operation equals the temperature during read disturb operation) degrades the read disturb characteristics in comparison with room temperature operation, as shown in Fig. 6.46. The steady-state leakage current after 125°C operation increases about three times in comparison with that at room temperature. Therefore, in the case of an accelerated test of read disturb, the high-temperature operation of the flash memory cell should be used. On the other hand, the boundary time decreases during high-temperature operation, while the initial threshold voltage shift is nearly constant

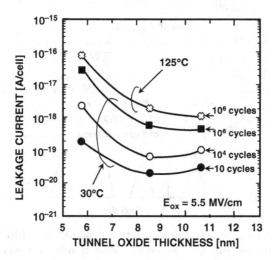

FIGURE 6.46   The stress-induced leakage current as a function of tunnel-oxide thickness.

(0.1 V). This indicates that, in the decay region, the trapping or detrapping of carriers in the oxide is accelerated by higher temperatures, while the amount of the charge depends very little on the operation temperature.

6.4.3   Read Disturb in NAND Flash Product

Raw bit error rate (RBER) of read disturb was reported for several company's NAND flash product [37].

Figure 6.47a [37] shows RBER as a function of the number of reads per page performed on devices, which have cycled program/erase 10K times. When a NAND cell is in read operation, a voltage of V_{passR} is applied to all unselected word lines in the block. V_{passR} must be higher than the highest V_t of the programmed cells so that the unselected cells do not block the current from the cell being read. The V_{passR} bias tends to disturb bits at high values of V_t either through SILC [5,8,27,38–40,44], which allows electrons to reach the floating gate, or through the filling of traps in the tunnel oxide.

Failure bit is mainly caused from L0 (erase state) due to higher electric field in tunnel oxide, as shown in Fig. 6.47b. This is as expected for the SILC mechanism, which is strongly field-dependent, because the lowest V_t state has the highest electric field in the tunnel oxide under read bias V_{passR}. The characteristics of the read disturb failures were studied by excluding the program errors and plotting only the incremental read-disturb errors, as shown in Fig. 6.48. The RBER increases as a power law in the number of reads (Fig. 6.48a) and in the number of P/E cycles (Fig. 6.48b), consistent again with SILC [38–40,59]. Failure rate is degraded about 2 orders of magnitude with increasing program/erase cycling from 1000 to 10,000. Failure rate of read disturb should be saved by ECC.

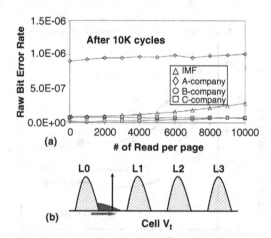

(a)

(b)

FIGURE 6.47   Read disturb characteristics in NAND flash product. (a) RBER versus number of read per page after 10K program/erase cycles. Failure is mainly caused in L0 failure. Bit failure rate increase as increasing number of read cycles. (b) SILC mechanism after P/E cycling.

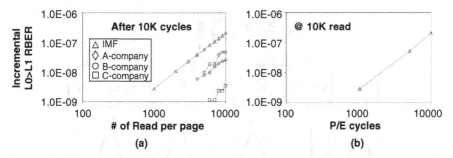

FIGURE 6.48   Incremental RBER in read disturb as measured by L0 bit failures, as a function of (a) number of reads per page for 10K-cycled blocks and (b) number of program/erase cycles before the 10K reads per page. Bit failure rate increases as an increasing number of read, and bit failure rate also increases as an increasing number of P/E cycling precondition.

6.4.4   Hot Carrier Injection Mechanism in Read Disturb

The other read disturb mechanism was reported [60]. It is called the "boosting hot-carrier injection effect." The hot carrier injection is occurred by unexpected boosting voltage induced by V_{pass_read} in unselected cells in a NAND string.

In order to investigate the read disturb mechanisms of "boosting hot-carrier injection effect," three different read voltages and four different cell data states (S0, S1, S2, and S3) were applied on the selected cell. Figure 6.49 shows (a) the operation condition and (b) waveform of SGS/SGD rising time shift scheme [61] for read-disturb evaluation. In the evaluation, the selected WLn was performed with more than 100K read cycles.

During the read operation, the channel potential in part of string (WL2–31 area) is boosting up by V_{pass_read} (V_{passR}) in unselected word lines, as shown in Fig. 6.50 bottom. This boosting potential generates hot electrons at a selected cell which has a large potential difference between source and drain (see Fig. 6.50a,b). Due to a large potential difference, some of the hot electrons are injected to the floating gate in a cell adjacent to the selected cell, as illustrated in Fig. 6.51.

Figure 6.50, upper portion, shows the measured results of WL2 V_{th} shift (i.e., read disturb failure) during select WL1 read disturb cycles with different WL1 voltages (V_{WL1}) and cell data states (WL1 = S0–S3; see Fig. 6.49a). From these data, a serious WL2 V_{th} shift can be observed in $V_{WL1} = 0.5$ V and $V_{WL1} = 1.8$ V after 1K read cycles. In Fig. 6.50a, upper portion, the magnitude of the V_{th} shift with the WL1 = S2 state is larger than that with WL1 = S3 state. However, obviously WL2 V_{th} shift can be found only when WL1 is at S3 state in Fig. 6.50b, upper portion. In Fig. 6.50c, upper portion, WL2 V_{th} is unchanged while V_{WL1} is set to 3.6 V.

To precisely analyze the phenomenon, TCAD simulation and analysis were carried out to clarify the mechanism of the read disturb failure. Based on simulation results of Fig. 6.50, the channel potential difference between selected WLn (e.g., WL1) and unselected WLn+1 (e.g., WL2) is related to cell data states (S0–S3) and the read voltage of the selected WL (V_{WL1}). Figure 6.50a, bottom portion, shows that the

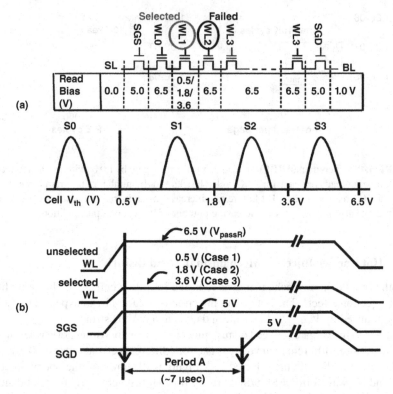

FIGURE 6.49   (a) Schematic diagram of read condition. The read cycling on WL1 was performed more than 100K. Four different cell data states (S0–3) are assigned for MLC NAND flash memory operation. (b) Waveform of read operation. There are three different voltages applied on the selected WL; these are represented as case 1 to case 3. In order to avoid SG–WL coupling noise, an SGS/SGD rising time shift scheme is used.

selected WL1 channel is tuned off and the channel potential of unselected WL2~31 is boosted to a high level when the WL1 cell data state is S2 or S3. Therefore, a sufficient potential difference between WLn and WLn+1 causes a high transverse electric field. When V_{WL1} is increased to 1.8 V as in Fig. 6.50b, bottom, a high programming cell state (S3) is required to support the potential boosting of unselected WL2–31. In addition, from Fig. 6.50c, bottom portion, and the case of WL1 = S2 in Fig. 6.50b, bottom portion, the large potential difference cannot be observed since the WL1 channel is turned on by high WL1 voltage. Therefore, the potential difference can be reduced by the turn-on effect of the selected cell. These simulation results are well corresponding with read disturb results of Fig. 6.50, top portion.

Electron current density is another factor to cause the V_{th} shift of WLn+1. From Fig. 6.49a, the current density of WL1 = S2 should be higher than that of WL = S3 since its V_{th} is lower. Consequently, the probability of impact ionization can be increased due to the high current density in the case of WL1 = S2. According to the

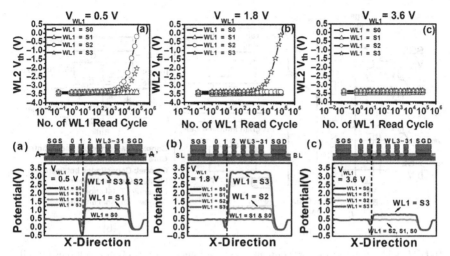

FIGURE 6.50 Measured results of WL2 V_{th} shift during WL1 read cycles. (a) WL1 voltage (V_{WL1}) is 0.5 V (case 1). The data shows serious WL2 V_{th} shift if WL1 cell was programmed (S2 and S3). (b) WL1 voltage (V_{WL1}) is 1.8 V (case 2). The data shows that the WL2 V_{th} shift occurred only with S3 state. (c) WL1 voltage (V_{WL1}) is 3.6 V (case 3). There is no obvious WL2 V_{th} shift in this condition.

model, the phenomenon of the serious WL2 V_{th} shift in the condition of WL1 = S2 rather than WL1 = S3 can be clearly explained.

Figure 6.51 shows the schematic diagram of the mechanism of boosting hot-carrier injection in MLC NAND flash memories. The transverse *E*-field can be enhanced by the channel potential difference and consequently make a high probability of impact ionization. As a result, electron–hole pairs are generated, and then electrons are injected into the adjacent cell (WL2) since the higher vertical field of V_{WL2}. Thus, the V_{th} of an adjacent cell is changed after 1K cycles with the repeating injecting of the hot electrons.

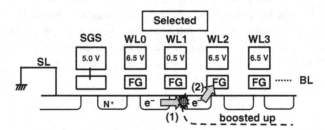

FIGURE 6.51 The schematic diagram of the mechanism of boosting hot-carrier injection (HCI) on read disturb in NAND flash memories. The probability of HCI could be enhanced by (1) high transverse E-field, and then (2) electrons injected into the floating gate due to the high vertical *E*-field.

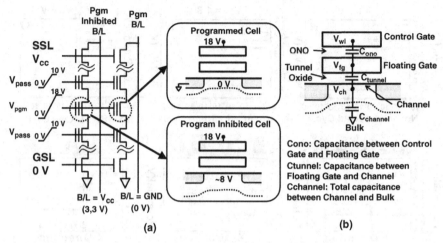

FIGURE 6.52 Self-boosted program inhibit voltage generation. (a) Bias conditions for self-boosting. (b) Capacitance model for coupling ratio calculation.

6.5   PROGRAM DISTURB

6.5.1   Model of Self-Boosting

The program self-boosting operation is used in an inhibit cell string of NAND flash memory cells, as shown in Section 2.2.4. The channel potential in inhibit strings are boosted up mainly by inhibit WL voltage (V_{pass}). The bias conditions for generating program inhibit boosting voltages to the channel of the inhibit NAND string is shown in Fig. 6.52a [62]. With the SSL transistors (drain side select transistor) turned on and the GSL transistors (source side select transistor) turned off, the bit-line voltages for cells to be programmed are set to 0 V, while the bit-line voltages for cells to be program inhibited are set to V_{cc}. In program-inhibited cells, the V_{cc} bit-line initially precharges the associated channel, which is normally V_{cc}-V_{tssl} (V_t of SSL transistor). When the word lines of the NAND string rise (selected word line to the program voltage of V_{pgm} and unselected word lines to the pass voltage of V_{pass}), the series capacitances through the control gate, floating gate, channel, and bulk are coupled and the channel potential is boosted automatically. Assuming a single boosted pass cell and using the model of Fig. 6.52b, the boosted channel voltage, V_{ch}, can be estimated as follows:

$$V_{ch} = V_{wl} * C_{ins}/(C_{ins} + C_{channel}) \qquad (6.7)$$

where C_{ins} is the total capacitance between control gate and channel (C_{ono} in series connection with C_{tunnel})

$$C_{ins} = C_{ono} * C_{tunnel}/(C_{ono} + C_{tunnel})$$

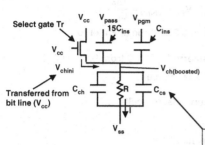

$$Cins = \frac{C_{ono} \cdot C_{tun}}{C_{ono} + C_{tun}}$$

Cell capacitance per 1 cell

C_{ono}: ONO capacitance
C_{tun}: Tunnel-oxide capacitance

C_{ch}: Channel capacitance per 16 cells between inversion layer and P-well, and between N-diffusion layer and P-well. C_{ch} strongly depends on the field implantation dose.

I : Channel leakage current including the junction leakage current and the field leakage current.

R : Resistance for the channel leakage current, I. This depends on field implantation dose, N-dose, P-well dose and isolation width. And this strongly depends on the operation temperature.

C_{cs}: Additional capacitance/16 cells caused by neighboring cells. C_{cs} increases in case of the neighboring bitline = 0 V (column stripe), and strongly depends on the field implantation dose.

FIGURE 6.53   The equivalent circuit for self-boosting model, including an additional capacitor, C_{cs}. In the column stripe pattern, C_{cs} becomes large and decreases the channel voltage, V_{ch}. As a result, the number of programmable cycles (NOP) is limited by the column stripe pattern.

In a program inhibit string, the coupled channel voltage rises from V_{cc}-V_{tssl}, to V_{ch} when the word-lines voltages rise. The SSL transistor shuts off under the conditions of the drain BL in V_{cc} and the source in V_{ch}, due to the body effect of the SSL transistor. The GSL transistor also shuts off by applying 0 V to the gate and applying V_{cc} to the source line (SL). Then the channel becomes a floating node. By calculating (6.7), it is determined that the floating channel voltage rises to approximately 80% of the gate voltage. Thus, channel voltages of program inhibited cells are boosted to approximately 8 V when program (15.5–20 V) and pass (~10 V) voltages are applied to the control gates. This high-boosted channel voltage prevents the FN tunneling current from being initiated in the program-inhibited cells.

The program-boosting mechanism and limitation had been investigated in detail [63] in the LOCOS cell. Figure 6.53 shows the equivalent circuit for the channel-boosting mechanism. For the "1" data program (program inhibit), which keeps the negative threshold voltage (V_{th}), the program inhibit channel voltage (V_{ch}) is raised by the capacitive coupling with the pass voltages (V_{pass}) and the program voltage (V_{pgm}). V_{ch} must be raised enough to reduce the tunnel-oxide electric field, because the difference between V_{pgm} and V_{ch} is an effective program voltage for program inhibit cells.

In measurement data, program inhibit performance has a neighbor string data dependence. In the case of the column stripe pattern (channel of neighbor string is 0 V during programming "0" data), the number of allowable programmable cycles (NOP) is decreased to about 2/3 of that in the case of the all "1" pattern (channel of neighbor string is boosting voltage V_{ch} for programming "1" data). This means that the program inhibit performance is degraded in the column stripe pattern. Therefore, NOP is limited by the column stripe pattern.

In a conventional model, the program disturbance in the column stripe pattern is explained by the field leakage current from the program inhibit channel voltage V_{ch}.

$$V_{ch} = V_{chini} + C_r(V_{pass} - V_t - V_{chini}) + C_r{}'(V_{pgm} - V_{th} - V_{chini}) - \frac{T_{pw}}{C_{tot}} \; I$$

[$V_{pgm} - V_{ch}$] is a potential difference between CG and the channel for "1" data programming cells.

Vchnin is the initial voltage, which is transferred by SGD from bitline (V_{cc})

$C_{tot} = 16C_{ins} + C_{ch} + C_{cs}$: Total channel capacitance

$C_r = \dfrac{15C_{ins}}{C_{tot}}$: Channel boost ratio for un-selected cells

$C_r{}' = \dfrac{C_{ins}}{C_{tot}}$: Channel boost ratio for selected cell

FIGURE 6.54   Channel voltage equation used in the simulation. V_{th} is the threshold voltage of un-selected cell (if D-type cells then $V_{th} = 0$ V). Cr, Cr' is the channel boost ratio. T_{pw} is the pulse width of V_{pass} and V_{pgm}.

On the other hand, in the new model the program disturbance is extended mainly by the additional capacitance (C_{cs}) between the active area and the neighboring cells, as shown in Fig. 6.53. The C_{cs} increases when the depletion area under LOCOS isolation is widened in column stripe pattern. As shown in Fig. 6.54, an increasing C_{cs} (increasing C_{tot}) decreases the channel boost ratio (C_r), and then decreases the V_{ch} in the column stripe pattern. Figure 6.55 shows the measured and simulated program disturbance characteristics. The C_{cs} is a fitting parameter. The simulated result is well matched with the measured result.

V_{pass} and V_{pgm} waveforms dependence on the program disturbance was also investigated for analysis of the channel leakage current, as shown in Fig. 6.56. The various pulse widths of T_{pw} are used, as shown in Fig. 6.56a. Figure 6.56b shows the V_{th} of the "1" program cell (program inhibit cell), as a function of T_{pw}. In a conventional model, V_{th} difference between the all "1" pattern and the column stripe pattern at $T_{pw} = 30$ μs

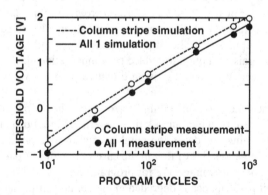

FIGURE 6.55   Simulated and measured program inhibit (self-boosting) characteristics for "1" data program cell. C_{cs} is a fitting parameter. Operation temperature is 85°C. Field implantation dose is 1 E14/cm^2. $V_{pass} = 10$ V, $V_{pgm} = 17$ V, $C_{cs} = 5$ E-16 farad/16 cells.

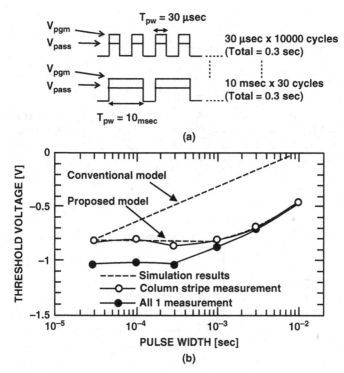

FIGURE 6.56 (a) V_{pgm} and V_{pass} waveform for analysis of the channel leakage current, "1". (b) Threshold voltage as a function of V_{pass} and V_{pgm} pulse width. The model well reproduces the measurement data. On the other hand, the conventional field leakage current model deviates from the measurement data. The threshold voltage shift at long T_{pw} (>1 ms) is caused by the junction leakage current.

is considered to be due to the field leakage current. However, the measurement data deviates from the simulation result as pulse width increases. On the other hand, in the proposed model, the simulation result can well reproduce the measurement data, where the V_{th} difference is calculated by the enlarged C_{cs} with $C_{cs} = 5E\text{-}16$ (farad/16 cells) derived from Fig. 6.55. The increase of V_{th} when the T_{pw} is longer than 1 ms is caused by the junction leakage current in the boosted channel.

A quantitative NAND string boosting model was investigated in a sub-30-nm NAND cell [64] to clarify the impacts of the channel capacitance, the channel leakage current, and the cell scaling on the program disturb. The model is including the channel boosting ratio (CBR) from capacitances network of 3-D technology computer-aided design (TCAD) simulations, the transient channel potential with the junction leakage (J/L) current, the band-to-band tunneling (BTBT) current, and the Fowler–Nordheim (FN) tunneling current of cells.

Figure 6.57a illustrates a schematic of NAND strings during programming, along with the various mechanisms that impact the program disturb. Typical program and

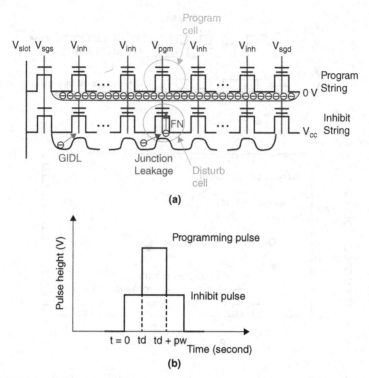

FIGURE 6.57 (a) Bias conditions of NAND strings during programming, along with the different program disturb mechanisms. (b) Program and inhibit-pulse waveforms. A td is a ramp-up delay between inhibit pulse (V_{inh}) and programming pulse (V_{pgm}).

inhibit-pulse waveforms are shown in Fig. 6.57b. Here, td is the time delay between ramp-up of the inhibit-pulse V_{inh} and ramp-up of the program-pulse V_{pgm}, and pw is the programming pulse width.

Figures 6.58a and 6.58b show the V_t of disturbed cell as a function of the inhibit voltage (V_{inh}, i.e., V_{pass}) for three different delay times td of 5, 100, and 500 μs. Simulation results and experimental data show a good fit over a series of the inhibit voltage V_{inh}. The model also shows the overall trend in both low and high channel boron concentration cases. In the case of the low channel boron concentration of Na (see Fig. 6.58a), the cell program disturb continues to improve with higher V_{inh} voltage values. This indicates that the channel potential is mainly determined by the channel boosting ratio (CBR). On the other hand, the cell with a higher channel boron concentration shows a different behavior. In the case of the higher boron dose (see Fig. 6.58b), the V_t of the disturbed cell starts to saturate at a V_{inh} value of around 7 V, which indicates that the channel boosting potential is limited by the channel leakage current. The difference between the delay times of 5 and 100 μs is almost the same as that between 100 and 500 μs, which suggests a much higher channel leakage current at the high channel boosting voltage. In Fig. 6.58b, the V_t of the disturbed

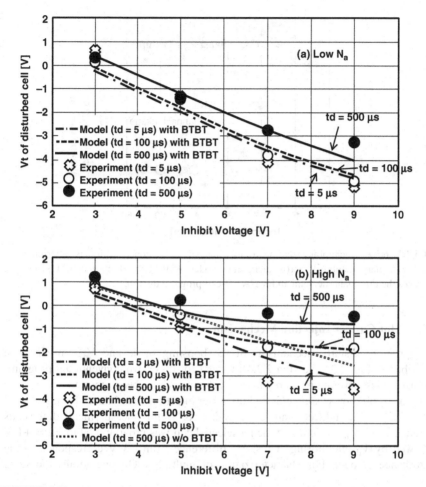

FIGURE 6.58   Disturbed cell V_t as a function of the inhibit voltage in the case of (a) low and (b) high channel boron concentrations of Na. $V_{seed} = 1$ V. Initial erased V_t is below -7 V.

cell without the BTBT current in model is also shown. Without the BTBT current, the simulation results do not match the experimental data. This result indicates that the dominant leakage mechanism for the program disturb is the BTBT current when the boron concentration is high.

As the NAND cell scales down further, a higher channel boron concentration is required to mitigate the short-channel effect. The leakage current of channel boosting node is increased. Figure 6.59 illustrates the boron concentration requirement (closed circles) and the resultant channel leakage current (open circles) during the boosting across generations [64]. The boron concentration is determined to maintain a charge neutral V_t across technology nodes. The BTBT current is expected to be a dominant program disturb mechanism for cells beyond 20 nm.

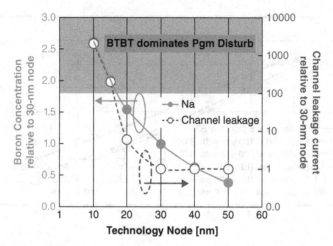

FIGURE 6.59 Technology scaling trend. Channel boron concentration is determined to maintain a charge-neutral V_t across technology nodes. At the sub-20-nm node, BTBT current becomes the dominant mechanism for NAND-cell program disturb.

6.5.2 Hot Carrier Injection Mechanism

Two program disturb mechanisms of the boosting mode and V_{pass} mode were described in Fig. 2.21 (Section 2.2.4). Except for these two conventional program-disturb modes, several program disturb mechanisms have been reported.

Figure 6.60 shows "source/drain hot-carrier injection disturbance" [65], so-called "SGS GIDL (gate-induced drain leakage) disturb." Before measurement, all the cells were erased to $V_{th} = -3$ V, and then a selected cell was programmed to $V_{th} = +1$ V followed by V_{th} monitoring. Thus, V_{th} difference from -3 V corresponds to the disturbance amount. For characterization of multiple NOP operations, the same

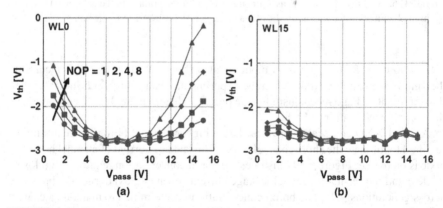

FIGURE 6.60 Programming disturbance characteristics of a NAND cell array measured at WL0 and WL15 during programming operation.

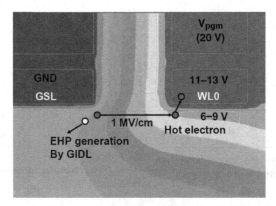

FIGURE 6.61 Simulation of the "hot-carrier disturbance" phenomenon, illustration of the "hot-carrier disturbance" model at WL0 cell.

programming operation cycles were repeated in the same word line with a repetition number of NOP. In Fig. 6.60, conventional V_{pgm} disturbance can be observed at low V_{pass} voltages ($V_{pass} < 6$ V) and the V_{pgm} disturbance is logarithmically proportional to NOP. The program disturbance characteristic at WL15 (Fig. 6.60b) is typical, which occurs only at low V_{pass} voltages. However, the program disturbance characteristic at WL0 (Fig. 6.60a) shows that another disturbance phenomenon can be observed at high V_{pass} voltages, and it is more severe at higher V_{pass} voltages. In contrast to typical V_{pgm} disturbance, the disturbance at high V_{pass} voltages is linearly proportional to NOP. As shown in Fig. 6.60, the new programming disturbance is worst at WL0, while negligible at other WLs.

The model is verified by device simulation of a NAND cell string. Figure 6.61 shows the potential profile across the simulation structure at GSL/WL0 during programming operation. The channel potential is raised to 8 V, and the lateral electric field at the GSL-WL0 space is around 1 MV/cm. The large hole current is generated at the GSL (SGS) edge due to the GIDL mechanism. Also, the large electron current is generated, and a part of the generated electron is injected to the floating gate of WL0 by accelerating with lateral electric field. The GIDL situation also occurred at the SSL (SGD; drain side select gate) edge; however, comparing with GSL bias conditions, the V_{gs} (voltage different between gate and source) at the SSL transistor is lowered as much as the applied voltage at the SSL gate. In addition, the lateral electric field at the SSL-WL31 space is lowered by the same amount with the same reason. Therefore, although the same phenomenon happens at the WL31 cell, the situation is even better than the WL0 cell.

By using simulation tools, a method to minimize the disturbance problem was obtained. This disturb phenomenon is strongly depended on WL0-SGS(GSL) space length, as shown in Fig. 6.62. The narrower space of less than 110 nm makes this program disturb worse. This suggests that "SGS GIDL disturb" becomes worse as a memory cell scaling (WL0-SGS (DSL) space scaling).

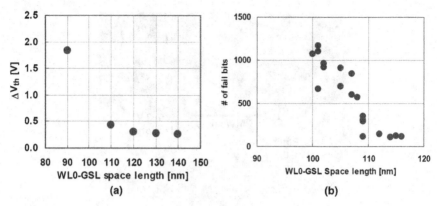

FIGURE 6.62 (a) Simulation result of the number of electrons injected to the WL0 cell for various WL0-GSL space. (b) Number of failure bits measured with 1-Mb block array at $V_{pass} = 10$ V.

The other mechanism was also reported for the same disturb phenomenon [66]. It was concluded that this program disturb is caused by hot electrons which are generated due to the generation-recombination center (GR-center) at the oxide–silicon interface of SSL, not generated by GIDL. The high electric field in space from SSL transistor to cell (WL0) accelerates generated electrons, and then hot electrons are injected to FG of WL0.

Another program disturb mechanism had been reported as a "DIBL-generated hot-electron injection" mechanism in 51-nm memory cell [67, 68], as described in Figs. 6.63–6.65. Due to DIBL (drain-induced barrier lowering) by the channel boosting voltage, punch-through between source and drain occurred at cut-off cells

• **Erase(E)/Program(P) Pattern**

Pattern	WL00–WL13	WL14	WL15	WL16–WL31
EP1		E	P1	
EP3	E	E	P3	E
PP1		P1	P1	
PP3		P3	P3	

• **BVdss Meas. condition**

Pattern	V_{csl} (source)	SSL Tr. & WL0–13	WL14	WL15	WL16–WL31 & GSL Tr.	V_{bl} (drain)
EP1			V_{pass}	0V		
EP3	0 V	V_{pass}	V_{pass}	0V	V_{pass}	0 – 8 V sweep
PP1			0V	0V		
PP3			0V	0V		

(left side labels: BL, GSL, WL31, WL16, WL15, WL14, WL00, GND, CSL)

FIGURE 6.63 Erase(E)/program(P) patterns and BV_{dss} measurement conditions. E: $V_{th} = -3.0$ V; P1: $V_{th} = +1.0$ V; and P3: $V_{th} = +3.0$ V.

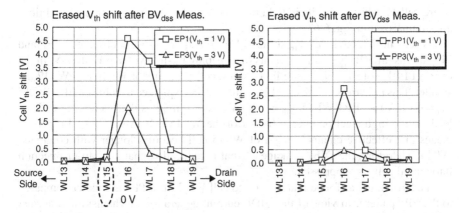

FIGURE 6.64 Erased cell V_{th} shifts from initial erased state after BV_{dss} measurements of EP1, EP3, PP1, PP3 patterns, respectively. EP1(WL14: erased E, WL15: programmed P1), EP3(E, P3), PP1(P1, P1), and PP3(P3, P3).

in the program self-boosting scheme. This punch-through generates the hot electrons, and a portion of hot electrons is injected near the floating gate, as shown in Fig. 6.65. The punch-through has a strong dependence of cell V_t. The lower V_t makes punch-through and this DIBL disturb worse.

To investigate a new program disturb induced by DIBL leakage, BV_{dss} curves of the selected cell in various erase/program data patterns are measured by sweeping bit-line voltage V_{bl} to 8 V. Before BV_{dss} measurements, all the cells were erased to $V_{th} = E(-3.0 \text{ V})$, and then one (WL15) or two (WL14, 15) selected cells in each array were programmed to $V_{th} = P1(+1.0 \text{ V})$ or P3(+3.0 V), followed by monitoring the

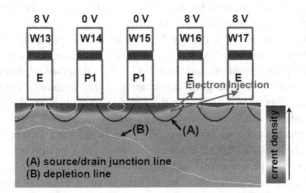

FIGURE 6.65 Device simulation is performed on the BV_{dss} measurement condition at PP1 pattern. Current density shows the punch-through of WL15 off-cell at $V_{bl} = 4$ V. DIBL (drain-induced barrier lowering) generates punch-through between source and drain in unselected cell. Electrons generated by punch-through are injected at nearby cells (disturb happens). Lower V_t cell shows larger DIBL.

entire cell V_{th} shifts from its initial erased states. Detailed measurement conditions are summarized in Fig. 6.63.

As seen in Fig. 6.64, cell V_{th} shifts of WL16 and WL17 were observed at EP and PP patterns after BV_{dss} measurements of sweeping bit-line voltage V_{bl} to 8 V. In the order of EP1 → PP1 → EP3 → PP3 patterns, cell V_{th} shifts of WL16 and WL17 are smaller. This pattern dependency of V_{th} shifts can be explained as follows. EP1(WL15 $V_{th} = +1.0$ V) → EP3(WL15 $V_{th} = +3.0$ V) pattern implies that increasing cell V_{th} leads to the high electron energy barrier of the WL15 cell. Also, EP → PP pattern means that effective channel length of WL15 cell becomes longer with combined WL14. Therefore, the DIBL leakage current is reduced so that hot carrier injection into erased cells is suppressed.

In addition, it was found that the erased cell V_{th} shift of PP3 is smaller compared to the PP1 pattern. In view of the GIDL current generation mechanism in a normal NMOS transistor, the greater the number of electrons that are stored in the floating-gate poly-Si (referred to as high programmed state in MLC flash operation), the greater the amount of GIDL-current that is generated. According to this assumption, PP1 should have a good immunity against a GIDL-induced hot carrier program disturb compared to EP3 pattern. However, the cell V_{th} shift of PP1 was larger cell V_t shift than PP3, as shown in Fig. 6.64. This indicates that hot carrier program disturb is mainly caused by DIBL, not by GIDL current in a 51-nm device. Therefore the short channel effect by DIBL should be controlled for an MLC NAND flash device beyond 51 nm. The simulation result also supports that the leakage source for a hot carrier is mainly originated by DIBL as seen in Fig. 6.65.

6.5.3   Channel Coupling

Channel boosting potential is decreasing as the cell dimension is scaled down. The boosted potential has a dependency on the neighboring string potential pattern. In the V_{cc}–V_{cc}–V_{cc} mode (condition of bit-line voltage order; see Fig. 6.66a), two adjacent active lines are under program-inhibit conditions when the center active line is in the program-inhibited condition. In the 0 V–V_{cc}–0 V mode, two adjacent active lines are under program operation. The 0 V–V_{cc}–V_{cc} mode means that only one adjacent active line is under program operation, and the other is under program inhibition.

Figure 6.66b shows the program disturbance characteristics of neighbor string potential pattern dependence [69]. The threshold voltage shifts of the program cell and program-inhibited cells in three neighbor data pattern modes are concurrently measured as selected cells are programmed by the incremental step pulse programming (ISPP). The boosted channel potential is derived from the difference of program voltage between the program cell and the inhibit cell to reach the same program threshold voltage (see Fig. 6.66b). The boosted channel potential appeared to be worst in the case of the 0 V–V_{cc}–0 V mode. From the measurement, adjacent channel potentials have a large impact on a boosted channel potential under program inhibit conditions.

To reveal a physical mechanism of neighbor data pattern dependence, TCAD simulation was performed [69]. In the 0 V–V_{cc}–0 V mode, the virtual sidewall transistor

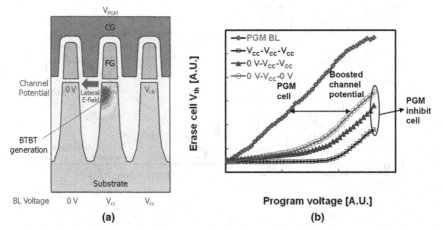

FIGURE 6.66   (a) Schematic illustration of the BTBT generation phenomenon by the lateral electric field. Channel boosting potential drops by BTBT generation at the side wall of the channel. (b) Program characteristics of the selected cell in program bit line, along with program disturbance characteristics of the unselected cell in program-inhibit bit line. For the V_{cc}–V_{cc}–V_{cc} mode, all the adjacent active lines are under program inhibition. For the 0 V–V_{cc}–V_{cc} mode, one of the adjacent active lines is under program operation, and the other is under program inhibition. For the 0 V–V_{cc}–0 V mode, all the adjacent active lines are under program operation.

is built up. When the neighboring channel is under the program by applying 0 V to the bit line, the channel potential is set to GND. The channel acts as a virtual gate of 0 V, and trench isolation dielectrics acts as a gate oxide; thus, program-inhibited channel potential is controlled by the neighboring channel virtual gate. The program-inhibited boosted channel near the Si surface below a tunnel oxide acts as the drain side of the virtual sidewall transistor. In the condition of the 0 V–V_{cc}–0 V mode, a large "gate-induced drain leakage" (GIDL) is generated in a booting channel. The GIDL current appears in the form of band-to-band tunneling (BTBT) leakage, and the BTBT leakage is source of losing boosting channel potential. The BTBT increases very abruptly over critical electric field, and thus the boosted channel potential has become saturated in spite of increasing pass voltage.

The BTBT generation mechanism at the sidewall of the channel is illustrated in Fig. 6.66a [69–71]. At the sidewall of the boosted channel facing a channel of GND, a large lateral electric field is built, thus BTBT electron–hole pair generation occurs at the sidewall. The generated electrons drop the boosted potential sooner than expected considering only the capacitive coupling effects.

To overcome the program disturb at 1X-nm node cell, the active air gap was developed to reduce the active channel coupling effect [69, 72]. Figure 6.67a shows a schematic active air gap [72]. And Fig. 5.25 (in Section 5.3.4) shows a SEM photograph of active air gap. The active air gap can improve program disturb in 0 V–V_{cc}–0 V (0 F inhibit cell), as shown in Fig 6.67b.

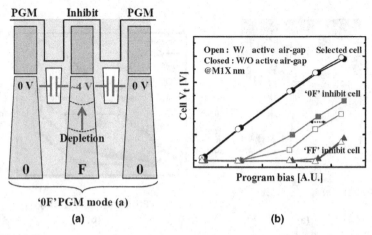

FIGURE 6.67 (a) The schematic of "0 F" disturbance mode. Boosted channel potential of inhibit BL is severely affected by grounded adjacent BL. (b) The improvement of "0 F" boosting level with active air gap.

6.6 ERRATIC OVER-PROGRAM

The erratic over-program is the phenomenon of unexpected large V_t shift during programming. Erratic over-program cells make the tail of V_t distribution in upper side distribution, as shown in Fig. 6.68. If the V_t of tail bit exceeds over read voltage in the case of L0, L1, and L2, it produces a single bit failure. However, in case of L3, if the V_t of tail bit exceeds (or closed to) V_{passR}, all cells of L0/L1/L2 in the NAND string become failure bits because the over-programmed cell is always OFF during read due to high V_t. Then the over-program in L3 makes a failure rate much worse than in other states.

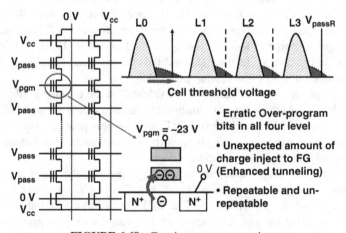

FIGURE 6.68 Erratic over-programming.

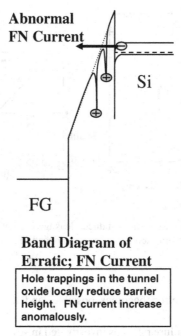

FIGURE 6.69   Schematic model of erratic over-program. Two hole traps in tunnel oxide induce an abnormal FN current.

The root cause of the erratic over-program is explained by an enhanced tunneling at two or more hole traps at close location in tunnel oxide. Figure 6.69 shows an image of an enhanced tunneling at a two-hole trap site. Hole traps locally reduce the barrier height of the tunnel oxide.

As another model of the erratic over-program, the neutral electron-assisted two-step tunneling is considered. It was observed as an anomalous increase of tunneling current of the stress-induced leakage current in the tunnel oxide at low electric field. Figure 6.70a shows an example of the stress-induced leakage current that suddenly increases during the gate stressing [63]. The increased stress-induced leakage current can fit with the FN tunneling current line, and the estimated barrier height is 0.57 eV. During the gate stressing, there is no hole generation, and hole capturing into the tunnel oxide could not happen because of the lower electric field. As a model for the stress-induced leakage current of the typical cell, neutral electron-assisted two-step tunneling is supported by several works [55–58]. Figure 6.70b shows a schematic model to explain the anomalous current increase phenomenon. Before the increase of the stress-induced leakage current, the two-step tunneling through a neutral electron trap located at the trap depth level Φ_{t1} occurs. Due to hole movement in the tunnel oxide by the electric field during the gate stress, another electron trap site which has the shallow barrier height Φ_{t2} is leveled down to the position where the electron in the conduction band of the substrate can tunnel. As a result, local

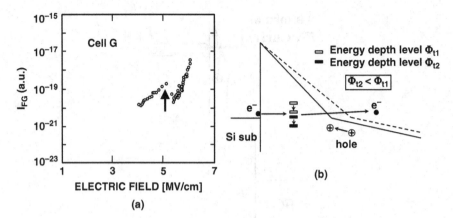

FIGURE 6.70 (a) Increase of the stress-induced leakage current during the gate stress. (b) Energy band diagram to explain increase of stress-induced leakage current during the gate stressing.

two-step tunneling of electron becomes possible, and the stress-induced leakage current increases.

Erratic over-program failure rate was investigated in several NAND flash products from different suppliers [37], as shown in Fig. 6.71. Bit failure rate is increased with P/E cycling increased. And the percentage of failure bits for each level is dependent on the product (suppliers). It would be dependent on program condition setting, read condition setting, and process difference.

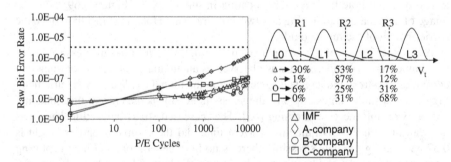

FIGURE 6.71 Erratic over-programming. RBER after programming data, as a function of prior program/erase cycles. Errors were logged at the cycle points noted by the symbols. The dashed line is the RBER where the instantaneous UBER (Uncorrectable Bit Error Rate) reaches 10^{-15}. Nonmonotonic curves result from small sample sizes and erratic RBER behavior. Schematic illustration of the dominant types of program errors and the percentage weighting of each type for each product. The RBER was dominated by cells with higher V_t's than intended; exceptions exist to some degree and result in percentages sometimes adding to less than 100%. RBER is gradually increased with P/E cycling increasing. The percentage of fail bits for each level is dependent on product (provider). Oxide trap-assisted tunneling is a major root cause.

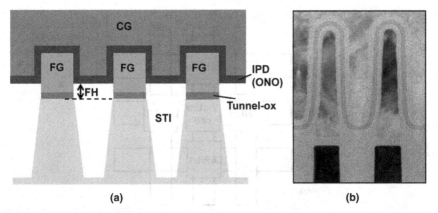

(a)   **(b)**

FIGURE 6.72   (a) The self-aligned shallow trench isolation cell (SA-STI cell) structure along word line (WL). (b) TEM micrograph of a 26-nm SA-STI cell [74]. A field oxide height (FH) is the distance between a channel Si and the top of an STI field oxide.

6.7   NEGATIVE V_t SHIFT PHENOMENA

6.7.1   Background and Experiment

The self-aligned shallow trench isolation cell (SA-STI cell) [10, 73] has been used for NAND flash memory products for a long time from 0.2-μm generation [74–76] to present middle-1X-nm generation [72, 77]. The structure of the SA-STI cell along word line (WL) is shown in Fig. 6.72. A floating gate (FG) is self-aligned patterned with STI to avoid overlap of FG on STI edge corner. In the SA-STI cell, the sidewall of FG is used for increasing a capacitance between FG and CG, to increase a coupling ratio. Then the field oxide height (FH), which is the distance between the channel Si and the top of the STI field oxide, has to be decreased as small as possible to increase a coupling ratio, as shown in Fig. 6.72a. The decreasing FH can also obtain a small FG–FG coupling interference [78] along the WL direction. However, in a small FH, high voltage (~20 V) is applied directly between substrate (channel) and CG during program and erase. This high electric field is a concern that could have an impact on reliability and performance of NAND flash memory.

Section 6.7 describes the "negative" V_t shift phenomena in program inhibit conditions of a 2X- to 3X-nm SA-STI NAND flash cell [11, 79]. The negative V_t shift occurs in the small FH case, thus it is one of the high field effects during programming. The negative V_t shift phenomena makes the V_t read window margin (RWM) worse for MLC/TLC due to widening V_t distribution width. Therefore, the negative shift phenomena could become a new potential obstacle of scaling NAND flash cells in the scaled 2X-nm NAND flash.

A 2X- and 3X-nm rule SA-STI cells with various FH were used for this experiment. The range of FH small/middle/large in experiments are 10–20 nm. And thickness of IPD (ONO) is around 12 nm. The cross-sectional TEM micrograph of 26-nm SA-STI cell [80] is shown in Fig. 6.72b.

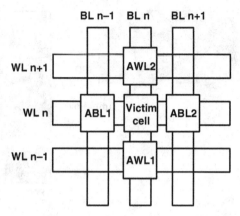

FIGURE 6.73   The cell arrangement for program-inhibited test. Attack cells of ABL1&2 are the adjacent cells along WL direction, and attack cells of AWL1&2 are adjacent cells along BL direction.

Figure 6.73 shows the cell arrangement for the program-inhibited test. Attack cells of ABL1&2 are adjacent cells along the WL direction, and attack cells of AWL1&2 are adjacent cells along the BL direction. The V_t of inhibit victim cells are monitored before and after programming the attack cells.

For an analysis of current flow in an SA-STI structure in the program condition in Section 6.7.4, a cell-structured capacitor is used. Terminals of CG, FG, source/drain junction, and substrate are independently connected to monitor the current.

6.7.2   Negative V_t Shift

Figure 6.74 shows the victim cell V_t shift during the programming of the attack cell. In the case of attack AWL2, the V_t of the victim cell monotonously increases with attack AWL2 cell V_t increased due to conventional FG–FG coupling interference [78] (see Section 5.3). However, in the case of ABL1&2, the V_t of the victim cell initially increases and then decreases as the attack ABL1&2 are programmed. V_t shift caused by programming a neighbor cell should be positive if it is caused by conventional FG–FG coupling interference. However, the V_t shift is showing a negative direction as attack cell V_t increases over $V_t > 7$. This phenomenon is called the "negative V_t shift."

Figure 6.75 shows the dependence of the negative V_t shift on FH. The negative V_t shift has a strong FH dependence. The negative V_t shift is larger when FH is small. In the region of attack cell $V_t < 6$, the slope of (victim cell V_t)/(attack cell V_t) is showing the conventional FG–FG coupling interference. In the case of FH low, the slope is smaller than the case of FH middle and large. This means that the FH low case has small FG–FG coupling interference due to the CG shield effect between FGs.

Figure 6.76 shows the victim cell V_t dependence on (a) attack cell: program and (b) attack cell: inhibit condition, which are illustrated in the right-hand side of Fig. 6.76.

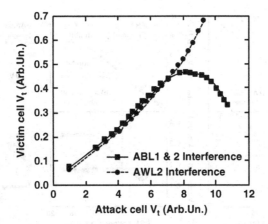

FIGURE 6.74 Victim cell V_t shift versus attack cell programmed V_t. Negative V_t shift phenomena are observed in the case of attack ABL1 and ABL2. Victim cell V_t is corresponding to L1 for MLC.

In (a) attack cell: program, attack cell V_t is monotonously increased. Victim cell V_t initially increases and then decreases as attack cell V_t increases. This V_t movement is the same as that in Fig. 6.74 and Fig. 6.75. However, in the case of (b) attack cell: inhibit, attack cell V_t initially increases, and when the attack cell V_t has reached to around 4 or 7, attack cell V_t stops increasing by changing channel voltage from 0 V to V_{boosting} (inhibit mode) during program pulse. This operation corresponds to the program verify operation [81] in product, such that when V_t has reached a certain V_t, programming is stopped by changing to the inhibit mode during next program pulse (channel voltage changes from 0 V to V_{boosting}). For victim cell V_t in (b) attack cell

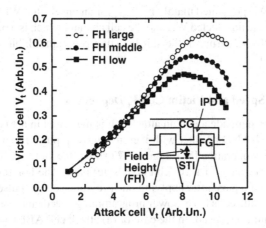

FIGURE 6.75 Field height (FH) dependence of "negative" V_t shift. Small FH has the larger "negative" shift. Victim cell V_t corresponds to L1 for MLC.

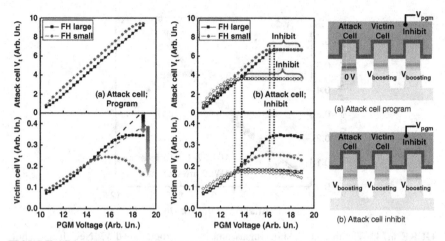

FIGURE 6.76 Victim cell V_t shift in (a) attack cell: program and (b) attack cell: Inhibit. In the case of (a) attack cell program, a large negative shift of victim cell is observed, however, in the case of (b) attack cell: inhibit, negative V_t shift is much smaller. Difference of bias conditions between (a) and (b) is a channel voltage of 0 V for program or V_{boost} (~8 V) for inhibit. Victim cell V_t corresponds to L1 for MLC.

inhibit, a negative shift is much smaller than case (a) attack cell: program, when attack cells are in the inhibit mode, even high program voltage (V_{pgm}) is applied. Because the same high V_{pgm} pulses are applied in both cases (a) and (b), the difference of bias condition between (a) attack cell: program and (b) attack cell: inhibit is only the channel voltage in attack cell, which are 0 V for program or $V_{boosting}$ for inhibit, as shown on the right-hand side of Fig. 6.76. Thus, the channel voltage of 0 V in attack cell produces a negative V_t shift in inhibit victim cell, especially in the case of small FH.

$V_{boosting}$ (~8 V) is generated mainly by V_{pass} for an unselected WL with capacitive coupling between the unselected WL and the cell channel which is isolated by a select transistor in the NAND string in the self-boosting program inhibit scheme [63, 82] (see Section 6.5.1).

6.7.3 Program Speed and Victim Cell V_t Dependence

The program speed dependence of an attack cell is measured in actual page program sequence of MLC NAND product with incremental step pulse program (ISPP) [82] and bit-by-bit verify operation [81]. Figure 6.77 shows the victim cell delta V_t versus attack cell program speed for a 16-Kbit (2-KByte) page. The horizontal axis shows the V_t distribution of a page after applying one program-voltage pulse (V_{pgm}), which means that the left-side cells have slower programming speed, and the right-side cells have faster programming speed. In the case of (a) attack cell ABL1 = L1 & ABL2 = L1 (programming to L1 (lower V_t); Erase → L1), victim cell delta V_t is larger in a slow attack cell and the smaller in a fast attack cell, due to conventional FG–FG

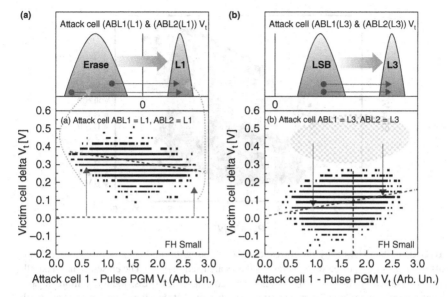

FIGURE 6.77   Victim cell delta V_t distribution of a 16-Kbit cell versus attack cell program speed (attack cell V_t distribution of a page after applying one program-voltage pulse). In the case of (a) attack cell ABL1 = L1 & ABL2 = L1, victim cell delta V_t shows dependence of conventional FG–FG coupling interference. However, in the case of (b) attack cell ABL1 = L3 & ABL2 = L3, victim cell delta V_t shows the larger negative V_t shift for slow attack cell. Victim cell V_t corresponds to L3 for MLC.

interference. The slower attack cells have the larger victim cell delta V_t, because an attack cell V_t change during programming (Erase → L1) is larger in a slow attack cell, as shown in Fig. 6.77a, upper portion.

In order to ensure the larger V_t change in a slow attack cell, the bit-by-bit V_t distribution after program and erase are measured. Figure 6.78 shows the bit-by-bit correspondence of program and erase V_t distributions after one program pulse and one erase pulse. We can see that the program cells on the left-hand side of programmed V_t distribution are also on the left-hand side of erased V_t distribution, and similarly the program cells on the right-hand side of the programmed V_t distribution are also on the right-hand side of the erased V_t distribution. Therefore, it is confirmed that the slow program cells are on the left-hand side of erased V_t distribution, as shown in Fig. 6.77a, upper portion.

On the other hand, in case of (b) attack cell ABL1 = L3 and ABL2 = L3 (programming to L3 (higher V_t); LSB →L3) in Fig. 6.77b, the victim cell delta V_t is smaller in a slow attack cell and larger in a fast attack cell, even if the V_t shift by a conventional FG–FG coupling should be the same between slow and fast cells due to the same attack cell V_t change during programming (LSB → L3), as shown in Fig. 6.77b, upper portion. This means the negative V_t shift is much larger (−0.2 to −0.4 V) in the case of a slow attack cell, compared with a fast attack cell.

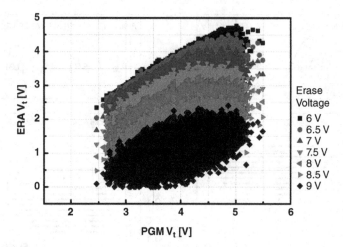

FIGURE 6.78 The bit-by-bit correspondence of program and erase V_t after one program pulse and one erase pulse. Several erase voltages are used (6–9 V). The erased V_t has a parallel shift as erase voltage increases. Then it is supposed that erased V_t distribution in more negative V_t has the same correspondence, so that it is confirmed that the slow program cells are on the left side of the erased V_t distribution.

The reason is supposed that in a slow attack cell, the larger number and higher voltage (V_{pgm}) of program pulses are subjected to the condition of channel voltage = 0 V (see Fig. 6.76a attack cell program). Then the negative shift becomes larger in a slow attack cell. Conversely, in a fast attack cell, the larger number and higher voltage of program pulse are shortly subjected to the condition of channel voltage = 0 V, because a fast attack cell becomes inhibit mode (channel voltage = $V_{boosting}$) earlier than a slow attack cell.

Furthermore, this new negative V_t shift results in wider placement V_t distribution. In the case of (a) attack cell ABL1 = L1 & ABL2 = L1, the victim cell delta V_t distribution width is 0.36 V; however, in the case of (b) attack cell ABL1 = L3 & ABL2 = L3, the victim cell delta V_t distribution width is 0.48 V. These V_t distribution widths have an impact on the read window margin of the MLC/TLC NAND flash product.

Figure 6.79 shows the victim cells dependence on the program state: (a) L1, (b) L2, and (c) L3. The victim cell delta V_t of L3 is smaller than that of L1 and L2, especially in the case of the attack cell ABL1 = L3 & ABL2 = L3. It means the negative V_t shift is larger in victim cell L3. This is considering the fact that if the FG of L3 is negatively charged, then it could gather more positive charge during attack cell programming.

Summarizing the results of the negative V_t shifts phenomena (Sections 6.7.2 and 6.7.3), the negative V_t shift is enhanced in the case of (1) neighbor cell along WL (ABL1,2) in programming (channel voltage = 0 V), (2) small FH, (3) higher V_{PGM}, (4) attack cell: L3, (5) attack cell: slow programming, and (6) victim cell: higher V_t (L3).

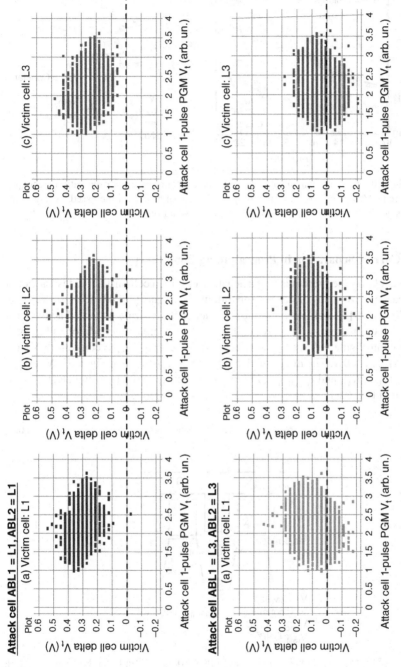

FIGURE 6.79 The dependence of victim cell delta V_t, on program states (L1, L2, L3), in attack cell ABL1 = L1 & ABL2 = L1 and Attack cell ABL1 = L3 & ABL2 = L3. Victim cell delta V_t, of (c) L3 has the larger negative shift than that of (a) L1 and (b) L2.

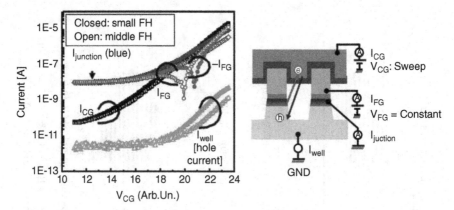

FIGURE 6.80 Current analysis of cell structured capacitor by carrier separation technique. Hole currents (I_{Well}) are observed, and they are increased as V_{CG} is increased. A large hole current is generated due to large I_{junction} in the case of small FH.

6.7.4 Carrier Separation in Programming Conditions

In order to clarify the mechanism of a negative V_t shift, a cell-structured test capacitor was measured by using a carrier separation technique [29, 83–86], as shown in Fig. 6.80 [11, 79]. A cell-structured test capacitor has a stripe patterned active area/STI and flat CG pattern, with source and drain. A measurement condition of Fig. 6.80 is that a control gate voltage (V_{CG}) sweeps while keeping V_{FG} = constant (8 V) and $V_{\text{well}} = V_{\text{junction}} = 0$ V. The image of electron flow is illustrated in Fig. 6.81. Measured current of I_{CG}, I_{FG}, and I_{junction} in Fig. 6.80 can be expressed by using electron flow as shown in Fig. 6.81.

$$I_{\text{CG}} = -I_{\text{CG\_Juction}} - I_{\text{CG\_FG}} \tag{6.8}$$

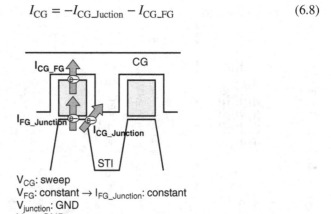

FIGURE 6.81 Electron flow in condition of Fig. 6.80. I_{junction} (in Fig. 6.80) = $I_{\text{CG\_Junction}}$ + $I_{\text{FG\_Junction}}$. Hole current (I_{well}) at region of $V_{\text{CG}} > 18$ (Fig. 6.80) is generated by $I_{\text{CG\_Junction}}$.

$$I_{FG} = -I_{FG\_Junction} + I_{CG\_FG} \qquad (6.9)$$

$$I_{Junction} = I_{CG\_Junction} + I_{FG\_Junction} \qquad (6.10)$$

$$I_{FG\_Junction} = \text{constant}$$

It was observed in Fig. 6.80 that the hole current (I_{well}) was increased as V_{CG} increased in a high $V_{CG} > 18$ region. In the region of $V_{CG} > 18$, $I_{CG\_Junction}$ and I_{CG_FG} are increased as V_{CG} increased, while $I_{FG\_Junction}$ are basically constant because of constant V_{FG}. At $V_{CG} > 18$ in Fig. 6.80, $I_{junction}$ of small FH is larger than $I_{junction}$ of middle FH, and also I_{well} of small FH is larger than I_{well} of middle FH. From these observations of $I_{Junction}$ and I_{well} in Fig. 6.80, I_{well} (hole current) is considered to be generated by $I_{CG\_Junction}$, not by I_{CG_FG}, as shown in Fig. 6.81. $I_{CG\_Junction}$ is the electron flow of FN injection from channel/junction to CG. $I_{CG\_Junction}$ may generate I_{well} (hole current) based on the anode hole injection model [29, 83, 89].

For I_{FG}, I_{FG} is almost constant at $V_{CG} < 20$ because $I_{FG\_Junction} = \text{constant}$ while I_{CG_FG} is small due to small $V_{CG} - V_{FG}$. As increased V_{CG} to $V_{CG} > 20$, I_{FG} polarity is changed, because I_{CG_FG} is increased by increasing $(V_{CG} - V_{FG})$ and become the dominant current of I_{FG}.

Figure 6.82 shows a current flow in constant V_{CG}. Even if a voltage between channel/junction and CG is constant, I_{CG} is increased as V_{FG} is increased. This means that I_{CG}, which is mainly a direct electron injection from channel/junction to CG, is strongly enhanced by V_{FG}, even if constant V_{CG} is applied. As a scaling down of memory cell size, a FG–FG space becomes narrow. Then I_{CG} will increase because V_{FG} could enhance I_{CG} intensively. It suggests that the negative V_t shift phenomena may be enhanced by scaling down of a memory cell.

The ratio of (substrate hole current) / (gate electron current) [$= I_{well}/I_{junction}$] in Fig. 6.82 is in the range of 10^{-3} over $V_{FG} = 3$–5. The substrate hole current (I_{well}) is

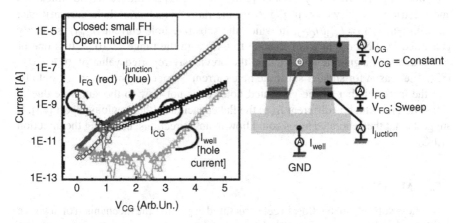

FIGURE 6.82 Current analysis of cell structured capacitance. Even if V_{CG} is constant, I_{CG} (direct electron injection from channel/junction to CG) is increased as V_{FG} is increased. This means I_{CG} is enhanced by FG potential.

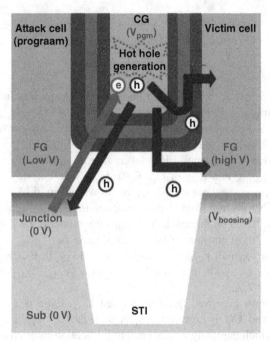

FIGURE 6.83 Suggested mechanism of negative V_t shift in victim cell. Electron injection from channel/Junction (0 V) to CG (V_{PGM}) could generate hot hole at CG, and parts of hot holes are injected to FG of victim cell through field dielectric and IPD. This hot hole injection causes a "negative V_t shift."

mainly generated by FN current through tunnel oxide ($I_{FG\_Junction}$). The value is in the same range as previously reported [29, 83, 84, 86, 89] for the same oxide thickness and electric field. However, in Fig. 6.80, the ratio of (substrate hole current)/(gate electron current) [$= I_{well}/I_{CG}$], in which the substrate hole current (I_{well}) is mainly generated by FN current from channel to CG, is in the range of 10^{-4}. It is one or two orders of magnitude smaller than the previously reported value of 10^{-3}–10^{-2} [83]. The reason for smaller substrate hole current is not clear; however, it would be that the large number of the generated holes does not flow to the substrate due to cell structure, which is different from the flat capacitor in the previous report [83]. It suggests that the generated holes could flow to any directions, including the direction of FG.

6.7.5 Model

From the results of current flow in cell structured capacitor, the mechanism of negative V_t shift is considered as illustrated in Fig. 6.83. During programming, electrons are injected from channel/junction (0 V) to CG (V_{PGM}) directly. Electron injection generates hot holes by impact ionization, and hot holes are injected to IPD on STI.

Parts of hot holes are injected into the FG of the victim cell through the field-oxide dielectric or IPD. Consequently, the V_t of the victim cell has shifted negatively.

This phenomenon would be accelerated with memory cell scaling since I_{CG} increases due to narrow FG–FG space field effect. Then the negative V_t shift phenomenon will be worse in future scaled memory cells. Then the negative V_t shift will be one of new scaling limitation factors to manage the V_t read window margin of 2 bits/cell and 3 bits/cell in 2X nm and beyond the NAND flash memory cell.

A novel program inhibit phenomenon of "negative" cell V_t shift had been presented in 2X- to 3X-nm self-aligned STI NAND flash memory cells. The negative V_t shift is caused in an inhibit cell when an along-WL adjacent cell is programming. The magnitude of the negative shift becomes larger in the case of higher program voltage (V_{PGM}), lower field oxide height (FH), slower program speed of the adjacent cell, and high V_t of the victim cell. The experimental results suggest that the mechanism of negative V_t shift is attributed to hot holes that are generated by FN electrons injection from channel/junction to the control gate (CG). Many reports had previously described the substrate hole current (I_{well}) in an MOS capacitor. However, this negative V_t shift phenomenon was a very rare case where a generated hole current could be directly observed in the device of flash memory cells as V_t shift.

6.8 SUMMARY

In Chapter 6, the phenomena of NAND flash memory reliability have been described.

In Section 6.2, the program/erase (P/E) cycling degradation and data retention characteristics were described. A uniform program and erase scheme, which uses uniform Fowler–Nordheim tunneling over the whole channel area both program and erase, guarantees a wide cell threshold voltage window even after 1 million program/erase cycles. The data retention characteristics could be also guaranteed by applying a uniform program and erase scheme. This uniform program/erase scheme has been used in NAND flash as de facto standard.

Several reliability aspects related to P/E cycling endurance and data retention are discussed in Section 6.3. The degradation by P/E cycling stress are mainly caused by electron/hole traps, stress-induced leakage current (SILC), and interface state generation. And by scaling memory cells, these degradation phenomena are becoming more severe.

In Section 6.4, read disturb characteristics were described. It was clarified experimentally that flash memory cell programmed and erased by Fowler–Nordheim tunneling (FN-t) has 10 times longer retention time than the conventional one, which is programmed by channel-hot-electron (CHE) injection and erased by FN-t. This difference of data retentivity between these two P/E schemes is due to decreasing the stress–induced leakage current (SILC) of thin gate oxide by bipolarity FN-t stress. Also, this improvement in data retention is more remarkable as the gate oxide thickness decreases. Therefore, a bipolarity FN-t P/E scheme, which enables a flash memory cell to scale down its oxide thickness, promisingly becomes the key technology to realize reliable flash memory.

Several analysis results of read disturb were also presented. The V_t shift of read disturb can be separated into two regions. Initially, the electron detrapping from tunnel oxide causes V_t shift in read disturb; after that, SILC causes a V_t shift. Also, the hot carrier injection phenomenon was presented in Section 6.4.4. During read operation, a hot carrier is generated by locally boosted node in NAND string and then is injected to a cell floating gate.

Program disturb was described in Section 6.5. By scaling down memory cells, it is becoming difficult to manage a program disturb failure because of the unexpected degradation mechanism of hot carrier injection and channel coupling. The hot carrier injection is caused in the high-field location between the self-boosting voltage (~8 V) and other (0 V), as described in Section 6.5.2. The channel coupling effect during programming was also described in Section 6.5.3. The neighbor channel voltage of 0 V has an impact on self-boosted inhibit channel voltage not only by capacitive coupling but also by band-to-band leakage in boosting node.

The erratic over program are described in Section 6.6. The erratic over-program is caused by excess electron injection through tunnel oxide during programming. Strong ECC can manage the erratic over-programming failure.

In Section 6.7, a novel program inhibit phenomena of "negative" cell V_t shift had been presented. The negative V_t shift is caused in an inhibit cell when an along-WL adjacent cell is programming. The magnitude of the negative shift becomes larger in the case of higher program voltage (V_{PGM}), lower field oxide height (FH), slower program speed of the adjacent cell, and high V_t of the victim cell. The experimental results suggest that the mechanism of negative V_t shift is attributed to hot holes that are generated by FN electrons injection from channel/junction to the control gate (CG).

Flash memory reliability and physical phenomena are summarized in Fig. 6.84 [90]. It is clarified that carrier traps in tunnel oxide, detrapping, SILC (stress–induced leakage current) are the major root causes of degradation of flash memory reliability.

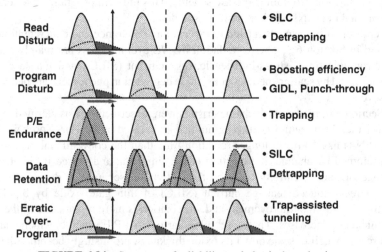

FIGURE 6.84   Summary of reliability and physical mechanism.

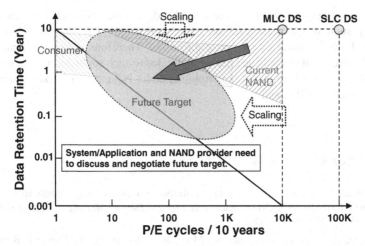

FIGURE 6.85 Prospect of data retention versus program/erase cycling. Due to encountering physical limitation, reliability in a future scaling device will be worse than that in the current device. System management would be essential.

The important reliability aspects of program/erase cycling endurance and data retention are trade-off relationship, as shown in Fig. 6.85 [90]. The future target of P/E cycling and data retention will be compromised as <1K P/E cycling and <1 year data retention even with system solutions.

Performance and reliability are also trade-off relationship [90], as shown in Fig. 6.86. If high-speed programming is required, the reliability (such as P/E cycling) will be degraded because the higher electrical field is applied to tunnel oxide in the

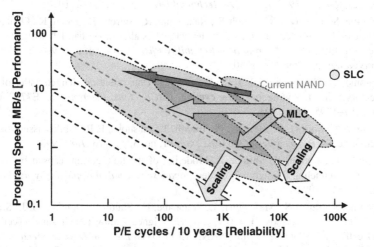

FIGURE 6.86 Prospect of performance versus reliability. Performance and reliability are "trade-off." As device scaling, both performance and reliability will be degraded naturally. By efforts of increasing page size, performance could be kept or be improved.

memory cell during program. On the other hand, if some application requires high reliability, such as >10K P/E cycling, >3 years retention, the performance of such high-speed programming should be compromised. Therefore, target specification of NAND flash reliability would be greatly subdivided to each application, such as memory cards, consumer application (smartphone, tablet PC, etc), high-end applications (enterprise server SSD), and so on.

REFERENCES

[1] Masuoka, F.; Momodomi, M.; Iwata, Y.; Shirota, R. New ultra high density EPROM and flash EEPROM with NAND structure cell, *Electron Devices Meeting, 1987 International*, vol. 33, pp. 552– 555, 1987.

[2] Shirota, R.; Itoh, Y.; Nakayama, R.; Momodomi, M.; Inoue, S.; Kirisawa, R.; Iwata, Y.; Chiba, M.; Masuoka, F. New NAND cell for ultra high density 5v-only EEPROMs, *Digest of Technical Papers—Symposium on VLSI Technology*, 1988, pp. 33–34.

[3] Momodomi, M.; Kirisawa, R.; Nakayama, R.; Aritome, S.; Endoh, T.; Itoh, Y.; Iwata, Y.; Oodaira, H.; Tanaka, T.; Chiba, M.; Shirota, R.; Masuoka, F. New device technologies for 5 V-only 4 Mb EEPROM with NAND structure cell, *Electron Devices Meeting, 1988. IEDM '88. Technical Digest, International*, pp. 412–415, 1988.

[4] Kirisawa, R.; Aritome, S.; Nakayama, R.; Endoh, T.; Shirota, R.; Masuoka, F. A NAND structured cell with a new programming technology for highly reliable 5 V-only flash EEPROM, *1990 Symposium on VLSI Technology. Digest of Technical Papers*, pp. 129–130, 1990.

[5] Aritome, S.; Shirota, R.; Kirisawa, R.; Endoh, T.; Nakayama, R.; Sakui, K.; Masuoka, F. A reliable bi-polarity write/erase technology in flash EEPROMs, *International Electron Devices Meeting, 1990. IEDM '90. Technical Digest*, pp. 111–114, 1990.

[6] Aritome, S.; Kirisawa, R.; Endoh, T.; Nakayama, R.; Shirota, R.; Sakui, K.; Ohuchi, K.; Masuoka, F. Extended data retention characteristics after more than 10^4 write and erase cycles in EEPROMs, *International Reliability Physics Symposium, 1990. 28th Annual Proceedings*, pp. 259–264, 1990.

[7] Aritome, S.; Shirota, R.; Sakui, K.; Masuoka, F. Data retention characteristics of flash memory cells after write and erase cycling, *IEICE Trans. Electron*, vol. E77-C, no. 8, pp. 1287–1295, Aug. 1994.

[8] Aritome, S.; Shirota, R.; Hemink, G.; Endoh, T.; Masuoka, F. Reliability issues of flash memory cells, *Proceedings of the IEEE*, vol. 81, no. 5, pp. 776–788, 1993.

[9] Hemink, G.; Endoh, T.; Shirota, R. Modeling of the hole current caused by fowler–nordheim tunneling through thin oxides, *Japanese Journal of Applied Physics*, vol. 33, pp. 546–549, 1994.

[10] Aritome, S.; Satoh, S.; Maruyama, T.; Watanabe, H.; Shuto, S.; Hemink, G. J.; Shirota, R.; Watanabe, S.; Masuoka, F. A 0.67 μm^2 self-aligned shallow trench isolation cell (SA-STI cell) for 3 V-only 256 Mbit NAND EEPROMs, *Electron Devices Meeting, 1994. IEDM '94. Technical Digest, International*, pp. 61–64, 11–14 Dec. 1994.

[11] Aritome, S.; Seo, S.; Kim, H.-S.; Park, S.-K.; Lee, S.-K.; Hong, S. Novel negative V_t shift phenomenon of program–inhibit cell in 2X–3X nm self-aligned STI NAND flash

memory, *Electron Devices, IEEE Transactions on*, vol. 59, no. 11, pp. 2950–2955, Nov. 2012.

[12] Masuoka, F.; Asano, M.; Iwahashi, H.; Komuro, T.; Tanaka, S. A new flash E^{2}PROM cell using triple polysilicon technology, *Electron Devices Meeting, 1984 International*, vol. 30, pp. 464–467, 1984.

[13] Tam, S.; Sachdev, S.; Chi, M.; Verma, G.; Ziller, J.; Tsau, G.; Lai, S.; Dham, V. *1988 Symposium on VLSI Technology, Technical Papers*, pp. 31–32, 1988.

[14] Kume, H.; Yamamoto, H.; Adachi, T.; Hagiwara, T.; Komori, K.; Nishimoto, T.; Koike, A.; Meguro, S.; Hayashida, T.; Tsukada, T. A flash-erase EEPROM cell with an asymmetric source and drain structure, *Electron Devices Meeting, 1987 International*, vol. 33, pp. 560–563, 1987.

[15] Ajika, N.; Ohi, M.; Arima, H.; Matsukawa, T.; Tsubouchi, N. A 5 volt only 16M bit flash EEPROM cell with a simple stacked gate structure, *Electron Devices Meeting, 1990. IEDM '90. Technical Digest, International*, pp. 115–118, 9–12 Dec. 1990.

[16] Onoda, H.; Kunori, Y.; Kobayashi, S.; Ohi, M.; Fukumoto, A.; Ajika, N.; Miyoshi, H. A novel cell structure suitable for a 3 volt operation, sector erase flash memory, *Electron Devices Meeting, 1992. IEDM '92. Technical Digest, International*, pp. 599–602, 13–16 Dec. 1992.

[17] Kodama, N.; Saitoh, K.; Shirai, H.; Okazawa, T.; Hokari, Y. A 5V only 16 Mbit flash EEPROM cell using highly reliable write/erase technologies, *VLSI Technology, 1991. Digest of Technical Papers, 1991 Symposium on*, pp. 75–76, 28–30 May 1991.

[18] Verma, G.; Mielke, N. Reliability performance of ETOX based flash memories, *Reliability Physics Symposium 1988. 26th Annual Proceedings, International*, pp. 158–166, 12–14 April 1988.

[19] Baglee, D. A.; Smayling, M. C. The effects of write/erase cycling on data loss in EEPROMs, *Electron Devices Meeting, 1985 International*, vol. 31, pp. 624–626, 1985.

[20] Witters, J. S.; Groeseneken, G.; Maes, H. E. Degradation of tunnel-oxide floating-gate EEPROM devices and the correlation with high field-current-induced degradation of thin gate oxides, *IEEE Transactions on Electron Devices*, vol. 36, no. 9, part 2, pp. 1663–1682, 1989.

[21] Haddad, S.; Chang, C.; Swaminathan, B.; Lien, J. Degradations due to hole trapping in flash memory cells, *Electron Device Letters, IEEE*, vol. 10, no. 3, pp. 117–119, March 1989.

[22] Mori, S.; Kaneko, Y.; Arai, N.; Ohshima, Y.; Araki, H.; Narita, K.; Sakagami, E.; Yoshikawa, K. Reliability study of thin inter-poly dielectrics for non-volatile memory application, *Reliability Physics Symposium, 1990. 28th Annual Proceedings, International*, pp. 132–144, 27–29 March 1990.

[23] Chen, J.; Chan, T.-Y.; Chen, I.-C.; Ko, P.-K.; Chenming, Hu. Subbreakdown drain leakage current in MOSFET, *Electron Device Letters, IEEE*, vol. 8, no. 11, pp. 515, 517, Nov. 1987.

[24] Chan, T.-Y.; Chen, J.; Ko, P.-K.; Hu, C. The impact of gate-induced drain leakage current on MOSFET scaling, *Electron Devices Meeting, 1987 International*, vol. 33, pp. 718, 721, 1987.

[25] Shirota, R.; Endoh, T.; Momodomi, M.; Nakayama, R.; Inoue, S.; Kirisawa, R.; Masuoka, F. An accurate model of subbreakdown due to band-to-band tunneling and its application,

Electron Devices Meeting, 1988. IEDM '88. Technical Digest, International, pp. 26, 29, 11–14 Dec. 1988.

[26] Olivo, P.; Nguyen, T.N.; Ricco, B. High-field-induced degradation in ultra-thin SiO_2 films, *Electron Devices, IEEE Transactions on*, vol. 35, no. 12, pp. 2259–2267, Dec. 1988.

[27] Naruke, K.; Taguchi, S.; Wada, M. Stress induced leakage current limiting to scale down EEPROM tunnel oxide thickness, *IEDM Technical Digest*, pp. 424–427, Dec. 1988.

[28] Hemink, G. J.; Shimizu, K.; Aritome, S.; Shirota, R. Trapped hole enhanced stress induced leakage currents in NAND EEPROM tunnel oxides, *IEEE International Reliability Physics Symposium, 1996. 34th Annual Proceedings*, pp. 117–121, 1996.

[29] Weinberg, Z. A.; Fischetti, M. V.; Nissan-Cohen, Y. SiO_2 induced substrate current and its relation to positive charge in field effect transistor, *Journal of Applied Physics*, vol. 59, no. 3, pp. 824–832, 1 Feb. 1986.

[30] Lee, J.-D.; Choi, J.-H.; Park, D.; Kim, K. Data retention characteristics of sub-100 nm NAND flash memory cells, *IEEE Electron Device Letters*, vol. 24, no. 12, pp. 748–750, 2003.

[31] Lee, J.-D.; Choi, J.-H.; Park, D.; Kim, K. Degradation of tunnel oxide by FN current stress and its effects on data retention characteristics of 90 nm NAND flash memory cells, *41st Annual. 2003 IEEE International Reliability Physics Symposium Proceedings*, pp. 497–501, 2003.

[32] Lee, J.-D.; Choi, J.-H.; Park, D.; Kim, K. Effects of interface trap generation and annihilation on the data retention characteristics of flash memory cells, *IEEE Transactions on Device and Materials Reliability*, vol. 4, no. 1, pp. 110–117, 2004.

[33] Fayrushin, A.; Seol, K.S.; Na, J.H.; Hur, S.H.; Choi, J.D.; Kim, K. The new program/erase cycling degradation mechanism of NAND flash memory devices, *2009 IEEE International Electron Devices Meeting (IEDM)*, pp. 822–826, 2009.

[34] Perniola, L.; Bernardini, S.; Iannaccone, G.; Masson, P.; De Salvo, B.; Ghibaudo, G.; Gerardi, C. Analytical model of the effects of a nonuniform distribution of stored charge on the electrical characteristics of discrete-trap nonvolatile memories, *Nanotechnology, IEEE Transactions on*, vol. 4, no. 3, pp. 360, 368, May 2005.

[35] Cappelletti, P.; Bez, R.; Modelli, A.; Visconti, A. What we have learned on flash memory reliability in the last ten years, IEEE International Electron Devices Meeting, 2004. IEDM Technical Digest. 2004, pp. 489–492.

[36] Arai, F.; Maruyama, T.; Shirota, R. Extended data retention process technology for highly reliable flash EEPROMs of 10^6 to 10^7 W/E cycles, *36th Annual 1998 IEEE International Reliability Physics Symposium Proceedings, 1998.* pp. 378–382, 1998.

[37] Mielke, N.; Marquart, T.; Ning, Wu; Kessenich, J.; Belgal, H.; Schares, E.; Trivedi, F.; Goodness, E.; Nevill, L.R. Bit error rate in NAND flash memories, *IEEE International Reliability Physics Symposium, 2008. IRPS 2008.* pp. 9–19, 2008.

[38] Satoh, S.; Hemink, G.; Hatakeyama, K.; Aritome, S. Stress-induced leakage current of tunnel oxide derived from flash memory read-disturb characteristics, *IEEE Transactions on Electron Devices*, vol. 45, no. 2, pp. 482–486, 1998.

[39] Satoh, S.; Hemink, G. J.; Hatakeyama, F.; Aritome, S. Stress induced leakage current of tunnel oxide derived from flash memory read-disturb characteristics, *Microelectronic Test Structures, 1995. ICMTS 1995. Proceedings of the 1995 International Conference on*, pp. 97–101, 22–25 March 1995.

[40] Kato, M.; Miyamoto, N.; Kume, H.; Satoh, A.; Adachi, T.; Ushiyama, M.; Kimura, K. Read-disturb degradation mechanism due to electron trapping in the tunnel oxide for low-voltage flash memories, *Technical Digest., International Electron Devices Meeting, 1994. IEDM '94.* pp. 45–48, 1994.

[41] Mielke, N.; Belgal, H.; Kalastirsky, I.; Kalavade, P.; Kurtz, A.; Meng, Q.; Righos, N.; Wu, J. Flash EEPROM threshold instabilities due to charge trapping during program/erase cycling, *IEEE Transactions on Device and Materials Reliability*, vol. 4, no. 3, pp. 335–344, 2004.

[42] Belgal, H. P.; Righos, N.; Kalastirsky, I.; Peterson, J. J.; Shiner, R.; Mielke, N. A new reliability model for post-cycling charge retention of flash memories, *Reliability Physics Symposium Proceedings, 2002. 40th Annual*, pp. 7–20, 2002.

[43] Mielke, N.; Belgal, H.P.; Fazio, A.; Meng, Q.; Righos, N. Recovery effects in the distributed cycling of flash memories, *IEEE International Reliability Physics Symposium Proceedings, 2006. 44th Annual*, pp. 29–35, 2006.

[44] Modelli, A.; Gilardoni, F.; Ielmini, D.; Spinelli, A. S. A new conduction mechanism for the anomalous cells in thin oxide flash EEPROMs, *Proceedings. 39th Annual. 2001 IEEE International Reliability Physics Symposium, 2001*. pp. 61–66, 2001.

[45] Compagnoni, C. M.; Miccoli, C.; Mottadelli, R.; Beltrami, S.; Ghidotti, M.; Lacaita, A. L.; Spinelli, A. S.; Visconti, A. Investigation of the threshold voltage instability after disturbuted cycling in nanoscale NAND flash memory array, *IEEE International Reliability Physics Symposium, 2010, IRPS 2010*, pp. 604–610, 2010.

[46] Ielmini, D.; Spinelli, A. S.; Lacaita, A. L.; Modelli, A. Equivalent cell approach for extraction of the SILC distribution in flash EEPROM cells, *IEEE Electron Device Letters*, vol. 23, no. 1, pp. 40–42, 2002.

[47] Ielmini, D.; Spinelli, A. S.; Lacaita, A. L.; Leone, R.; Visconti, A. Localization of SILC in flash memories after program/erase cycling, *40th Annual Reliability Physics Symposium Proceedings*, pp. 1–6, 2002.

[48] Ielmini, D.; Spinelli, A. S.; Lacaita, A. L.; van Duuren, M. J. Correlated defect generation in thin oxides and its impact on flash reliability, *Digest International Electron Devices Meeting, 2002. IEDM '02*, pp. 143–146, 2002.

[49] Ielmini, D.; Spinelli, A. S.; Lacaita, A. L.; Modelli, A. A statistical model for SILC in flash memories, *IEEE Transactions on Electron Devices*, vol. 49, no. 11, pp. 1955–1961, 2002.

[50] Ielmini, D.; Spinelli, A. S.; Lacaita, A. L.; Visconti, A. Statistical profiling of SILC spot in flash memories, *Electron Devices, IEEE Transactions* on, vol. 49, no. 10, pp. 1723–1728, 2002.

[51] Ghidini, G.; Sebastiani, A.; Brazzelli, D. Stress induced leakage current and bulk oxide trapping: Temperature evolution, *40th Annual Reliability Physics Symposium Proceedings*, pp. 415–416, 2002.

[52] Scott, R. S.; Dumin, N. A.; Hughes, T. W.; Dumin, D. J.; Moore, B. T. Properties of high-voltage stress generated traps in thin silicon oxide, *IEEE Transactions on Electron Devices*, vol. 43, no. 7, pp. 1133–1143, 1996.

[53] Kurata, H.; Otsuga, K.; Kotabe, A.; Kajiyama, S.; Osabe, T.; Sasago, Y.; Narumi, S.; Tokami, K.; Kamohara, S.; Tsuchiya, O. The impact of random telegraph signals on the scaling of multilevel flash memories, *VLSI Circuits, 2006. Digest of Technical Papers. 2006 Symposium on*, pp. 112–113.

[54] Yamada, S.; Amemiya, K.; Yamane, T.; Hazama, H.; Hashimoto, K. Non-uniform current flow through thin oxide after Fowler–Nordheim current stress, *Reliability Physics Symposium, 1996. 34th Annual Proceedings, IEEE International*, pp. 108–112, 30 Apr–2 May 1996.

[55] Kimura, M.; Koyama, H. Stress-induced low-level leakage mechanism in ultrathin silicon dioxide films caused by neutral oxide trap generation, *Reliability Physics Symposium, 1994. 32nd Annual Proceedings, IEEE International*, pp. 167–172, 11–14 April 1994.

[56] Dumin, D. J.; Maddux, J. R. Correlation of stress-induced leakage current in thin oxides with trap generation inside the oxides, *Electron Devices, IEEE Transactions on*, vol. 40, no. 5, pp. 986–993, May 1993.

[57] Rofan, R.; Hu, C. Stress-induced oxide leakage, *Electron Device Letters, IEEE*, vol. 12, no. 11, pp. 632–634, Nov. 1991.

[58] Takagi, S.; Yasuda, N.; Toriumi, A. A new I–V model for stress-induced leakage current including inelastic tunneling, *Electron Devices, IEEE Transactions on*, vol. 46, no. 2, pp. 348–354, Feb. 1999.

[59] Belgal, H.P.; Righos, N.; Kalastirsky, I.; Peterson, J.J.; Shiner, R.; Mielke, N. A new reliability model for post-cycling charge retention of flash memories, *Reliability Physics Symposium Proceedings, 2002. 40th Annual*, pp. 7–20, 2002.

[60] Wang, H.-H.; Shieh, P.-S.; Huang, C.-T.; Tokami, K.; Kuo, R.; Chen, Shin-Hsien.; Wei, Houng-Chi.; Pittikoun, S.; Aritome, S. A new read-disturb failure mechanism caused by boosting hot-carrier injection effect in MLC NAND flash memory, *Memory Workshop, 2009. IMW '09. IEEE International*, pp. 1–2, 10–14 May 2009.

[61] Takeuchi, K.; Kameda, Y.; Fujimura, S.; Otake, H.; Hosono, K.; Shiga, H.; Watanabe, Y.; Futatsuyama, T.; Shindo, Y.; Kojima, M.; Iwai, M.; Shirakawa, M.; Ichige, M.; Hatakeyama, K.; Tanaka, S.; Kamei, T.; Fu, J.Y.; Cernea, A.; Li, Y.; Higashitani, M.; Hemink, G.; Sato, S.; Oowada, K.; Shih-Chung Lee; Hayashida, N.; Wan, J.; Lutze, J.; Tsao, S.; Mofidi, M.; Sakurai, K.; Tokiwa, N.; Waki, H.; Nozawa, Y.; Kanazawa, K.; Ohshima, S. A 56 nm CMOS 99 mm² 8 Gb Multi-level NAND flash memory with 10 MB/s Program Throughput, *Solid-State Circuits Conference, 2006. ISSCC 2006. Digest of Technical Papers. IEEE International*, pp. 507–516, 6–9 Feb. 2006.

[62] Suh, K.-D.; Suh, B.-H.; Lim, Y.-H.; Kim, J.-K.; Choi, Y.-J.; Koh, Y.-N.; Lee, S.-S.; Kwon, S.-C.; Choi, B.-S.; Yum, J.-S.; Choi, J.-H.; Kim, J.-R.; Lim, H.-K. A 3.3 V 32 Mb NAND flash memory with incremental step pulse programming scheme, *Solid-State Circuits, IEEE Journal of*, vol. 30, no. 11, pp. 1149–1156, Nov. 1995.

[63] Satoh, S.; Hagiwara, H.; Tanzawa, T.; Takeuchi, K.; Shirota, R. A novel isolation-scaling technology for NAND EEPROMs with the minimized program disturbance, *Electron Devices Meeting, 1997. IEDM '97. Technical Digest, International*, pp. 291–294, 7–10 Dec. 1997.

[64] Torsi, A.; Yijie Zhao; Haitao Liu; Tanzawa, T.; Goda, A.; Kalavade, P.; Parat, K. A program disturb model and channel leakage current study for sub-20 nm NAND flash cells, *Electron Devices, IEEE Transactions on*, vol. 58, no. 1, pp. 11,16, Jan. 2011.

[65] Lee, J. D.; Lee, C. K.; Lee, M. W.; Kim, H.S.; Park, K. C.; Lee, W. S. A new programming disturbance phenomenon in NAND flash memory by source/drain Hot-electrons generated by GIDL current, *NVSMW*, pp. 31–33, 2006.

[66] Joo, S. J.; Yang, H. J.; Noh, K. H.; Lee, H. G.; Woo, W. S.; Lee, J. Y.; Lee, M. K.; Choi, W. Y.; Hwang, K. P.; Kim, H. S.; Sim, S. Y.; Kim, S. K.; Chang, H. H.; Bae,

G. H. Abnormal disturbance mechanism of Sub-100 nm NAND flash memory, *Japanese Journal of Applied Physics*, 45, pp. 6210–6215, 2006.

[67] Oh, D.; Lee, S.; Lee, C.; Song, J.; Lee, W.; Choi, J. Program disturb phenomenon by DIBL in MLC NAND Flash Device, *NVSMW*, pp. 5–7, 2008.

[68] Oh, D.; Lee, C.; Lee, S.; Kim, T. K.; Song, J.; Choi, J. New self-boosting phenomenon by source/drain depletion cut-off in NAND flash memory, *NVSMW*, pp. 39–41, 2007.

[69] Lee, C.; Hwang, J.; Fayrushin, A.; Kim, H.; Son, B.; Park, Y.; Jin, G.; Jung, E. S. Channel coupling phenomenon as scaling barrier of NAND flash memory beyond 20 nm node, *Memory Workshop (IMW), 2013 5th IEEE International*, pp. 72,75, 26–29 May 2013.

[70] Park, Y.; Lee, J. Device considerations of planar NAND flash memory for extending towards sub-20 nm regime, *Memory Workshop (IMW), 2013 5th IEEE International*, pp. 1,4, 26–29 May 2013.

[71] Park, Y.; Lee, J.; Cho, S. S.; Jin, G.; Jung, EunSeung. Scaling and reliability of NAND flash devices, *Reliability Physics Symposium, 2014 IEEE International*, pp. 2E.1.1, 2E.1.4, 1–5 June 2014.

[72] Seo, J.; Han, K.; Youn, T.; Heo, H.-E.; Jang, S.; Kim, J.; Yoo, H.; Hwang, J.; Yang, C.; Lee, H.; Kim, B.; Choi, E.; Noh, K.; Lee, B.; Lee, B.; Chang, H.; Park, S.; Ahn, K.; Lee, S.; Kim, J.; Lee, S. Highly reliable M1X MLC NAND flash memory cell with novel active air-gap and p+ poly process integration technologies, *Electron Devices Meeting (IEDM), 2013 IEEE International*, pp. 3.6.1, 3.6.4, 9–11 Dec. 2013.

[73] Aritome, S. Advanced flash memory technology and trends for file storage application, *Electron Devices Meeting, 2000. IEDM Technical Digest. International*, pp. 763–766, 2000.

[74] Shimizu, K.; Narita, K.; Watanabe, H.; Kamiya, E.; Takeuchi, Y.; Yaegashi, T.; Aritome, S.; Watanabe, T. A novel high-density $5F^2$ NAND STI cell technology suitable for 256 Mbit and 1 Gbit flash memories, *Electron Devices Meeting, 1997. IEDM '97. Technical Digest, International*, pp. 271–274, 7–10 Dec. 1997.

[75] Takeuchi, Y.; Shimizu, K.; Narita, K.; Kamiya, E.; Yaegashi, T.; Amemiya, K.; Aritome, S. A self-aligned STI process integration for low cost and highly reliable 1 Gbit flash memories, *VLSI Technology, 1998. Digest of Technical Papers. 1998 Symposium on*, pp. 102–103, 9–11 June 1998.

[76] Imamiya, K.; Sugiura, Y.; Nakamura, H.; Himeno, T.; Takeuchi, K.; Ikehashi, T.; Kanda, K.; Hosono, K.; Shirota, R.; Aritome, S.; Shimizu, K.; Hatakeyama, K.; Sakui, K. A 130 mm^2 256 Mb NAND flash with shallow trench isolation technology, *Solid-State Circuits Conference, 1999. Digest of Technical Papers. ISSCC. 1999 IEEE International*, pp. 112–113, 1999.

[77] Hwang, J.; Seo, J.; Lee, Y.; Park, S.; Leem, J.; Kim, J.; Hong, T.; Jeong, S.; Lee, K.; Heo, H.; Lee, H.; Jang, P.; Park, K.; Lee, M.; Baik, S.; Kim, J.; Kkang, H.; Jang, M.; Lee, J.; Cho, G.; Lee, J.; Lee, B.; Jang, H.; Park, S.; Kim, J.; Lee, S.; Aritome, S.; Hong, S.; Park, S. A middle-1X nm NAND flash memory cell (M1X-NAND) with highly manufacturable integration technologies, *Electron Devices Meeting (IEDM), 2011 IEEE International*, pp. 199–202, Dec. 2011.

[78] Lee, J.-D.; Hur, S.-H.; Choi, J.-D. Effects of floating-gate interference on NAND flash memory cell operation, *Electron Device Letters, IEEE*, vol. 23, no. 5, pp. 264–266, May 2002.

[79] Seo, S.; Kim, H.; Park, S.; Lee, S.; Aritome, S.; Hong, S. Novel negative V_t shift program disturb phenomena in 2X–3X nm NAND flash memory cells, *Reliability Physics Symposium (IRPS), 2011 IEEE International*, pp. 6B.2.1–6B.2.4, 10–14 April 2011.

[80] Shim, H.; Lee, S.-S.; Kim, B.; Lee, N.; Kim, D.; Kim, H.; Ahn, B.; Hwang, Y.; Lee, H.; Kim, J.; Lee, Y.; Lee, H.; Lee, J.; Chang, S.; Yang, J.; Park, S.; Aritome, S.; Lee, S.; Ahn, K.-O.; Bae, G.; Yang, Y. Highly reliable 26 nm 64 Gb MLC E2NAND (embedded-ECC & enhanced-efficiency) flash memory with MSP (Memory Signal Processing) controller, *VLSI Technology (VLSIT), 2011 Symposium on*, pp. 216–217, 14–16 June 2011.

[81] Tanaka, T.; Tanaka, Y.; Nakamura, H.; Oodaira, H.; Aritome, S.; Shirota, R.; Masuoka, F. A quick intelligent program architecture for 3 V-only NAND-EEPROMs, *VLSI Circuits, 1992. Digest of Technical Papers, 1992 Symposium on*, pp. 20–21, 4–6 June 1992.

[82] Suh, K.-D.; Suh, B.-H.; Lim, Y.-H.; Kim, J.-K.; Choi, Y.-J.; Koh, Y.-N.; Lee, Sung-Soo; Kwon, Suk-Chon.; Choi, B.-S.; Yum, J.-S.; Choi, J.-H.; Kim, J.-R.; Lim, H.-K. A 3.3 V 32 Mb NAND flash memory with incremental step pulse programming scheme, *Solid-State Circuits, IEEE Journal of*, vol. 30, no. 11, pp. 1149–1156, Nov. 1995.

[83] Chen, I. C.; Holland, S.; Hu, C. Oxide breakdown dependence on thickness and hole current—enhanced reliability of ultra thin oxides, *Electron Devices Meeting, 1986 International*, vol. 32, pp. 660–663, 1986.

[84] Schuegraf, K. F.; Hu, C. Hole injection SiO breakdown model for very low voltage lifetime extrapolation, *IEEE Trans. Electron Devices*, pp. 761–767, April 1994.

[85] Esseni, D.; Bude, J. D.; Selmi, L. On interface and oxide degradation in VLSI MOSFETs. II. Fowler–Nordheim stress regime, *Electron Devices, IEEE Transactions on*, vol. 49, no. 2, pp. 254–263, Feb. 2002.

[86] Chang, C.; Hu, C.; Robert, W. Brodersen. Quantum yield of electron impact ionization in silicon, *Journal of Applied Physics*, vol. 57, p. 302, 1985.

[87] Fischetti, M. V.; Weinberg, Z. A.; Calise, J. A. The effect of gate metal and SiO_2 thickness on the generation of donor states at the $Si–SiO_2$ interface, *Journal of Applied Physics*, vol. 57, no. 2, pp. 418–425, 1985.

[88] DiMaria, D. J.; Cartier, E. Mechanism for stress-induced leakage currents in thin silicon dioxide films, *Journal of Applied Physics*, vol. 78, pp. 3883–3894, 1995.

[89] Hemink, G.; Endoh, T.; Shirota, R. Modeling of the hole current caused by Fowler–Nordheim tunneling through thin oxides, *Japanese Journal of Applied Physics*, 33 pp. 546–549, 1994.

[90] Aritome, S. NAND flash memory reliability, in *International Solid-State Circuits Conference 2009 (ISSCC 2009)*, at Forum "SSD Memory Subsystem Innovation."

7

THREE-DIMENSIONAL NAND FLASH CELL

7.1 BACKGROUND OF THREE-DIMENSIONAL NAND CELLS

The demand of NAND flash memory [1] greatly increases with expanding applications, such as solid-state drive (SSD), because a bit cost greatly decreases by
aggressive scaling of two-dimensional (2D) NAND flash memory cell, as shown in
Fig. 3.1 [2] in Chapter 3. However, beyond the 20-nm technology node, the scaling
of 2D NAND flash memory cell is facing several serious physical limitations, such
as floating-gate capacitive (FG–FG) coupling interference, random telegraph signal
noise (RTN), and so on, as described in Chapter 5.

In order to further scale down the memory cell size of NAND flash memory,
several three-dimensional (3D) NAND flash cells had been proposed before 2006
[3–9], as shown in Fig. 7.1 [2]. One is the stacked NAND cell [3–5], as shown in
Fig. 7.2. The NAND strings are fabricated on each stacked silicon layers in vertical
and are connected to common bit line and source line. The channel Si-layers and gates
have to be fabricated for each duplicated layer. Then a fabrication cost is increased
due to increasing process steps. The other is the stacked-surrounding gate transistor
cell (S-SGT cell) [7–9], as shown in Fig. 7.3. The surrounding gate transistor cells
with a floating gate are vertically connected in series to fabricate the NAND cell
strings. However, a fabrication process was very complicate, and also the cell size
was very large due to the step structure of silicon substrate for each cell, as shown
in Fig. 7.3. Therefore, before 2006, these 3D cells could not effectively decrease a

Nand Flash Memory Technologies, First Edition. Seiichi Aritome.
© 2016 The Institute of Electrical and Electronics Engineers, Inc. Published 2016 by John Wiley & Sons, Inc.

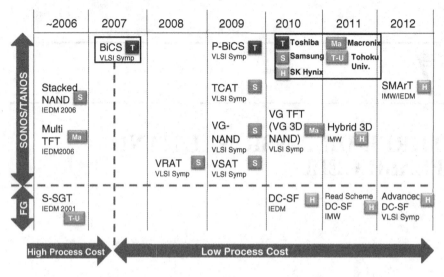

FIGURE 7.1   History of three-dimensional NAND flash.

- **Stacked Si-layer for NAND strings**
- **TANOS charge trap cell**
- **Increased process cost due to increasing number of process steps for each duplicated layers.**

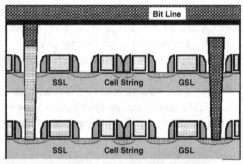

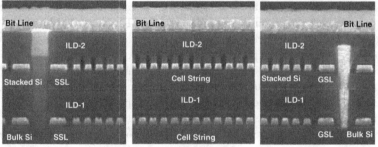

FIGURE 7.2   Stacked NAND cell.

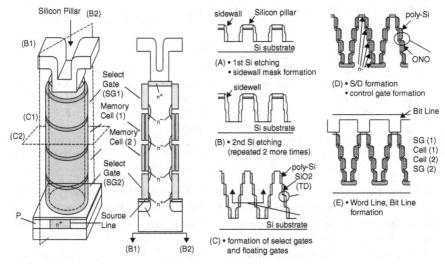

• Structure Concept of 3D NAND
• Complex fabrication process⟶Cost increased
• Increased Cell size due to step structure⟶Cost increased

FIGURE 7.3   Stacked-surrounding gate transistor cell (S-SGT cell).

bit cost because of a complicate fabrication process, an increased process steps, and large unit cell size.

In 2007, the BiCS cell (bit cost scalable cell) technology had been proposed [10], as shown in Fig. 7.1. The BiCS cell has a new structure of the stacked control gate layers and vertical poly-Si channel, as described in detail in Section 7.2. The new process concept is shown in Fig. 7.4 [11, 12]. The multi-stacked layers of gate (plate) and dielectric are deposited ("Stack"), and then the through-hole is fabricated through

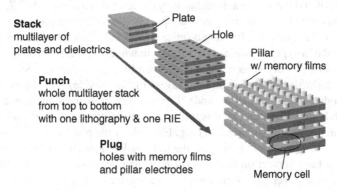

FIGURE 7.4   Basic concept of BiCS (bit cost scalable) technology.

TABLE 7.1   Comparison of major 3D NAND cells.

	BICS/ P-BiCS	TCAT (V-NAND)	SMArT	VG-NAND	DC-SF
Channel	Vertical	Vertical	Vertical	Horizontal	Vertical
Gate structure	Surrounding (GAA)	Surrounding (GAA)	Surrounding (GAA)	Dual	Surrounding (GAA)
Charge storage	SONOS	SONOS (TANOS)	SONOS (TANOS)	SONOS (TANOS)	FG
Gate process	Gate first	Gate last	Gate last	Gate last	Gate last

the multi-stacked layers ("Punch"). After that, the through-holes are filled by memory film (ONO) and channel poly-Si (pillar electrodes) ("Plug"). Due to this new process concept, the fabrication process became very simple and low cost, and also it was expected that very small effective cell size could be achieved.

After introducing BiCS cell in 2007 [10], several 3D NAND cells were proposed, such as an advanced BiCS [13, 14], P-BiCS [15–17], VRAT [18], TCAT [19], VG-NAND [20], VSAT [21], VG-TFT [22,23], DC-SF [24–28], SMArT [29], and so on, as shown in Fig. 7.1.

Table 7.1 shows the comparison of major 3D cells. 3D cells are categorized by channel structure (vertical or horizontal), gate structure (surrounding gate (GAA) or dual gate), charge storage (SONOS/TANOS or FG), and gate process (gate first or gate last). Each 3D cell has both advantage and disadvantage of structure, process, operation, and so on. In Chapter 7, major 3D cells are introduced and discussed.

7.2   BiCS (BIT COST SCALABLE TECHNOLOGY) / P-BiCS (PIPE-SHAPE BiCS)

7.2.1   Concept of BiCS

Concept of BiCS (bit cost scalable technology) is described in Figs. 7.5–7.7. All of the stacked electrode plates (control gates) are punched through and plugged with poly-silicon channel at one time, forming a series of vertical FETs which act as a NAND string of SONOS-type memories [10, 13], as shown in Fig. 7.5. The single memory cell has a vertical poly-silicon surrounding by both the ONO dielectrics (silicon dioxide/SiN of charge storage layer/silicon dioxide) and the surrounding gate electrode (GAA; gate all around). The memory cells work in depletion mode with the body poly-silicon, which is undoped or lightly n-doped uniformly without source/drain n-type diffusion within plug. Each electrode plate acts as a control gate except the lowest plate, functioning as the lower select gate (lower SG). A single bit is located in the intersection of a control gate plate and plugged poly-silicon. The control gates and lower SG are commonly connected in each layer in block, as shown in Fig. 7.7. The string is selected by a bit line and an upper select gate (upper SG), as shown in Fig. 7.7. As shown in Fig. 7.6, the control gates and upper/lower SG

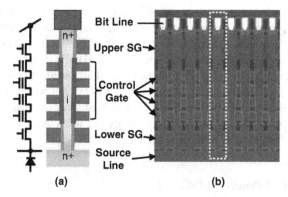

FIGURE 7.5 BiCS (bit cost scalable) flash memory. (a) The memory string. (b) Cross-sectional SEM image of BiCS flash memory array.

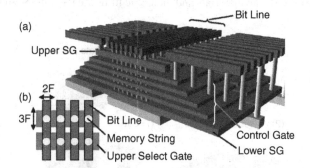

FIGURE 7.6 (a) Bird's-eye view of BiCS flash memory. (b) Top-down view of BiCS flash memory array.

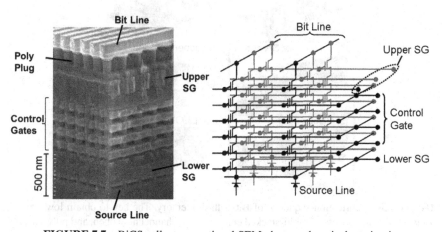

FIGURE 7.7 BiCS cell cross-sectional SEM photo and equivalent circuit.

are connected to the metal layers at stair-like gate structure, which is fabricated by resist slimming process, described in Section 7.2.2. The bottom of memory string is connected to the common source diffusion formed on the silicon substrate.

The surrounding gate transistor (SGT), which is used in BiCS, had been previously proposed for logic CMOS in IEDM 1988 because of an excellent current drivability and body effect [30]. The vertical channel SGT EPROM cell was proposed in 1993 [31–33]. Also, the stacked-SGT cell for NAND flash was proposed in 2001 [7]. Therefore, the special feature of BiCS technology is a low-cost process to fabricate many stacked cells all at one process sequence.

7.2.2 Fabrication Process of BiCS

Figure 7.8 shows the fabrication sequence of BiCS flash memory cell [10]. Lower select gate transistors ((2) and (3) in Fig. 7.8), series-connected memory cells ((4)–(7) in Fig. 7.8), and upper select gate transistors ((8) and (9) in Fig. 7.8) are fabricated sequentially. Stacked gate materials and dielectric films are P+ poly-Si and Si dioxide

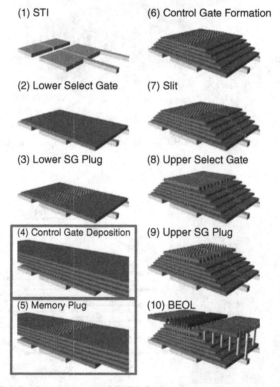

FIGURE 7.8   Fabrication sequence of BiCS flash memory. The key to obtain low process cost is a one-time process of multi-stacked control gates, channel hole open, and poly-channel plug.

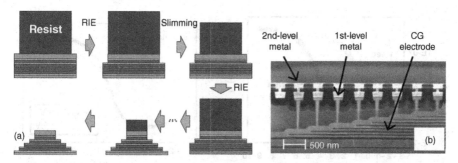

FIGURE 7.9 (a) Fabrication sequence of edge of control gates (control gate pick-up area) into stair-like structure. (b) Cross-sectional SEM image of edge of control gates.

(SiO_2), respectively. Holes for transistor channel or memory plug are punched through by RIE. And then LPCVD TEOS film or ONO films are deposited for lower/upper SG and for memory cells, respectively. The bottom of dielectric films are removed by RIE and plugged by amorphous Si to connect Si substrate. Arsenic is implanted and activated for drain and also source of the upper device. ONO films are deposited in the opposite order as compared to the conventional SONOS device—that is, from LPCVD TEOS film as top block oxide (5 nm), LPCVD SiN film (11 nm), and LPCVD TEOS film as tunnel oxide (2.5 nm).

Edges of control gate are processed into stair-like structure by repeating of RIE and resist sliming, as shown in Fig. 7.9a. Figure 7.9b shows the cross-sectional SEM image of control gates stair-like structure and contact area in the edge of control gates. The control gates and SGs can be connected to metal layer by the control gates stair-like structure and the contacts.

For minimizing program disturb and read disturb, all stacked control gates and lower select lines have to be separated by a slit which separates a block of memory plugs from each other, as shown in (7) of Fig. 7.8. Only upper select gate is cut into line pattern to work as row address selector, as shown in (8) of Fig. 7.8 and Fig. 7.7. Via contact hole and BL metal layer are fabricated on the array and peripheral circuit simultaneously, as shown in (10) of Fig. 7.8. ·

7.2.3 Electrical Characteristics

Figure 7.10 shows the I_d–V_g characteristics of (a) the SONOS memory cell [10] and (b) the select gate transistor [13] of BiCS technology. In the SONOS memory cell, a good I_{on}/I_{off} ratio of more than 6 orders of magnitude is obtained in both erase and program states, as shown in Fig. 7.10a. For the select transistor, the gate dielectric is 7 nm SiN which is deposited by LPCVD. A good subthreshold slope of ~190 m V/dec is obtained. The program/erase characteristics are shown in Fig. 7.11a [13]. Low-voltage program and erase operations are confirmed with surrounding gate transistors. The channel hole diameter dependence of the program and erase (P/E) window is measured on the test structure of cylindrical capacitors in Fig. 7.11b [15].

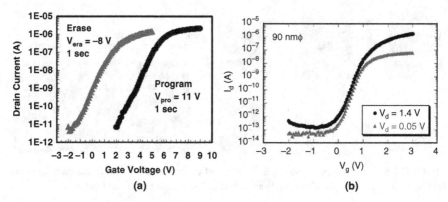

FIGURE 7.10 I_d–V_g characteristics of (a) the SONOS memory cell and (b) the select gate vertical transistor with SiN-gate in BiCS technology.

The P/E window can be enlarged by the smaller channel hole diameter, because the electric field strength of the tunnel film is enhanced by the curvature effect.

The field enhancement effect of channel hole curvature plays a significant role in programming/erasing, since the BiCS cell is a surrounding gate device. When channel radius (R1) is decreased and comparable to the ONO thickness, the field enhancement effect become very large, as shown in Fig. 7.12 [22, 34]. The field in tunnel oxide (bottom oxide) is enhanced as the radius of the channel is decreased. Meanwhile, the field in the blocking oxide (top oxide) of ONO is simultaneously decreased.

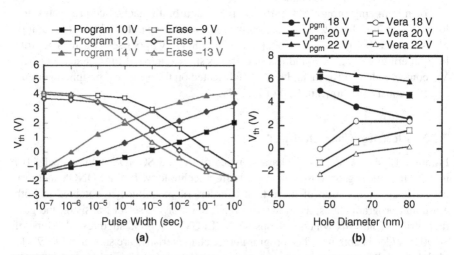

FIGURE 7.11   (a) Program/erase (P/E) characteristics of vertical SONS memory in BiCS technology. (b) Hole-diameter dependence of P/E characteristics in BiCS technology. The P/E window is enhanced by a smaller hole.

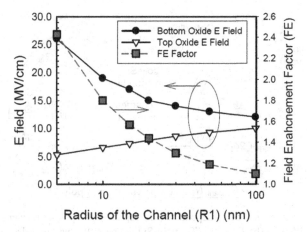

FIGURE 7.12 Calculated bottom oxide and top oxide electrical field for the surrounding gate (gate all around; GAA) nano wire SONOS device. ONO = 5/8/7 nm. The radius R1 is half of the poly diameter. The applied voltage is +18 V in this calculation. Field enhancement factor is shown in the right axis. The FE factor is defined as the ratio of the bottom oxide field (with radius $R1$) over the capacitor (with $R1 \rightarrow \infty$).

Conventional erase operation in two-dimensional NAND flash cell with applying erase voltage to substrate (p-well) cannot be used in BiCS cell because erase p-well voltage cannot be directly transferred to channel plug poly-silicon. Then, erase operation is executed by raising the potential of channel poly-Si pillars of NAND string with injection of holes which are generated by GIDL (gate-induced drain leakage) at the junction side edge of the select gate, as shown in Fig. 7.13 [13, 14].

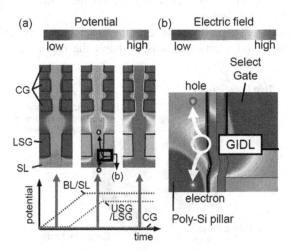

FIGURE 7.13 BiCS erase operation.

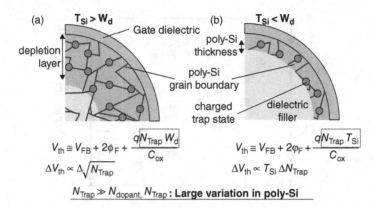

$$V_{th} \cong V_{FB} + 2\phi_F + \frac{qN_{Trap}W_d}{C_{ox}}$$

$$\Delta V_{th} \propto \Delta\sqrt{N_{Trap}}$$

$$V_{th} \cong V_{FB} + 2\varphi_F + \frac{qN_{Trap}T_{Si}}{C_{ox}}$$

$$\Delta V_{th} \propto T_{Si}\Delta N_{Trap}$$

$$N_{Trap} \gg N_{dopant}, N_{Trap} : \textbf{Large variation in poly-Si}$$

FIGURE 7.14   Schematics of V_{th} dependence on trap density at poly-silicon grain boundaries when (a) $T_{Si} > W_d$, and (b) $T_{Si} < W_d$ (T_{Si}, poly-Si thickness; W_d, width of depletion layer). If poly-Si thickness (T_{Si}) is thinner than depletion layer width (W_d), ΔV_{th} becomes dependent on the total number of traps, and thus ΔV_{th} becomes smaller with thinner body thickness.

There are many traps in grain boundary of channel poly silicon in BiCS cell, as described detail in Section 8.6. Then, subthreshold characteristics of poly-silicon channel transistors have quite a large variation, and it is difficult to control to ensure a tight distribution. Based on the model described in Fig. 7.14 [13], the approach to achieve better controllability of the V_t distribution is making the body silicon much thinner than the depletion width (W_d), in order to reduce the volume of poly-silicon and total number of traps and to make threshold voltage less sensitive to the trap density fluctuation. The concept of the "macaroni" body vertical transistor is illustrated in Fig. 7.15 [13]. Very thin poly-silicon is deposited on the gate dielectric to form a macaroni-shaped body. The center of the body is filled with dielectric film to make process integration easier. Thinner body thickness makes channel potential better controlled by gate electrode.

Figure 7.16a shows I_d–V_g characteristics of the "macaroni" body vertical transistor [13]. The macaroni body vertical transistor shows the much better subthreshold

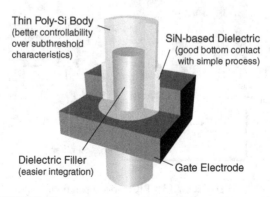

FIGURE 7.15   Concept of 'macaroni' body vertical transistor.

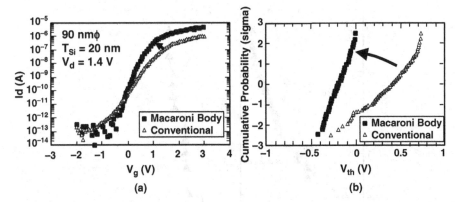

FIGURE 7.16   (a) Typical I_d–V_g characteristics of conventional vertical transistor and "macaroni" body vertical transistor. (b) V_{th} distribution of conventional and macaroni body vertical transistor.

characteristics as well as the better drive current as compared with a conventional channel transistor. V_{th} variation can be well reduced by the macaroni body vertical transistor, as shown in Fig. 7.16b [13].

The macaroni channel transistor was compared with the full channel transistor in detail [35]. Figure 7.17 compares the statistical distributions of threshold voltage V_{th} (Fig. 7.17a) and subthreshold swing (STS) (Fig. 7.17b) of full channel and macaroni channel devices (Si channel thickness d_{Si} of 7, 10, and 13 nm). From both plots, it is clarified that the distributions of both V_{th} and STS are tighter in the case of macaroni.

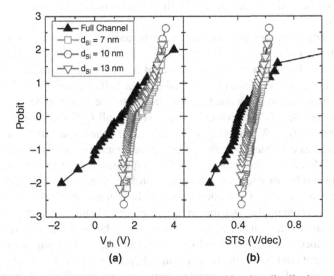

FIGURE 7.17   (a) Threshold voltage and (b) subthreshold swing distributions of full channel and macaroni devices with different channel thicknesses; the memory hole diameter is $\varphi = 80$ nm.

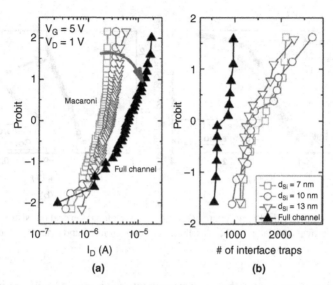

FIGURE 7.18 (a) Statistical distributions of drain current at read condition and (b) distribution of number of interface traps for macaroni (different d_{Si}) and full channel devices.

This is because the macaroni channel can achieve a better control of the conduction by reducing the volume of the poly-Si channel. The thinner poly-Si channel can obtain the easier electrostatic control performed by the gate electrode over the thinner body and the smaller poly-Si grains which reduce the impact on the channel current of statistical variation of smaller grain size configuration [36]. Furthermore, no relevant difference is observed between devices with different channel thickness in this range of thickness.

Figure 7.18a compares the statistical distributions of the drain current (I_D) for macaroni channel devices with different channel poly-Si thicknesses and for the full channel device. The drain current is measured by biasing the gate at 5 V and the drain voltage V_D at 1 V. It can be observed that the full channel device have higher average I_D than macaroni channel device; however, the full channel device has a larger variation. Especially in lower I_D tail, a very small I_D device can be observed, contrast to no tail I_D in Macaroni channel device. And the drain current in macaroni channel increases consistently with increasing the channel poly-Si thickness.

Figure 7.18b reports the number of interface traps measured by the charge pumping method. The number of interface defects is twofold higher in macaroni channel than in full channel devices. And it is independent on channel poly-Si thickness. It suggests that the macaroni channel device has an interface trap at both interfaces of gate oxide side and the filler side, thus resulting in the much higher number of defects. These results, combined with the I_D trend of Fig. 7.18a, are a clear indication that the conduction is not confined to the tunnel oxide/channel interface, but involves the full channel thickness of macaroni channel. Interface traps at the filler side interface are also responsible for the higher V_{th} values observed in Fig. 7.17a.

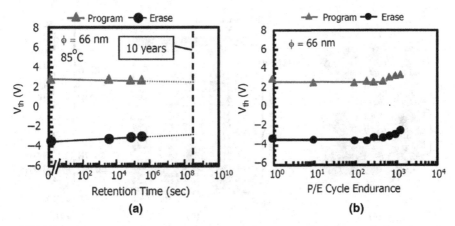

FIGURE 7.19   (a) Data retention characteristics of P-BiCS cells. It has no obvious degradation for 10-year-long data retention. (b) Program/erase endurance characteristics.

Figure 7.19 shows (a) data retention characteristics and (b) program/erase cycling endurance characteristics of the BiCS cell [17]. They show good enough characteristics to implement a NAND flash product. Figure 7.20 shows V_{th} distribution of three programmed levels of storage data for MLC [17]. Tight V_t distribution widths of MLC cells were obtained.

7.2.4   Pipe-Shaped BiCS

BiCS technology evolved to the pipe-shape BiCS (P-BiCS). Figure 7.21 shows the schematic of pipe-shaped BiCS flash memory [15]–[17]. Two adjacent NAND strings are connected at the bottoms by pipe-connection (PC) which is gated by the bottom electrode (pipe gate). One of the terminals for the U-shaped pipe is connected to the bit line (BL), and the other is connected to the source line (SL). The SL consists

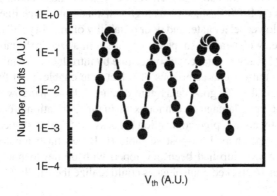

FIGURE 7.20 V_{th} distribution of a fabricated test chip of a BiCS cell.

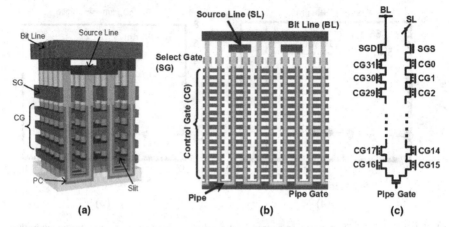

FIGURE 7.21   (a) Schematic of pipe-shaped-BiCS (P-BiCS) flash memory. (b) Cross-sectional view of the P-BiCS. (c) An equivalent circuit of P-BiCS.

of the meshed wiring of the third level metal (not shown in Fig) and accessed by the first and the second level metal like a conventional planar (2D) NAND flash cell technology. Therefore the resistance of the SL is sufficiently low. The both of the SG transistors are placed on the stacked control gates (CG). The control gate (CG) is isolated by the slits and faces to each other as a couple of combs pattern.

The fabrication process of the P-BiCS NAND strings is described in Fig. 7.22 [15]. The pipe connection (PC) is filled by a sacrificial-film and connected to a sacrificial film in the stacked control gates after the memory hole formation. And the sacrificial films are removed after SG-hole formation. Then, the gate dielectric ONO (memory films) and the channel poly-silicon film (silicon-body) can be sequentially deposited, because the gate dielectric etching process is not required before channel poly silicon deposition, compared that the gate dielectric etching is required before channel poly silicon deposition in conventional straight-shaped BiCS process to connect channel poly-Si with substrate. Therefore, P-BiCS flash could improve some critical problems on BiCS flash, such as high resistance of source line, cut-off characteristics of the lower select gate, and poor reliability of memory cells. Three advantages are compared with straight-shaped BiCS, as summarized in the table in Fig. 7.23. P-BiCS realizes (1) Low resistance of source line by introducing metal wiring and (2) tightly controlled cut-off characteristics of select gate enable good functionality of memory array. And also, (3) good data retention and wide V_{th} window can be realized because of less process damage on tunnel oxide in the fabrication process. P-BiCS is easily fabricated by adding pipe connection process to straight- shaped BiCS.

By using the pipe-shaped bit-cost scalable (P-BiCS) flash memory cell, a 16-Gb flash memory test chip had been developed with 60 nm technology [16]. The three-dimensional 16 stacked control gates could realize the small effective 1-bit cell size of 0.00082 μm^2.

In the original BiCS flash memory, the control gates are shared by several neighboring NAND cell strings to minimize cell size. In P-BiCS flash memory cell, the

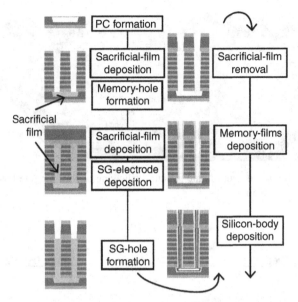

FIGURE 7.22 Fabrication process of P-BiCS flash memory. PC is formed on the first-level gate-conductor, which is also used for the support devices.

branched control gate configuration was adopted. Control gates are shared and connected by fork-shaped plates with four branches, as shown in Fig. 7.24. Each branch controls cells of two pages. The block in P-BiCS is formed by vertically stacked 16 pairs of control gates. Each pair of control gate plates is arranged in a staggered layout. (See CG0 and CG31 in Fig. 7.24 as an example.) Thanks to the high boost efficiency cell, program disturbances are not a serious concern in BiCS [14] and P-BiCS.

Unlike control gates, select gates are individually separated for the selectivity of cell strings. Figure 7.25 shows a schematic of row decoders. Two row decoders are placed on both sides of the cell array, one side for CG0 to CG15 and the other side

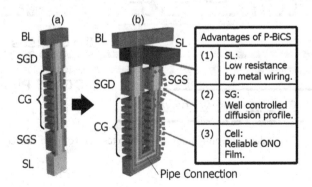

FIGURE 7.23 Advantages of pipe-shaped-BiCS.

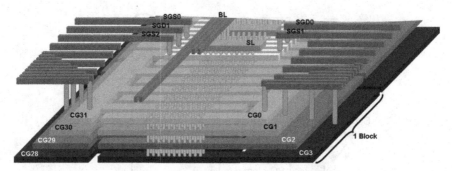

FIGURE 7.24 Branched control gate structure (showing upper four gate layers) of 16-Gb pipe-shaped BiCS test chip.

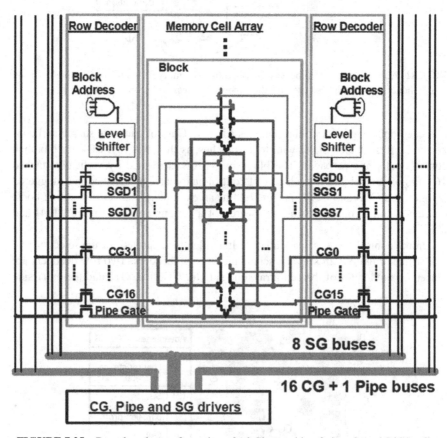

FIGURE 7.25 Row decoder configuration of 16-Gb test chip of pipe-shaped-BiCS cell.

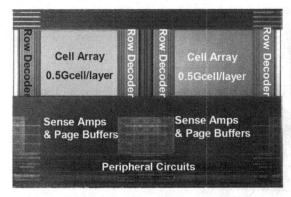

Density	1Gcell/layer
Cell Size	0.00082 μm²
Number of Layers	16 layers
Block Size	2M Byte(SLC) 4M Byte(MLC)
Page Size	8K Byte
Die Size	10.11x15.52 mm²

FIGURE 7.26   Image of 16-Gb pipe-shaped BiCS test chip.

for CG16 to CG31. Row decoders only decode the block address. The selectivity of eight select gates in a block is ensured by eight buses and drivers of select gates. Since the pipe connection (PC) forms a transistor, the pipe gate driver is needed.

Figure 7.26 shows a micrograph of the 16-Gb test chip. The number of cells in one string is 32 in 16 stacked control gate layers. Every layer has 1G memory cells. The chip contains 1K blocks with a page size of 8K byte. A commercial 64-Gb P-BiCS MLC NAND flash memory die size is estimated to be 10.5 mm × 12.3 mm by using the same configurations.

7.3   TCAT (TERABIT CELL ARRAY TRANSISTOR)/V-NAND (VERTICAL-NAND)

7.3.1   Structure and Fabrication Process of TCAT

The schematic structure of TCAT (terabit cell array transistor) is shown in Fig. 7.27 [19]. TCAT has a similar structure of BiCS, with a vertical poly-silicon channel, stacked word lines, and a silicon nitride (SiN) charge storage layer of the surrounding gate SONOS cell.

Figure 7.28 shows a process sequence of a TCAT flash memory cell. The points of process and structural differences with BiCS flash are (i) oxide(SiO_2)/nitride(SiN) multilayer stack for both memory cells and select gates, (ii) line-type "word line (W/L) cut" etched through the whole stack between the each row array of channel poly plug (see Fig. 7.28d), (iii) line-type CSL formed by an implant through the "W/L cut," (iv) replaced metal gate lines from SiN to tungsten (W), and (v) select transistors of GSL/SSL fabricated simultaneously with memory cell process. The most unique process is "gate replacement" to achieve the metal gate SONOS structure and low-resistance word line. Figure 7.29 shows the detail process of gate replacement from SiN to tungsten W [19]. After "W/L cut" dry etch and wet removal of sacrificial nitride layer (see Fig. 7.29b), gate dielectric layers (including tunnel oxide, charge

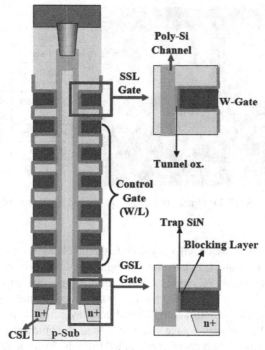

FIGURE 7.27   Schematic structures of TCAT flash cell string. Details of selection transistors are shown.

trap SiN and blocking dielectric) and gate metal are deposited in the conventional order (Fig. 7.29c). This is the gate last process, not "gate first" process as for BiCS flash [13]. The conventional gate last process is one of the advantages of a TCAT flash cell. And then, separation of each gate node is followed by etch processes (Fig. 7.29d).

The TEM cross-sectional view of the unit cell in Fig. 7.30a shows a damascened tungsten W metal gate SONOS structure in the vertical NAND flash cell string [19]. The electric field induced between the adjacent cells programmed in different state accelerates the charge losses by charge spreading mechanism at the hot temperature storage (HTS) test [37]. However, as shown in Fig. 7.30b,c [38], the TCAT structure of the gate last process has a biconcave structure which contributes to prevent from lateral charge losses.

An equivalent circuit of TCAT cell and cross-sectional SEM image are shown in Fig. 7.31 [19, 39, 40]. This configuration has 24 stacked WL layers, two dummy WL (DWL) of Dummy 0/1, and two string select gate layers of SSL and GSL. The string is selected by a bit line and an upper select gate (SSL). The control gates (WLs) and SSL/GSL are connected to the metal layers at stair-like gate structure (Fig. 7.28f). The bottom of memory string is connected to the common source line (CSL) diffusion formed by a CSL implant on the silicon substrate.

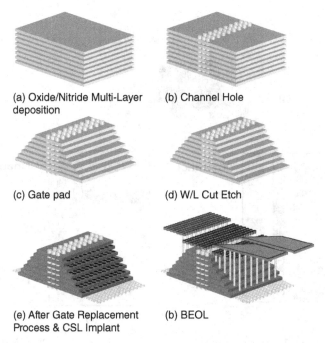

(a) Oxide/Nitride Multi-Layer deposition

(b) Channel Hole

(c) Gate pad

(d) W/L Cut Etch

(e) After Gate Replacement Process & CSL Implant

(b) BEOL

FIGURE 7.28 Process sequence of TCAT flash memory.

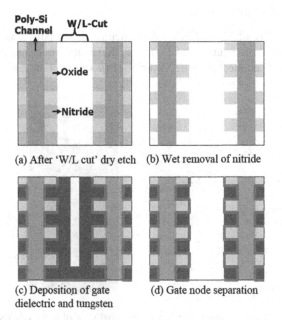

(a) After 'W/L cut' dry etch

(b) Wet removal of nitride

(c) Deposition of gate dielectric and tungsten

(d) Gate node separation

FIGURE 7.29 Concept of the process flow with "gate replacement."

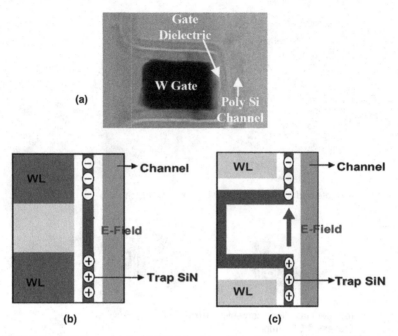

FIGURE 7.30 (a) Cross-sectional SEM images of a cell in the vertical NAND string of TCAT flash memory. Trap layer structure of (b) BiCS and (c) TCAT.

7.3.2   Electrical Characteristics

An important feature of TCAT NAND flash memory is the bulk erase operation. As shown in the schematic structure of Fig. 7.27, the channel poly plug in TCAT flash structure is directly connected to the Si substrate (p-sub), not the $n+$ common source

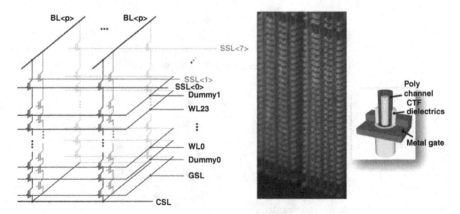

FIGURE 7.31 Schematic diagram and cross-sectional view of a 3D V-NAND array (TCAT flash memory array).

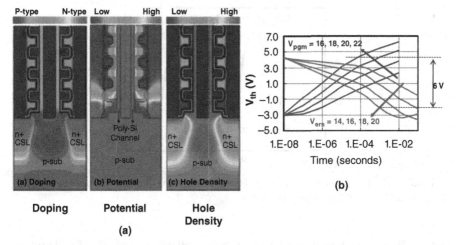

FIGURE 7.32   (a) Simulated profiles of doping, potential, and hole density during a bulk-erase operation of TCAT flash memory cells. (b) Program and erase characteristics.

diffusion layer as in the BiCS flash cell. Therefore, conventional bulk erase operation can be performed as described in the simulated profiles of Fig. 7.32a. Thanks to the bulk erase operation, the major peripheral circuits do not require to be changed from conventional 2D NAND flash memory to implement a TCAT NAND flash product [39, 40]. Figure 7.32b shows the program and erase (P/E) characteristics of the memory cell transistor [38]. A wide V_{th} window of ~6 V is obtained.

Figure 7.33 shows (a) program/erase (P/E) cycling endurance characteristics and (b) data retention characteristics. The V_t shift by 1K and 10K P/E cycling is kept less than 0.5 V and 1.5 V, respectively, which are small enough for mass production. And also, data retention characteristics show the wide V_t window even after 10 years of

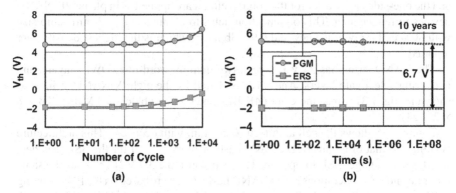

FIGURE 7.33   (a) Program/erase cycling endurance characteristics of TCAT cell. (b) Long-term data retention characteristics of a TCAT cell.

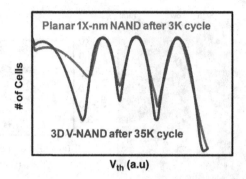

FIGURE 7.34   (a) Comparison of measured V_{th} distributions of planar 1X-nm NAND after 3K cycles and 3D V-NAND after 35K cycles.

lifetime. From these results, the performance and reliability of the TCAT cell proved to be appropriate for multilevel cell (MLC) operation.

Figure 7.34 shows a 2-bit/cell (MLC) V_{th} distribution of the 3D V-NAND (same as TCAT) after 35K program-erase cycles compared with that of a planar (2D) 1X-nm NAND after 3K cycles [39, 40]. The 3D V-NAND has a very excellent MLC V_{th} distribution, compared to that of the 1X-nm NAND, even with the tenfold greater number of program/erase cycles.

7.3.3   128-Gb MLC V-NAND Flash Memory

Based on the TCAT cell [19, 38, 41], the first three-dimensional NAND flash product with the V-NAND cell was implemented [39, 40]. Figure 7.35 shows the die micrograph of the 128-Gb MLC (2-bit/cell) 3D V-NAND flash memory device as the first generation of the V-NAND cell array technology. This memory chip has 24 stacked WL layers and consists of two planes each containing 64-Gb arrays (2732 of 3-MB blocks with 8-KB pages). The shared-WL-block scheme for the row circuit and the one-side page buffer for the column circuit are applied as in planar 2D NAND devices to reduce area. This organization helps to obtain small 133-mm^2 chip size with 80% cell array efficiency and 0.96 Gb/mm^2, which is the highest density ever reported.

The NAND string has 24 WL layers, two dummy word-line (DWL) layers, and two string select gate layers. And a BL is shared by the eight V-NAND strings, as shown in Fig. 7.31. Since there are 64K BLs, the size of a block is 3 MB (8 KB × 8 × 2 × 24).

Figure 7.36 shows the cell architecture diagram with decoders. This architecture is similar to that of a planar 2D NAND flash memory, except for the SSL decoders which select a target SSL to operate. This is because the 3D V-NAND has the same program and erase operations as 2D NAND flash, which are based on the FN tunneling mechanism. In particular, the bulk erase feature can provide better characteristics such as higher erase speed, lower power, and more reliability, compared with other 3D NAND cells based on the hole generation by the GIDL mechanism. So, conventional

Bits per cell	2
Density	128 Gb
Technology	Three-dimensional vertical NAND, 3-metal layer
Organization	8KB x 384 pages x 5464 blocks x 8 strings
Program performance	50 MB/s for embedded application 36 MB/s for enterprise SSD application
Data interface speed	667 Mbps@Mono, 533 Mbps@8-stack
Power Supply	V_{cc} = 3.3 V / V_{ccq} = 1.8 V

FIGURE 7.35   Die micrograph and main feature of 128-Gb 2-bit/cell 3D V-NAND flash. Die size is 133 mm^2.

operations used in a planar 2D NAND can also be applied to the 3D V-NAND with simple modifications.

This chip has over 2.5 billion channel holes. This configuration can be seen in Fig. 7.31, which shows a simplified schematic diagram, an SEM image of the fabricated V-NAND array, and a cross-sectional diagram of a unit cell. In order to ensure high performance, a damascened metal-gate structure was adopted. And, as a basic cell structure, the surrounding gate (gate all-around structure) charge trap flash (CTF) cell (SONOS-type cell) is used with advanced barrier engineering material in a damascened metal-gate structure.

The chip accomplishes 50-MB/s write throughput with 3K endurance for typical embedded applications. Also, extended endurance of 35K is achieved with 36 MB/s of write throughput for data center and enterprise SSD applications.

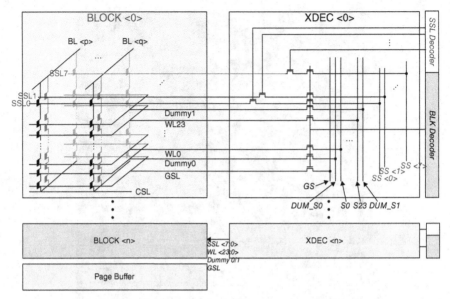

FIGURE 7.36 Block and X-decoder schematic diagram of 3D V-NAND flash memory.

7.3.4 128-Gb TLC V-NAND Flash Memory

The second generation three-dimensional V-NAND flash product with the 32 stacked cells was presented in ISSCC 2015 [42]. Figure 7.37 shows (a) the die micrograph of the 128-Gb TLC (3-bit/cell) 3D V-NAND flash memory device and (b) bit density. The die size is surprisingly reduced from the previous 133 mm^2 of 128-Gb MLC with 24 stacked WLs [39, 40] to 68.9 mm^2 of 128-Gb TLC with 32 stacked WLs [42]. This die size is smaller than 15-nm 64-Gb MLC 2D NAND flash memory of 75 mm^2, which was presented in the same conference [43], even higher density of 128 Gb.

In order to reduce die size, a new bit-line architecture was developed. In the new bit-line architecture, two bit lines are placed for one channel hole, as shown in Fig. 7.38b. Thanks to the new bit-line architecture, a two-cell plane could be changed to a one-cell plane. And also, a one-side page buffer, a shared block decoding scheme, and an MIM capacitor are adopted. Bit density is increased to 93% and exceeds over magnetic HDD (hard disk drive), as shown in Fig. 7.37b. A plane contains 2732 main blocks and additional 64 spare blocks. Each block has 6 MB in size with 384 16-KB pages. A device summary is shown in Fig. 7.39.

A new high-speed program (HSP) algorithm for TLC V-NAND was also developed. In a 2D FG cell, a three-step reprogram scheme (see Section 4.3 and 4.4) has been widely used for TLC programming to decrease the effect of floating-gate capacitive coupling interference (see Section 5.3). However, in a 3D V-NAND cell, floating-gate capacitive coupling interference is negligible due to the charge trap cell. Therefore, page programming algorithm could be simplified to the single program

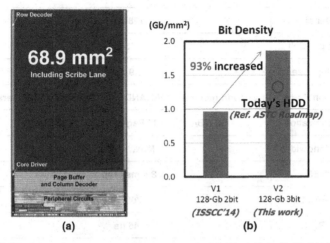

FIGURE 7.37   Die micrograph and bit density of 128-Gb 3-bit/cell 3D V-NAND flash. Small die size of 68.9 mm^2 can be achieved by using a 32-cell stacked V-NAND process, TLC (3 bit/cell), and 1-plane architecture. Bit density is increased 93% from former 128-Gb 2-bit/cell 3D V-NAND flash.

step of HSP (high speed program), as shown in Fig. 7.40. A page buffer receives three pages of data before 3b/cell programming operation, and it completes programming in a single sequence.

By using HSP, it is possible to improve programming performance and reduce programming power consumption, as shown in Fig. 7.41. The page program time of tPROG can be 200% faster than a typical planar 1X-nm 2D TLC NAND flash with the conventional reprogram algorithm, as shown in Fig. 7.41a. And also, power consumption during programming can be reduced 40%, as shown in Fig. 7.41b.

Even in fast page programming speed, memory cell V_t distribution is good enough to guarantee the reliability of product, as shown in Fig. 7.42. The V_t distribution is shown for initial and 5K P/E cycle cases. P/E cycling is performed at the ambient temperature of 55°C. After 5K P/E cycles, the cell V_t distribution is still good. Fast

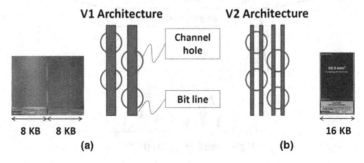

FIGURE 7.38   Bit-line architecture. Two bit lines in a channel hole pitch.

Density	128 Gb
Bit per Cell	3
Chip Size	68.9 mm²
Technology	Second generation V-NAND with 32 stacked WL layers
Organization	16 KB/Page, 384 Pages/Block, 2732 Blocks
I/O Bandwidth	Max. 1 GB/s
tBERS	3.5 ms (Typ.)
tPROG	700 µs
tR_4K	45 µs

FIGURE 7.39   Device summary of a 128-Gb 3-bit/cell 3D V-NAND flash.

tPROG of 700 µs and good endurance of over 5K P/E cycles make it suitable for both client and data-center SSD applications.

7.4   SMArT (STACKED MEMORY ARRAY TRANSISTOR)

7.4.1   Structural Advantage of SMArT

The SMArT (stacked memory array transistor) cell was presented as shown in Fig. 7.43 [29]. The SMArT cell has a similar structure and process with BiCS and TCAT cells. The fabrication process of charge storage ONO and metal gate is different from BiCS and TCAT. Figure 7.44 shows the structure comparison of BiCS, TCAT,

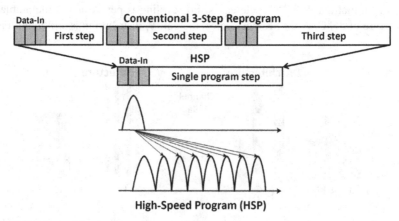

FIGURE 7.40   High-speed program (HSP) scheme. The single program step of HSP can be used in V-NAND cell because cell-to-cell interference is small in a charge trap cell.

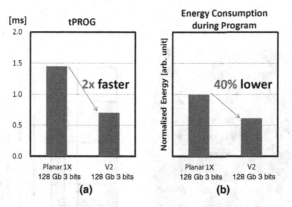

FIGURE 7.41 The program time and energy consumption during programming. Program time could be improved 200% (×2), and energy consumption can be reduced 40%.

and SMArT [44]. BiCS cell uses a poly-Si as word lines (WL), which is fabricated by SiO2/poly-Si multi-stacked layers, and does not use the WL replacement process by tungsten (W) metal as a low-resistance word line. Therefore, WL resistance of BiCS is higher than tungsten metal WL of TCAT and SMArT cell. However, the stacked height of SiO2/poly-Si multi-stacked layers can be lower than that of the W replacement process due to no ONO deposition in W replacement process. In the TCAT cell, WL resistance is much lower than that of the poly-Si WL of BiCS cell. However, the stacked height becomes high due to ONO deposition in the W replacement process. The SMArT cell can achieve both low stack height and low WL resistance, because of no ONO deposition in the W replacement process, as shown in Fig. 7.44. The ONO films are deposited before the W replacement process.

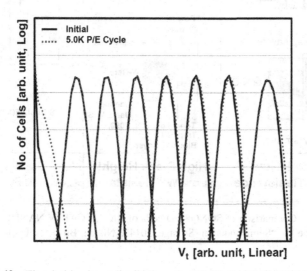

FIGURE 7.42 Threshold voltage distribution of a 128-Gb 3-bit/cell 3D V-NAND flash.

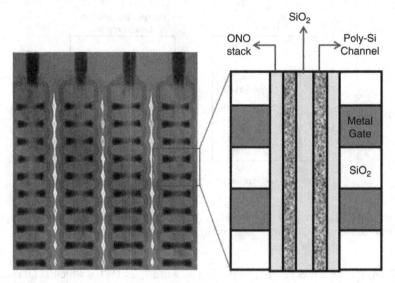

FIGURE 7.43   TEM cross-sectional image of SMArT (stacked memory array transistor) cell string and schematic drawing of a SMArT unit cell.

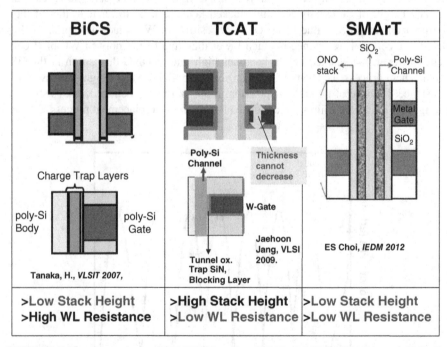

FIGURE 7.44   Comparison of SONOS 3D cells of BiCS, TCAT (V-NAND), and SMArT. First appearance in "Semiconductor Storage 2014", Nikkei Business Publications, Inc., 2013/07/31.

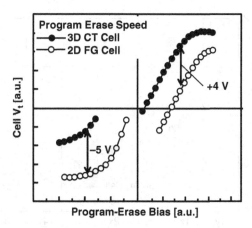

FIGURE 7.45 Comparison of program-erase characteristics of 2D 2y-nm node FG cell and 3D CTD SMArT cell.

Low stack height is very important for 3D NAND flash, because an aspect ratio of the multi-stack etching process can be reduced. Low aspect ratio and low stack height are the key challenge of 3D NAND flash fabrication, and it will be serious in the case of increasing number of stacked layers, as described in Section 8.7. The stack height is minimized by using the SMArT cell process (inserting ONO layer in the plug) with low resistive tungsten (W). Low WL resistance is fabricated by the gate replacement process. Then the SMArT cell has the "gate-last" process, providing better reliability.

7.4.2 Electrical Characteristics

Program and erase (P/E) speed of 3D charge trap (CT) SMArT cell is compared with 2D FG cell in Fig. 7.45 [29]. The program speed of 3D CT cell is much faster than 2D FG cell, but the erase speed is much slower in spite of the field enhancement effect of the surrounding-gate structure (gate all-around structure). Despite the smaller program/erase window, the NAND cell operation window is quite sufficient to MLC and TLC, because the program saturation V_t of a 3D CT cell is much larger than that of a 2D floating-gate cell, as shown in Fig. 7.45.

Figure 7.46 shows the cell-to-cell interference (floating-gate capacitive coupling interference) of a 2D floating-gate cell and a 3D SMArT cell [29]. As expected in the charge-trap-type (SONOS-type) cell of a 3D SMArT cell, the cell-to-cell interference is negligibly small in comparison with a 2D floating-gate cell. This means that the major scaling limiter of the cell-to-cell interference in 2D cell, described in Chapter 5, is not the problem in 3D charge trap (CT) cell.

The V_t distributions before and after program/erase cycling endurance of a 3D SMArT cell are shown in Fig. 7.47, compared with a two-dimensional (2D) 2y-nm generation floating-gate cell [29]. In a 2D 2y-nm generation cell, the V_t distribution width becomes wider after 3K or 5K program/erase cycling. However, in (a) 3D SMArT cell, the cell V_t distribution does not become wider up to 5K cycles. This

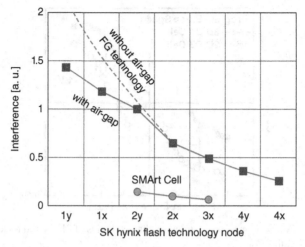

FIGURE 7.46   Comparison of total interference of a 2D FG cell and a 3D SMArT cell.

difference is known to be originated from the thin tunneling oxide of the charge trap cell where the interface traps are less generated than the floating-gate cell with thicker tunnel oxide.

7.5   VG-NAND (VERTICAL GATE NAND CELL)

7.5.1   Structure and Fabrication Process of VG-NAND

The vertical gate NAND (VG-NAND) flash array with horizontal multi-active layers had been proposed [20]. Figure 7.48 shows the structure and schematics of the VG-NAND array.

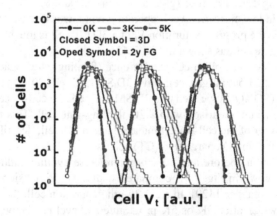

FIGURE 7.47   Comparison of V_t widening during program/erase cycling.

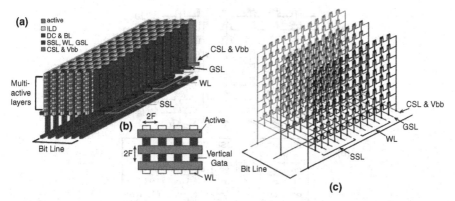

FIGURE 7.48   (a) Structure of vertical gate (VG)-NAND. (b) Top-down view of a VG-NAND showing $4*F^2$ cell size/layer. (c) Schematics of VG-NAND array with source and body tied to CSL (common source line). Each BL contains multi-actives, common vertical gate, and vertical plugs between multi-actives.

The VG-NAND flash has horizontal multilayer active strings and vertical gates (VG) for SSL, WL, and GSL. A charge trap layer is located between the active layer and the vertical gate, forming dual gate structure. Active layers are connected to a bit line (BL) and a source line (CSL) at the end of NAND strung. In the proposed array [20], word line (WL) and a BL are formed at the beginning of fabrication before the cell array, making an interconnect between WL, BL, and decoder easier. However, WL and BL can also be formed after making a memory cell array as a conventional 2D NAND flash memory cell. The source and active body (V_{bb}) are electrically connected to CSL (common source line) for enabling body erase operation. A positive bias is applied to CSL during erase. Array schematic of each layer is identical to the planar 2D NAND flash, except for SSL, as shown in Fig. 7.48c. VG-NAND requires 6 SSLs for 8 active layers and 8 SSLs for 16 active layers. Required number of SSL is expressed by

$$(\text{Number of SSL}) = 2 * \log_2(\text{Number of active layers})$$

A reason for the multi-SSLs is to select data from a chosen layer out of multilayers since VG-NAND cell uses common BL and common WL between multi-active layers. Figure 7.49 represents SSL schematics of VG-NAND with eight active layers and its operation table for a specific layer selection during read and program. A transistor with shade always turns on regardless of applied voltage on SSL while a transistor without shade only turns on under applied proper voltage.

Figure 7.50 describes a process sequence of VG-NAND cell. A process sequence is based on simple patterning and plugging. BL with $n+$ poly-Si is fabricated first, and then $n+$ poly-Si WL is formed on top of it (Fig. 7.50(1)). Multi-active layers with p-type poly-Si are formed with n-type ion implants for SSL layer selection, and alternated interlayer dielectrics are inserted between active layers. Then patterning is

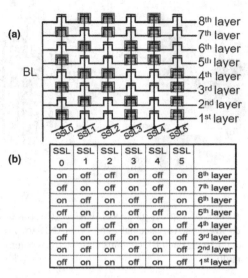

(a)

(b)

SSL 0	SSL 1	SSL 2	SSL 3	SSL 4	SSL 5	
on	off	off	on	off	on	8th layer
off	on	off	on	off	on	7th layer
on	off	on	off	off	on	6th layer
off	on	on	off	off	on	5th layer
on	off	off	on	on	off	4th layer
off	on	off	on	on	off	3rd layer
on	off	on	off	on	off	2nd layer
off	on	on	off	on	off	1st layer

FIGURE 7.49 (a) Schematics of SSL including depletion and enhancement transistors (b) operation table of SSL for a layer selection during read and program.

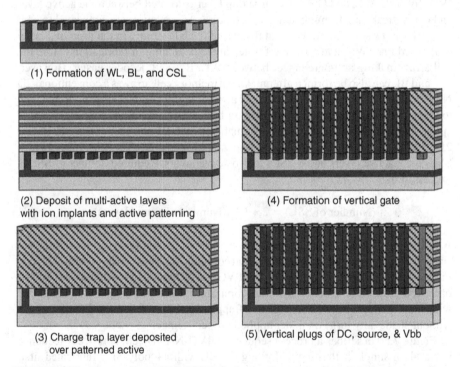

(1) Formation of WL, BL, and CSL

(2) Deposit of multi-active layers with ion implants and active patterning

(3) Charge trap layer deposited over patterned active

(4) Formation of vertical gate

(5) Vertical plugs of DC, source, & Vbb

FIGURE 7.50 Process flow of VG-NAND flash array Integration based on simple patterning and plugging.

carried on the multi-active layers (Fig. 7.50 (2)), and charge trap layers (ONO) are deposited over the patterned actives (Fig. 7.50 (3)). Consecutively, VG is formed by high aspect-ratio and high selective RIE, and connected to WL (Fig. 7.50 (4)). And as a final step, vertical plugs of DC and source-V_{bb} are connected to BL and CSL after contact ion implants (Fig. 7.50 (5)). N+ doped source and p-type active are electrically tied to CSL.

There are several challenges in VG-NAND for considering mass production. One is a VG patterning of high aspect and high selective RIE. As shown in Fig. 7.50 (4), a VG patterning has to remove WL poly silicon at the bottom of space between WLs and between active layers. This patterning has a very high aspect ratio (>30 @ 16 layers) etching with high selectivity for VG pattern material (SiO_2/SiN or resist).

The other challenge is the increased cost by the SSL formation. As described above, a large occupied area of SSL and many process steps of SSL formation are needed to fabricate the VG-NAND cell. The SSL occupied area is increased in accordance with an increasing number of stacked active layers. The effective cell size including the SSL area is getting larger due to large SSL occupied area. And, for each active layer, ion implantation process and mask lithography steps are needed to make an n-type channel. These process steps are also increased in accordance with an increasing number of stacked active layers.

In order to minimize the SSL area, several new schemes were proposed, such as the surrounding-gate transistor SSL scheme [45], the LSM (layer selection by multilevel scheme) with multi-state SSL V_t and multi-bias SSL [46], the island-gate SSL scheme [47–51], and so on.

As one example, Fig. 7.51 shows the island-gate SSL scheme of a VG NAND. Each channel layer is separately decoded by one island-gate SSL device. In one unit (contains $2*N$ channel layers, where N is number of stacked memory layers), all channel layers are grouped together for each memory layer and are connected to the metal-3 global BL through the "staircase" BL contacts formed at the BL pad region. All island-gate SSLs are connected by the interconnection of CONT/ML1/VIA1/ML2 toward SSL decoder. A common source line (CSL) is used to share the source of all memory layers. Figure 7.51b illustrates the detail layout in the case of $N = 4$. All $2*N$ (= 8) channel BLs are grouped into one unit, sharing the same BL pad. In the BL pad, "staircase" contacts are fabricated, where each contact corresponds to one memory layer, as shown in the inset. The staircase CONTs are then connected by ML3 BLs toward the page buffer for memory sensing. Each channel BL has own island-gate SSL for the selection/decoding, where SSL devices are all connected by ML2 lines (parallel to WL's) toward SSL decoder.

7.5.2 Electrical Characteristics

I_d–V_g characteristics of a VG-NAND cell is shown in Fig. 7.52 [20]. The conventional program and body-erase schemes have been performed. The V_t window of about 3.7 V is obtained. In the VG-NAND cell structure, a double gate is located on both sides of the active layer. Then gate controllability to channel is worse than the surrounding

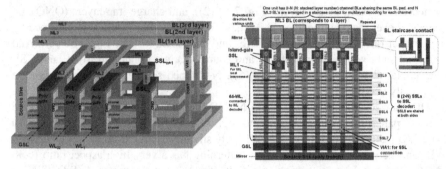

FIGURE 7.51 (a) Schematic diagram of a 3D VG NAND. WLs of the vertical layers are shared together. Every channel bit line is separately decoded by one island-gate SSL device. In one unit (containing 2∗N channel BLs), all channel BLs are grouped together for each layer and are connected to the metal 3 BL through the staircase BL contacts formed at the BL pad region. All island-gate SSL devices are connected by the interconnection of CONT/ML1/VIA1/ML2 routings toward the SSL decoder. A common source (CSL) is used to share the source planes of all memory layers. (b) Layout schematic for N (stack number) = 4. All 2∗N (= 8) channel BLs are grouped into one unit, sharing the same BL pad. In the BL pad, a staircase contact is fabricated, where each contact corresponds to one memory layer, as shown in the inset. The staircase CONT is connected by ML3 BLs toward page buffer for memory sensing. Each channel BL has its own island-gate SSL for the selection. The SSLs are connected through CONT/ML1/VIA1/ML2 toward the SSL decoder. A common source is fabricated to connect source lines of all memory layers.

gate cell such as BiCS, TCAT, and SMArT cells. The subthreshold slope would not be enhanced in case the of scaling cells due to using a thinner channel poly silicon layer as scaling.

Figure 7.53a shows the program/erase cycling endurance characteristics of a VG-NAND cell [20]. The window narrowing can be observed up to 1K cycles of

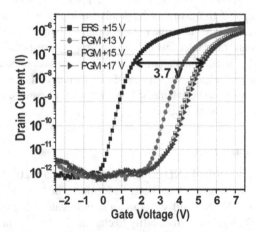

FIGURE 7.52 Program and body-erase window of a VG-NAND cell.

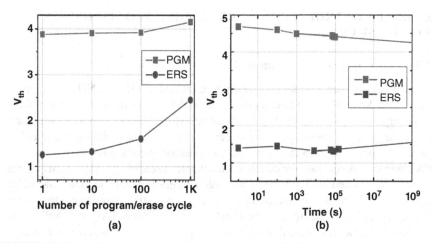

FIGURE 7.53   (a) Program and erase cycling endurance. (b) Data retention characteristics of a VG-NAND cell.

program/erase. Data retention characteristics at room temperature show the 10-year data retention capability in Fig. 7.53b.

In VG-NAND, a new interference mode had been reported [22, 47]. It comes from the channel potential interference in the proximity in the Z-direction (vertical direction; thickness of the buried oxide between stacked active channel layers), namely "Z-interference." Figure 7.54 shows that the interference in cell C may exceed 450 mV after programming the vertical adjacent cell (cell A) when the buried oxide thickness is scaled below 20 nm. Therefore, the Z-interference limits the Z-direction scaling. The thickness of buried oxide should be greater than 30 nm to avoid serious Z-interference.

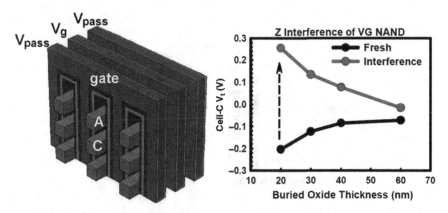

FIGURE 7.54   "Z interference" of VG-NAND when buried oxide thickness (F_Z) is scaled. In this calculation, $F = 60$ nm and $V_{pass} = 7$ V are used. Cell A is programmed with $e^- = 2E19$ cm^{-3} while cell C is the measured Z interference. When buried oxide thickness is only 20 nm, the Z interference may exceed 450 mV.

7.6 DUAL CONTROL GATE—SURROUNDING FLOATING GATE CELL (DC-SF CELL)

7.6.1 Concern for Charge Trap 3D Cell

There has been tremendous attention on three-dimensional (3D) NAND flash memory [10, 13–23]. Most 3D NAND cells use a SONOS/TANOS device structure with a charge trap nitride as a storage layer. However, it is well known that these structures with a charge trap nitride layer have suffered from inherent problems, such as low erase speed, poor retention characteristics, and charge spreading issues [37] between cells along the charge trap nitride layer. The charge spreading problem in 3D charge trap cell is illustrated in Fig. 7.55a. The stored charges in Si nitride move toward the neighbor cells through a connected nitride layer, because the charge trap nitride layer is physically connected from top to bottom CGs in 3D SONOS/TANOS NAND flash. As a result, this would cause degradation of data-retention characteristics and poor V_t distribution of cell state in the 3D SONOS cell. As these problems are related to a charge trap nitride, the floating-gate-type 3D NAND flash is required to be used instead of charge trap nitride. However, applying a conventional two-dimensional (2D) floating-gate structure without schematic change is not suitable for 3D NAND flash because the lateral space occupation of the floating gate is large, resulting in larger cell size.

Here, a dual control gate with a surrounding floating-gate (DC-SF) cell for 3D NAND flash memory had been proposed [24–26]. This structure allowed us to apply a floating gate to a 3D stacked cell structure with minimal cell size and high coupling ratio. The DC-SF cell and 3D SONOS cell [10, 13–17] are compared in a vertical schematic as shown in Fig. 7.55. The surrounding FG in the DC-SF cell is completely isolated by IPD and tunnel oxide as shown in Fig. 7.55b. This implies that significant improvement of data retention is expected for the DC-SF cell due to the absence of a physical leakage path.

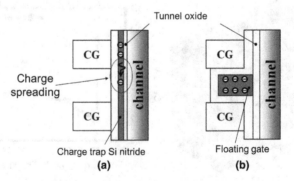

FIGURE 7.55 Comparison of 3D NAND flash cell structures. (a) SONOS cell (BiCS, etc.). (b) DC-SF cell. In the case of a SONOS cell, charge spreading problem is caused by a connected charge trap Si nitride layer.

7.6.2   DC-SF NAND Flash Cells

A. Concept.   Figure 7.56 shows the cross-sectional schematic of the DC-SF NAND flash cell. A surrounding FG is located in between the two control gates (CGs), which is a new approach for 3D NAND flash memory. The detailed schematic of the cell structure is illustrated in Fig. 7.56b. The surrounding FG is covered by inter-poly dielectric (IPD) and tunnel oxide. Therefore, two CGs are vertically capacitive coupled with FG. The tunnel oxide is only formed in between the channel poly and FG, while IPD is added on the side-wall of the CG, resulting in thicker dielectric layer formed between the channel poly and CG. This means that during program and erase, the charges can tunnel only through the tunnel oxide between the channel poly and FG without any tunneling between the channel poly and CGs.

There are several significant advantages of the DC-SF structure. The first advantage is that the floating gate, which is a proven and predictable technology in 2D NAND flash memory, can be used as a charge storage node so that many issues related to charge trap (SONOS and TANOS) cell can be eliminated. The second is the significant improvement in the coupling ratio because a new concept of functionality is implemented to 3D structure—one surrounding FG is controlled by two neighboring control gates. As a result, enlargement of the surface area between FG and two CGs can be achieved. Therefore, it can be attributed to low bias cell operation for program and erase. The third is to shrink the unit cell size in the horizontal direction, because FG is not positioned between the CG and channel poly in a horizontal direction, but is positioned in between two CGs in a vertical direction, implying that it is suitable for

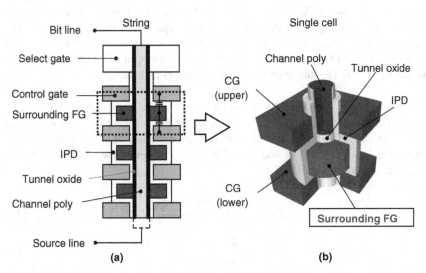

FIGURE 7.56   (a) Cross-sectional schematic of the DC-SF (dual control gate–surrounding floating gate) NAND flash cell. Two control gate (CGs) are vertically capacitive coupled with a floating gate (FG). (b) Bird's-eye view of the DC-SF NAND flash cell. The surrounding FG is capacitive, coupled with both upper and lower CGs.

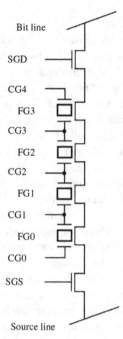

FIGURE 7.57   Equivalent circuit of the DC-SF NAND cell string.

fabrication of 3D NAND flash structure. The fourth is very small FG–FG coupling interference, because a CG positioned in two FGs plays a role of the electrical shield. Therefore, there is negligible capacitive coupled capacitance in between FGs. As a result, the DC-SF cell allows wide program/erase (P/E) window and low bias cell operation for 3D NAND flash device. The equivalent circuit of DC-SF cell string is given in Fig. 7.57. The single cell consists of one FG and two CGs.

B. Coupling Ratio.   Figure 7.58 shows the top and cross-sectional view of the DC-SF cell. The capacitive coupled capacitance of the FG is estimated by two different formulas in the DC-SF structure. In the vertical direction of the FG, the coupled capacitance between the FG and two CGs (C_{IPD}) is determined by parallel plate capacitance, as shown in Eq. (7.1).

$$C_{\text{IPD}} = \frac{2\varepsilon_r\varepsilon_0\pi(a_3^2 - a_2^2)}{d} \tag{7.1}$$

$$C_{\text{Tox}} = \frac{2\pi\varepsilon_r\varepsilon_0 h}{\ln(a_2/a_1)} \tag{7.2}$$

On the other hand, the capacitive-coupled capacitance of the FG with tunnel oxide (C_{Tox}) in the horizontal direction is extracted by coaxial cable capacitance as shown in Eq. (7.2), because the FG is covered on the cylindrical channel poly. The calculation

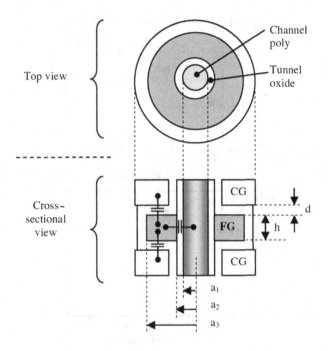

a_1: radius of channel poly
a_2: radius of channel poly + tunnel oxide
a_3: radius of channel poly + tunnel oxide + FG
d: IPD thickness
h: FG poly-Si height

FIGURE 7.58   Floating-gate capacitance of the DC-SF cell.

of coupling ratio is plotted as a function of the FG width, FG height, and the radius of channel poly in case of $T_{ox} = 8$ nm and $T_{IPD} = 12$ nm thick, as shown in Fig. 7.59. The high coupling ratio of 0.68 can be obtained, for example, in the structure of $a_1 = 20$ nm, $a_2 = 28$ nm, $a_3 = 68$ nm, $h = 40$ nm, and $d = 12$ nm; that is, FG width = $(a_3 - a_2) = 40$ nm, FG height = $h = 40$ nm, and radius of channel = $a_1 = 20$ nm.

The coupling ratio decreases with decreasing FG width $(a_3 - a_2)$, as shown in Fig. 7.59a, because the capacitor area of FG and CG decreases. On the contrary, the coupling ratio increases as the FG height (Fig. 7.59b) and radius of the channel (Fig. 7.59c) poly decrease, implying that the cell size in the horizontal direction can be reduced by increasing the coupling ratio in this structure. This result indicates that a high coupling ratio of about 0.7 can be maintained even though the cell size decreases, because the coupled capacitance of FG is compensated in both directions. Though the conventional planar 2D NAND flash memory suffers from low coupling ratio as the cell size decreases, the DC-SF cell structure has an advantage of maintaining a high coupling ratio even at the smaller cell size. With this structure, the coupling ratio of about 0.7 can be achieved.

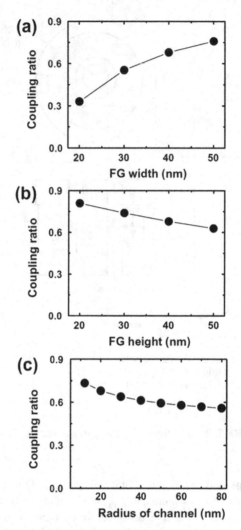

FIGURE 7.59 Coupling ratio of the DC-SF cell in case of T_{ox} = 8 nm and T_{IPD} = 12 nm. (a) Coupling ratio of the DC-SF cell as a function of FG width ($a_3 - a_2$) (FG height: 40 nm and radius of channel poly: 20 nm). (b) Coupling ratio of the DC-SF cell as a function of FG height (h) (FG width, 40 nm; radius of channel poly, 20 nm). (c) Coupling ratio of DC-SF cell as a function of radius of channel poly (a_1) (FG height, 40 nm; FG width, 40 nm).

C. Device Fabrication.   The process sequence of the DC-SF cell is shown in Fig. 7.60. First, in situ thermal CVD SiO_2 and poly Si are deposited in sequence to make multi-stacked layers. Thus, holes are formed by etch process through the entire SiO_2/poly-Si stacked layers (Fig. 7.60a). To make a space for IPD and FG, oxide recess is carried out in the horizontal direction (Fig. 7.60b). IPD deposition is followed (Fig. 7.60c), and the space is filled with FG poly-Si deposition overall

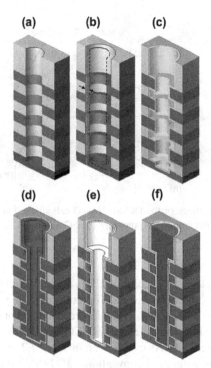

FIGURE 7.60 Process sequence of a DC-SF NAND flash cell. (a) Oxide/poly deposition and hole formation by etch process. (b) Oxide recess is carried out to make a space for IPD and FG. (c) IPD deposition. (d) FG poly-Si deposition. (e) Isotropic etch of FG and tunnel-oxide deposition. (f) Channel poly deposition.

inside the holes (Fig. 7.60d). To define complete FGs in the hole, the isotropic etch process of FG is performed. FGs of poly-Si are separated into each recess position. Tunnel oxide is deposited (Fig. 7.60e), and then the first channel poly-Si is deposited to cover tunnel oxide, and then the first channel poly-Si and tunnel oxide at bottom of holes are removed by RIE. And then the second channel poly-Si is deposited to fill hole, as shown in Fig. 7.60f. The second channel poly-Si is electrically connected to substrate. The cross-sectional TEM image of the DC-SF cell arrays is shown in Fig. 7.61a. It can be clearly seen that the surrounding FGs and CGs are well fabricated along the channel, as shown in Fig. 7.61b.

7.6.3 Results and Discussions

The operation conditions of the DC-SF cell are listed in Table 7.2. For the erase operation, an erase bias of -11 V is applied to all the CGs. Cell V_t values decrease to negative. In the program and read condition, FG2 between CG2 and CG3 is selected. The program bias (V_{pgm}: 15 V) is applied to both CG2 and CG3 simultaneously. Two different V_{pass} biases are used to prevent the program disturb. In the neighboring word

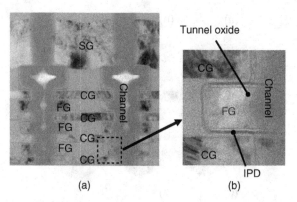

FIGURE 7.61 (a) TEM image of the DC-SF NAND cell string. FGs and CGs are stacked. (b) Detail TEM image of the single cell, showing the FG and two CGs with tunnel oxide and IPD on the channel.

line of CG1 and CG4, a lower V_{pass} bias of 2 V is applied. And the normal V_{pass} bias of 4 V is applied to the other CGs during programming. The potentials of FG1 and FG3 are compensated by a lower bias of 2 V in order to prevent program disturb. For the read operation, zero bias is applied to both CG2 and CG3 with a read pass bias (V_{read}) of 4 V to the other CGs. As can be seen in Table 7.2, all the operation biases are significantly lower than those of conventional 3D NAND flash memory based on charge trap nitride. This is due to the well-designed cell structure of the DC-SF and high coupling ratio.

The I_d–V_g characteristics of the DC-SF cell is plotted with different erase times in Fig. 7.62. The figure shows that cell V_t decreases as the erase time increases, implying that the erase cell operates effectively. As a result, a wide program/erase window of about 9.2 V is obtained.

Figure 7.63a shows the program characteristics. It shows that the DC-SF cell is well programmed even at a low program bias of 15 V. The coupling ratio of this cell

TABLE 7.2 Operation conditions for the DC-SF NAND cells^a

Bias	Erase (V)	Program (V)	Read (V)
BL	0	0/Vcc	1
SGD	4.5	4.5	4.5
CG4	V_{erase}:−11	V_{pass2}:2	V_{read}:4
CG3	V_{erase}:−11	V_{pgm}:15	0
CG2	V_{erase}:−11	V_{pgm}:15	0
CG1	V_{erase}:−11	V_{pass2}:2	V_{read}:4
CG0	V_{erase}:−11	V_{pass1}:4	V_{read}:4
SGS	4.5	0	4.5
SL	0	Vcc	0

^{a}FG2 between CG2 and CG3 is selected in program and read.

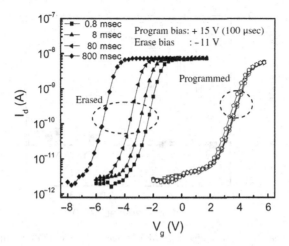

FIGURE 7.62   The I_d–V_g characteristics of the DC-SF NAND flash cell. V_g is VCG2 (= VCG3) bias. V_{read} for unselected control gate are VCG0 = VCG1 = VCG4 = 4 V. And bit-line voltage is V_d = 1 V.

is estimated to 0.71. Figure 7.63b shows the erase characteristics. The DC-SF cell can be erased well at low erase bias of -11 V for 1 ms. These operation voltages are significantly lower than those of conventional 3D SONOS NAND flash memory structure and planar 2D NAND flash cell. This implies that cell operation in the DC-SF structure is considerably effective because of a high coupling ratio.

The FG–FG coupling interference (floating-gate capacitive coupling interference) between a programmed cell and an adjacent cell was studied, as shown in Fig. 7.64. The ΔV_t of the adjacent cell has been measured as a function of programmed cell V_t from 2.0 to 3.6 V. Very small capacitive coupling interference of 12 mV/V is observed due to CG electric shield effect between FGs. And the FG–FG couplings

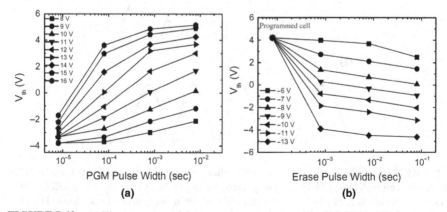

FIGURE 7.63   (a) The program and (b) erase characteristics of the DC-SF NAND flash cells.

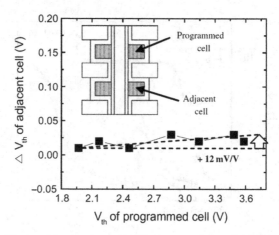

FIGURE 7.64   The interference characteristics between FG and FG (V_{th} difference of the adjacent FG versus V_{th} of the programmed cell). Very small FG–FG coupling interference value of 12 mV/V is obtained.

of x-direction (along WL) and y-direction (along BL) for one side are estimated to 1.1% and 0.7% of total FG capacitance, respectively, based on scaled cell size as shown in Fig. 7.67c. Then total FG–FG coupling of x- and y-directions is 3.6%. It is much smaller value in comparison with the conventional 2D FG cell. With this small FG–FG interference result, it is expected that V_t distribution setting could be acceptable for multilevel cell (MLC) or triple-level cell (TLC) as shown in Fig. 7.65.

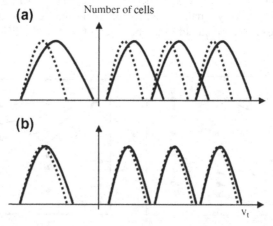

FIGURE 7.65   Cell threshold voltage (V_t) setting comparison between: (a) conventional planar FG cell with large FG–FG coupling interference and (b) DC-SF cell with small FG–FG coupling interference. The DC-SF cell has a wider V_t setting margin due to negligible FG–FG coupling.

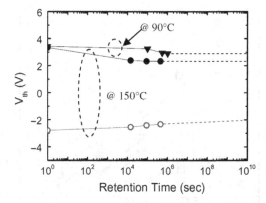

FIGURE 7.66   The data-retention characteristics of the DC-SF NAND flash cells.

Figure 7.66 shows the data-retention characteristics at two different temperatures (90°C and 150°C). For the high-temperature condition, the program and erase charge losses are 0.9 V and 0.2 V for 126 h, respectively.

7.6.4   Scaling Capability

In order to evaluate the scaling capability of the DC-SF cell, the effective cell size of the DC-SF structure and conventional 3D SONOS had been compared, as shown in Fig. 7.67. The assumption of physical unit cell size is that x- and y-pitch of BiCS [10, 13–17]/TCAT [19] are 100 and 160 nm, respectively. On the other hand, the x- and y-pitch of the DC-SF cell are 130 and 190 nm. The physical cell size of the DC-SF cell is larger than that of BiCS/TCAT, because space margin between FG and slit edge (gate edge) is needed. The feature size here is assumed to be 40 nm for both DC-SF and BiCS/TCAT. And, in this design rule, the high coupling ratio of 0.60 can still be obtained in the structure of $a_1 = 15$ nm, $a_2 = 23$ nm, $a_3 = 50$ nm, $h = 30$ nm, and $d = 12$ nm; that is, FG width = $(a_3 - a_2) = 27$ nm, FG height = $h = 30$ nm, and radius of channel = $a_1 = 15$ nm. Even if the physical cell size of the DC-SF cell is larger than that of conventional BiCS/TCAT, the effective cell size of the DC-SF can be comparable with BiCS/TCAT, because multilevel cells (2 bits/cell, 3 bits/cell, and 4 bits/cell) are available due to wide cell V_t window and negligible FG–FG coupling interference.

7.7   ADVANCED DC-SF CELL

7.7.1   Improvement on DC-SF Cell

Floating-gate (FG)-type DC-SF 3D NAND flash memory cell was proposed [24, 26], as shown in Fig. 7.56. (Section 7.6) to overcome the intrinsic disadvantages of the charge-trap-type 3D cell. However, in the DC-SF cell process [24, 26], there were

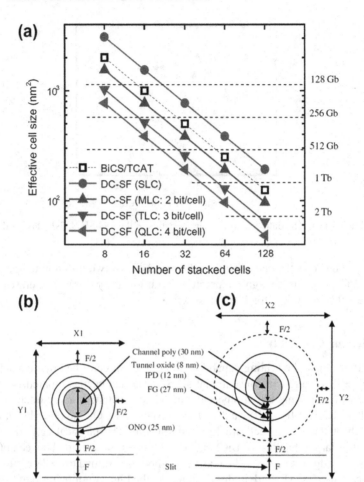

FIGURE 7.67 (a) Effective cell sizes for the DC-SF NAND flash cell. The DC-SF cell can be realized for 1 Tb with 3 bits/cell + 64 cells stacked, 2 Tb with 3 bits/cell + 128 cells stacked. (b) Assumption of cell size for BiCS ($F = 40$ nm, cell-to-cell distance: $F/2$, slit distance: F, $X1 * Y1 = 100 * 160$ nm^2). (c) Assumption of cell size for DC-SF ($F = 40$ nm, cell-to-cell distance: $F/2$, slit distance: F, $X2 * Y2 = 130 * 190$ nm^2).

still several critical problems, as shown in Fig. 7.68b, namely (1) high word-line resistance of the poly gate, (2) damage on IPD ONO by the FG separation process, and (3) field confinement at the FG edge during programming due to the horn shape FG. And also, read and program operations of the DC-SF cell had not been optimized yet, resulting in causing several disturb problems.

In Section 7.7, a novel metal control gate last process (MCGL process) [27, 28], new read scheme [25, 28], and new program schemes [28] for the DC-SF cell are introduced. Excellent performance and reliability of the DC-SF cell could be realized.

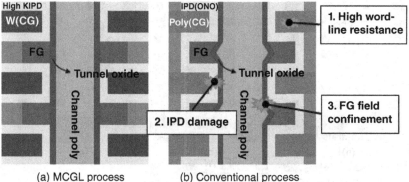

(a) MCGL process   (b) Conventional process

FIGURE 7.68 The DC-SF cell profile comparison. (a) MCGL (metal control gate last) process. (b) Conventional process. In (b) conventional process, there are several problems, such as (1) high word-line resistance, (2) IPD damage during FG separation process, and (3) FG field confinement due to the horn shape FG. The new MCGL process can solve all of these problems.

7.7.2 MCGL Process

The new MCGL (metal control gate last) process sequence of the DC-SF cell is described in Fig. 7.69 [27]. First, the multiple silicon oxide/nitride layers are deposited on N+/p-Si substrate. Next, the channel hole is patterned, and FGs are formed at oxide recess portion by isotropic poly etching (Fig. 7.69a). Tunnel oxide is deposited, and the channel contact hole is formed by etching through N+ layers to connect substrate and channel poly-Si directly (Fig. 7.69b). After channel poly-Si deposition, gates are patterned (Fig. 7.69c). Then stacked silicon nitride has recessed (Fig. 7.69d), and high-k IPD films are deposited on the FGs. After that, tungsten (W) film is deposited and separated to each stacked layers, as shown in Fig. 7.69e. Figure 7.70 shows the cross-sectional TEM image of the DC-SF cell arrays, fabricated by the MCGL process.

In this MCGL process, (1) low word-line resistance can be obtained by gate replacement process (SiN→W). And (2) IPD damage of FG separation process can be avoided due to IPD deposited after FG/channel-poly formation, in contrast with conventional process of ONO IPD before FG separation [24, 26]. And also, (3) FG field confinement at FG edge during programming can be suppressed due to better FG shape by no IPD deposition before FG formation, as shown in Fig. 7.70b.

7.7.3 New Read Scheme

Conventional read operation of the DC-SF NAND flash string is to apply the read voltage V_R to two neighbor CGs of the selected FG, while $V_{pass-read}$ is applied to unselected CGs [24, 26]. In order to investigate read operation issues in conventional read,

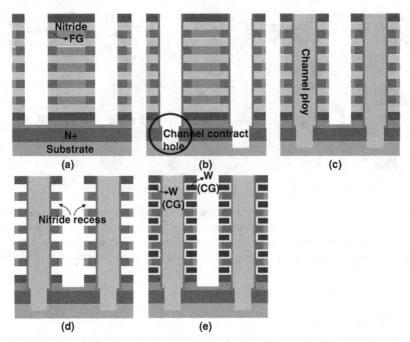

FIGURE 7.69 MCGL process sequence of the DC-SF NAND flash cell. (a) FG formation, (b) channel contact formation, (c) gate patterning, (d) nitride recess, (e) high-k IPD deposition and tungsten W (CG) formation.

the FG1 stored charge dependence has been investigated in several read conditions by TCAD simulation [25], as shown in Fig. 7.71.

Two unexpected characteristics are observed, as shown in Fig. 7.72. One is that selected cell V_t (FG0 and FG2 read) have shifted up when neighbor FG1 charge is negatively increased. This is because a transconductance is degraded due to increasing channel resistance under FG1, which stored negative charges, as shown in Fig. 7.73.

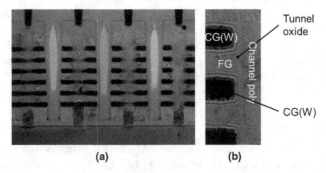

FIGURE 7.70 TEM image of (a) the MCGL process DC-SF cell array and (b) the FG/CG shape. The cell structure is showing a preferable vertical FG shape.

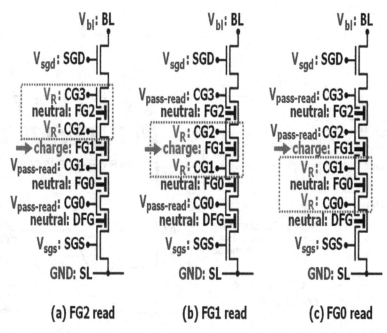

(a) FG2 read   **(b) FG1 read**   **(c) FG0 read**

FIGURE 7.71   Conventional read operations. (a) FG2 read operation (CG2&3 VR), (b) FG1 read operation (CG1&2 VR), and (c) FG0 read operation (CG0&1 VR). Charges in FG1 are set to be various amounts to derive the FG charge dependence.

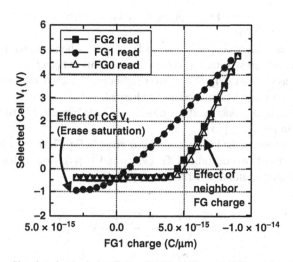

FIGURE 7.72   Simulated selected cell V_t of FG0, FG1, and FG2 read under various FG1 charges, as shown in Fig. 7.71.

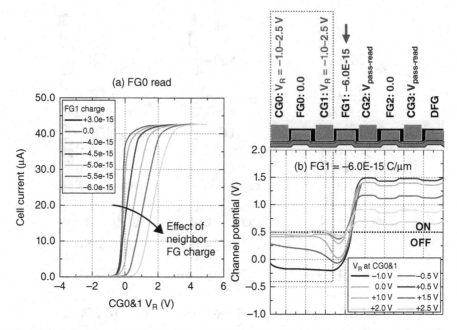

FIGURE 7.73   (a) Cell I_d–V_g of FG0 read under various FG1 charge. (b) Channel potential at FG1 = −6.0E-15 C/μm. When FG1 has a negative charge, the transconductance is degraded due to increasing channel resistance under FG1.

The selected cell V_t (FG0 or FG2) cannot be read correctly when FG1 is negatively charged (>−5 × 10^{-15}C/μm).

The other is that selected cell V_t (FG1 read) saturates when selected FG1 charge is positively increased (erase saturation phenomena). This is because CG V_t limits the selected cell V_t, as shown in Fig. 7.74. Two neighbors CG1 and CG2 turn the channel "off" directly, even when FG1 is positively charged.

In order to derive a proper read operation, a simple capacitor network model is used, as shown in Fig. 7.75 [25].

Analytic expression of the FG1 potential is derived in Eq. (7.3). V_{FG1} is determined by the average bias of two neighbor CGs (V_{CG1} and V_{CG2}), stored charge (σ_{FG1}) in FG1, and the coupling ratio α as written in Eq. (7.4).

$$V_{FG1} = \alpha \left(\frac{V_{CG1} + V_{CG2}}{2} - V_{T,FG1} \right) \tag{7.3}$$

$$\alpha = \frac{2C_1}{2C_1 + C_2} \tag{7.4}$$

$$V_{T,FG1} = -\frac{\sigma_{FG1}}{2C_1} \tag{7.5}$$

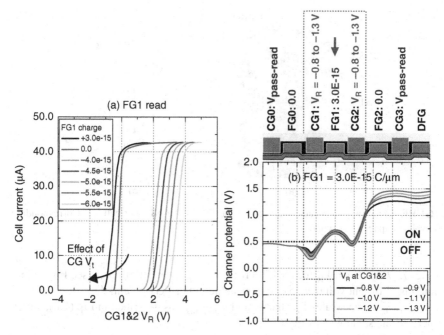

FIGURE 7.74 (a) Cell I_d–V_g of FG1 read under various FG1 charge. (b) Channel potential at FG1 = 3.0E-15 C/μm. The erase V_t saturation is caused by channel off state under CG1 and CG2.

The FG1 potential V_{FG1} during FG0 read operation can be described as Eq. (7.6) by substituting V_{CG1} and V_{CG2} with V_R and $V_{pass-read}$, respectively.

$$V_{FG1} = \alpha \left(\frac{V_R + V_{pass-read}}{2} - V_{T,FG1} \right) \qquad (7.6)$$

Because V_R is predetermined value for cell read operation, the potential decrease of the FG1 due to stored negative charge ($\Delta V_{T,FG1} = -\sigma_{FG1}/2C_1$) have to be compensated by increasing $V_{pass-read}$ to keep the FG1 cell "ON" as pass transistor. Therefore, increasing $\Delta V_{pass-read}$ is determined quantitatively as written in Eq. (7.7). σ_{FG1_max} is the allowed maximum negative charge (allowed highest V_t).

$$\Delta V_{pass-read} = 2\Delta V_{T,FG1} = -\frac{\sigma_{FG1-max}}{C_1} \qquad (7.7)$$

Equation (7.7) had confirmed $V_{pass-read}$ dependence in various FG1 charges (various $V_{T,FG1}$), as shown in Fig. 7.76. V_{pass_read2} have to increase 4 V to compensate a 2-V V_t increase of the neighbor FG1 cell. This strongly corresponds to Eq. (7.7).

Most of NAND devices are currently adopting multilevel cell (MLC) operation, which is using three read voltages (V_Rs) to identify each program levels, as shown

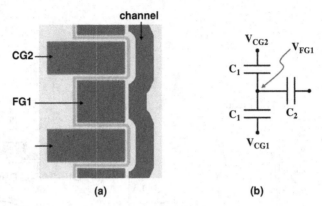

(a)   **(b)**

FIGURE 7.75   (a) Cross-sectional view of the DC-SF NAND unit cell. (b) Corresponding equivalent capacitor network.

in Fig. 7.77. For each V_R for PV1,2,3 read, $V_{\text{pass-read2}}$ and $V_{\text{pass-read1}}$ (see Fig. 7.76a) should be different voltages to compensate neighbor FG1 potential.

$V_{\text{pass-read2}}$ compensates V_R change and $\sigma_{\text{FG1\_max}}$, as described in Eq. (7.8), which is derived the same way as Eq. (7.7).

$$\Delta V_{\text{pass-read2}} = -\Delta V_R - \frac{\sigma_{\text{FG-max}}}{C_1} \tag{7.8}$$

And also, in order to maintain that all unselected FG cells are "ON" even if the cells are in high V_t (PV3) as the worst case, $V_{\text{pass-read1}}$ has to be equal to V_R (Eq. (7.9)).

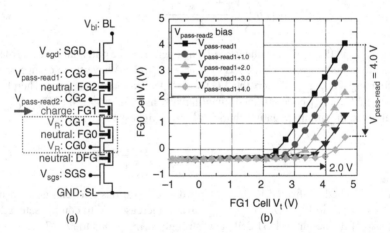

(a)   **(b)**

FIGURE 7.76   (a) FG0 read operation. (b) FG0 cell V_t under several $V_{\text{pass-read2}}$ conditions. $V_{\text{pass-read1}} = 5.0$ V.

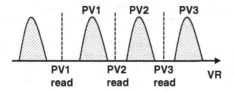

FIGURE 7.77   Multilevel cell (MLC) read operation.

This is because $V_R + V_{\text{pass-read2}}$ for FG1 should be equal to $V_{\text{pass-read2}} + V_{\text{pass-read1}}$ for FG2 (see Fig. 7.76a).

$$V_{\text{pass-read1}} = V_R \qquad (7.9)$$

Figure 7.78 shows the new read condition of the MLC DC-SF NAND cell, which is from Eqs. (7.8) and (7.9). $V_{\text{pass-read2}}$ has to be decreased as V_R is increased; conversely, $V_{\text{pass-read1}}$ has to be increased as V_R is increased. And in region of $V_R = 0\text{--}1$ V, $V_{\text{pass-read1}}$ is a used fixed voltage of 1.0 V, because the CG voltage has to turn channel "ON," as described in Fig. 7.74. The new multilevel read operation condition is summarized in Table 7.3.

7.7.4   New Programming Scheme

The programming scheme of the DC-SF cell has to be optimized to avoid program disturb problems. Figure 7.79 describes program disturb modes (inhibit modes) of DC-SF cell. There are two inhibit modes; mode (A), electron injection mode; and

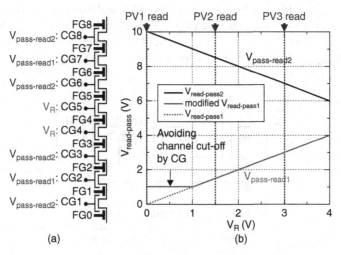

FIGURE 7.78   (a) New read operation of DC-SF NAND string. (b) $V_{\text{pass-read1\&2}}$ dependence on V_R, as shown in Eqs. (7.8) and (7.9).

TABLE 7.3   The new read scheme of DC-SF NAND string

Electrode		PV1 read $V_R = 0.0$ V	PV2 read $V_R = 1.5$ V	PV3 read $V_R = 3.0$ V
CG8	$V_{pass-read2}$	10.0 V	8.5 V	7.0 V
CG7	$V_{pass-read1}$	1.0 V	1.5 V	3.0 V
CG6	$V_{pass-read2}$	10.0 V	8.5 V	7.0 V
CG5 CG4	V_R	0.0 V	1.5 V	3.0 V
CG3	$V_{pass-read2}$	10.0 V	8.5 V	7.0 V
CG2	$V_{pass-read1}$	1.0 V	1.5 V	3.0 V
CG1	$V_{pass-read2}$	10.0 V	8.5 V	7.0 V

mode (B), charge loss mode. Mode (A) is a conventional program inhibit mode, which has a weak electron injection stress, caused by a high field in tunnel oxide due to FG coupled with two CGs (e.g., V_{pass_n+2} and V_{pgm_n+1}). Mode (A) becomes severe in the case of the lower V_t (e.g., erase state) due to a high field of tunnel oxide. On the other hand, mode (B) is a new charge loss mode that is unique in the DC-SF cell. Electrons in FG are ejected to CG by a high field in IPD. Mode (B) become severe in the case of high cell V_t (e.g., PV3 state, such as $V_t = 4$ V) and low V_{pass_n-2}. In order to minimize program disturb, V_{pass_n-2} and V_{pass_n+2} have to be optimized.

Figure 7.80 shows the measurement results of mode (B), which is accelerated by thin IPD (oxide equivalent 12 nm thick). V_t is decreased as V_{pgm_n-1} is increased and as V_{pass_n-2} is decreased. From this data, the maximum allowed electric field in IPD is estimated to 8.3 MV/cm from conditions of $V_{pgm\_n-1} = 15$ V, $V_{pass\_n-2} = 5$ V, and $V_t = 4$ V.

Figure 7.81 shows the potential differences among FG/CG/substrate during programming ($V_{pgm} = 15$ V) in the case of (a) $V_t = -1$ V and (b) $V_t = 4$ V. The potential difference between FG and substrate (V_{fg_sub}) becomes large as V_{pass_n-2}, V_{pass_n+2} are increased. In (a) $V_t = -1$ V, V_{fg_sub} reaches to the mode (A) limitation (criteria) of 7 V at V_{pass_n-2}, $V_{pass\_n+2} = 1$ V. Then, in (a) $V_t = -1$ V, V_{pass_n-2} and V_{pass_n+2}

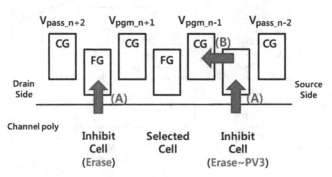

FIGURE 7.79   Two program inhibit modes of DC-SF cell. Mode (A), electron injection mode; weak electron injection from substrate to FG due to high field on tunnel oxide. Mode (B), charge loss mode; electron emission from FG to CG due to high field on IPD.

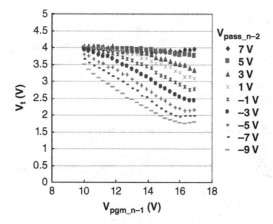

FIGURE 7.80   Measured charge loss of the mode (B) during programming of DC-SF cell. Charge loss (electron emission) from FG to CG is increased as increasing V_{pgm_n-1} and decreasing V_{pass_n-2}.

have to be less than 1 V to keep $V_{fg\_sub} < 7$ V. And in (b) $V_t = 4$ V, the potential difference between CG and FG (V_{cg1_fg}) reach the mode (B) limitation of 12.5 V at $V_{pass\_n-2} = -2$ V. The limitation is determined by the maximum allowed electric field of 8.3 MV/cm in mode (B) in the case of IPD thickness = 15 nm. Then, in (b) $V_t = 4$ V, V_{pass_n-2} has to be more than -2 V to keep $V_{cg1\_fg} < 12$ V. Mode (B) is located only in source side inhibit cell (right-side inhibit cell in Fig. 7.79). Therefore, V_{pass_n-2} has to be in the range of -2 V to 1 V, and V_{pass_n+2} has to be less than 1 V, because the drain-side inhibit cell (left-side inhibit cell in Fig. 7.79) is only in the case of low V_t (erase state).

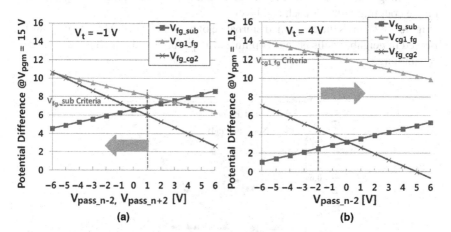

FIGURE 7.81   The potential difference among FG/CG/substrate during programming in the case of (a) $V_t = -1$ V and (b) $V_t = 4$ V. In (a) $V_t = -1$ V, maximum $V_{pass\_n-2}/V_{pass\_n+2}$ is determined to be 1 V by the criteria of mode (A). And, in (b) $V_t = 4$ V, minimum $V_{pass\_n-2}/V_{pass\_n+2}$ is determined to be -2 V by the criteria of mode (B).

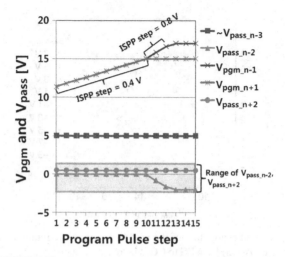

FIGURE 7.82   The proposed program scheme of the DC-SF cell.

Figure 7.82 shows one example of a new ISPP program scheme for the DC-SF cell. V_{pgm}s (V_{pgm_n-1} and V_{pgm_n+1}) are incrementally stepped up (ISPP) using 0.4-V steps until reaching 15 V. And if more program pulses are needed, V_{pgm_n-1} is stepped up using 0.8-V steps and V_{pass_n-2} has stepped down using −0.8-V steps in order to prevent mode (A) at the right-side inhibit cell in Fig. 7.79. V_{pgm_n+1} maintains 15 V, and then V_{pass_n+2} can be kept at a positive voltage of 0.5–1.0 V to transfer bit-line voltage (0 V) for programming cells.

The coupling ratio of DC-SF cell is sensitive with the channel diameters [26], which are normally large at top-side cells and small at bottom-side cells. Then, the programming voltage and inhibit voltage would need to be optimized by matching with the coupling ratio for each cell layer.

Figure 7.83 shows the program disturbance of neighbor cells. Almost no V_t shift is observed in neighbor cells because optimized V_{pass} for neighbor cell are used during programming.

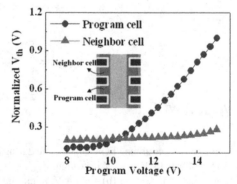

FIGURE 7.83   Program disturbance of neighbor cell. Low $V_{pass\_n-2} = 1$ V is used.

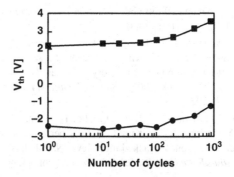

FIGURE 7.84 Program/erase cycling endurance characteristics of the DC-SF cell.

7.7.5 Reliability

Program/erase cycling endurance characteristics are shown in Fig. 7.84. The V_{th} shift after cycling is small, less than 1.3 V even after 1K cycles. Data retention characteristic is also evaluated. The PV3 V_{th} shift of 60 mV after 250°C 120 min is small, which is comparable with conventional planar 2D FG NAND flash characteristics. Therefore, any serious process damages on tunnel oxide and IPD in DC-SF cells have not been observed in the optimizing MCGL process.

Advanced DC-SF (dual control gate with surrounding floating gate) cell process and operation schemes had been introduced. In order to improve performance and reliability of the DC-SF cell, the new metal control gate last (MCGL) process had been developed. The MCGL process could realize a low resistive tungsten (W) metal word line, a low damage on tunnel oxide/IPD (inter-poly dielectric), and a preferable FG shape. Also, new read and program operation schemes had been developed. In the new read operation, the higher and lower $V_{pass-read}$ are alternately applied to unselected control gates (CGs) to compensate lowering FG potential to be a pass transistor. And in the new program scheme, the optimized V_{pass} are applied to neighbor WL of selected WL to prevent program disturb and charge loss through IPD. Thus, by using the MCGL process and the new read/program schemes, high performance and high reliability of the DC-SF cell could be realized for 3D NAND flash memories.

REFERENCES

[1] Masuoka, F.; Momodomi, M.; Iwata, Y.; Shirota, R. New ultra high density EPROM and flash EEPROM with NAND structure cell, *Electron Devices Meeting, 1987 International*, vol. 33, pp. 552–555, 1987.

[2] S. Aritome, 3D Flash Memories, International Memory Workshop 2011 (IMW 2011), short course.

[3] Jung, S.-M.; Jang, J.; Cho, W.; Cho, H.; Jeong, J.; Chang, Y.; Kim, J.; Rah, Y.; Son, Y.; Park, J.; Song, M.-S.; Kim, K.-H.; Lim, J.-S.; Kim, K. Three dimensionally stacked NAND flash memory technology using stacking single crystal Si layers on ILD and

TANOS structure for beyond 30 nm node, *Electron Devices Meeting, 2006. IEDM '06. International*, pp. 1–4, 11–13 Dec. 2006.

[4] Park, K.-T.; Kim, D.; Hwang, S.; Kang, M.; Cho, H.; Jeong, Y.; Seo, Y.-l.; Jang, J.; Kim, H.-S.; Jung, S.-M.; Lee, Y.-T.; Kim, C.; Lee, W.-S., A 45 nm 4 Gb 3-dimensional double-stacked multi-level NAND flash memory with shared bitline structure, *Solid-State Circuits Conference, 2008. ISSCC 2008. Digest of Technical Papers. IEEE International*, pp. 510, 632, 3–7 Feb. 2008.

[5] Park, K.-T.; Kang, M.; Hwang, S.; Kim, D.; Cho, H.; Jeong, Y.; Seo, Y.-l.; Jang, J.; Kim, H.-S.; Lee, Y.-T.; Jung, S.-M.; Kim, C. A fully performance compatible 45 nm 4-gigabit three dimensional double-stacked multi-level NAND flash memory with shared bit-line structure, *Solid-State Circuits, IEEE Journal of*, vol. 44, no. 1, pp. 208, 216, Jan. 2009.

[6] Lai, E.-K.; Lue, H.-T.; Hsiao, Y.-H.; Hsieh, J.-Y.; Lu, C.-P.; Wang, S.-Y.; Yang, L.-W.; Yang, T.; Chen, K.-C.; Gong, J.; Hsieh, K.-Y.; Liu, R.; Lu, C.-Y. A multi-layer stackable thin-film transistor (TFT) NAND-type flash memory, *Electron Devices Meeting, 2006. IEDM '06. International*, pp. 1–4, 11–13 Dec. 2006.

[7] Endoh, T.; Kinoshita, K.; Tanigami, T.; Wada, Y.; Sato, K.; Yamada, K.; Yokoyama, T.; Takeuchi, N.; Tanaka, K.; Awaya, N.; Sakiyama, K.; Masuoka, F. Novel ultra high density flash memory with a stacked-surrounding gate transistor (S-SGT) structured cell, *Electron Devices Meeting, 2001. IEDM '01. Technical Digest International*, pp. 2.3.1–2.3.4, 2001.

[8] Endoh, T.; Kinoshita, K.; Tanigami, T.; Wada, Y.; Sato, K.; Yamada, K.; Yokoyama, T.; Takeuchi, N.; Tanaka, K.; Awaya, N.; Sakiyama, K.; Masuoka, F. Novel ultrahigh-density flash memory with a stacked-surrounding gate transistor (S-SGT) structured cell, *Electron Devices, IEEE Transactions on*, vol. 50, no. 4, pp. 945, 951, April 2003.

[9] F. Masuoka, Forbes Global, *Forbes Magazine*, New York, 24 June 2002, 26 pp.

[10] Tanaka, H.; Kido, M.; Yahashi, K.; Oomura, M.; Katsumata, R.; Kito, M.; Fukuzumi, Y.; Sato, M.; Nagata, Y.; Matsuoka, Y.; Iwata, Y.; Aochi, H.; Nitayama, A. Bit cost scalable technology with punch and plug process for ultra high density flash memory, *VLSI Symposium Technical Digest, 2007*, pp. 14–15.

[11] Nitayama, A.; Aochi, H., Bit cost scalable (BiCS) technology for future ultra high density memories, *VLSI Technology, Systems, and Applications (VLSI-TSA), 2013 International Symposium on*, pp. 1, 2, 22–24 April 2013.

[12] Nitayama, A.; Aochi, H. Bit cost scalable (BiCS) technology for future ultra high density storage memories, *VLSI Technology (VLSIT), 2013 Symposium on*, pp. T60, T61, 11–13 June 2013.

[13] Fukuzumi, Y.; Katsumata, R.; Kito, M.; Kido, M.; Sato, M.; Tanaka, H.; Nagata, Y.; Matsuoka, Y.; Iwata, Y.; Aochi, H.; Nitayama, A. Optimal integration and characteristics of vertical array devices for ultra-high density, bit-cost scalable flash memory, *IEEE IEDM Technical Digest*, pp. 449–452, 2007.

[14] Komori, Y.; Kido, M.; Katsumata, R.; Fukuzumi, Y.; Tanaka, H.; Nagata, Y.; Ishiduki, M.; Aochi, H., Nitayama, A., Disturbless flash memory due to high boost efficiency on BiCS structure and optimal memory film stack for ultra high density storage device, *IEEE IEDM Technical Digest*, pp. 851–854, 2008.

[15] Katsumata, R.; Kito, M.; Fukuzumi, Y.; Kido, M.; Tanaka, H.; Komori, Y.; Ishiduki, M.; Matsunami, J.; Fujiwara, T.; Nagata, Y.; Zhang, L.; Iwata, Y.; Kirisawa, R.; Aochi, H.,

Nitayama, A. Pipe-shaped BiCS flash memory with 16 stacked layers and multi-level-cell operation for ultra high density storage devices, *VLSI Symposium Technical Digest, 2009*, pp. 136–137.

[16] Maeda, T.; Itagaki, K.; Hishida, T.; Katsumata, R.; Kito, M.; Fukuzumi, Y.; Kido, M.; Tanaka, H.; Komori, Y.; Ishiduki, M.; Matsunami, J.; Fujiwara, T.; Aochi, H.; Iwata, Y., Watanabe, Y., Multi-stacked 1G cell/layer pipe-shaped BiCS flash memory, *VLSI Symposium Technical Digest, 2009*, pp. 22–23.

[17] Ishiduki, M.; Fukuzumi, Y.; Katsumata, R.; Kito, M.; Kido, M.; Tanaka, H.; Komori, Y.; Nagata, Y.; Fujiwara, T.; Maeda, T.; Mikajiri, Y.; Oota, S.; Honda, M.; Iwata, Y.; Kirisawa, R.; Aochi, H., Nitayama, A., Optimal device structure for pipe-shaped BiCS flash memory for ultra high density storage device with excellent performance and reliability, *IEEE IEDM Technical Digest*, pp. 625–628, 2009.

[18] Kim, J.; Hong, A. J.; Ogawa, M.; Ma, S.; Song, E. B.; Lin, Y.-S.; Han, J.; Chung, U.-I.; Wang, K. L. Novel 3-D structure for ultra high density flash memory with VRAT (vertical-recess-array-transistor) and PIPE (planarized integration on the same plane), *VLSI Symposium Technical Digest, 2008*, pp. 122–123.

[19] Jang, J. H.; Kim, H.-S.; Cho, W.; Cho, H.; Kim, J.; Shim, S. I.; Jang, Y.; Jeong, J.-H.; Son, B.-K.; Kim, D. W.; Kim, K.; Shim, J.-J.; Lim, J. S.; Kim, K.-H.; Yi, S. Y.; Lim, J.-Y.; Chung, D.; Moon, H.-C.; Hwang, S.; Lee, J.-W.; Son, Y.-H.; Chung, U.-I., Lee, W.-S. Vertical cell array using TCAT (terabit cell array transistor) technology for ultra high density NAND flash memory, *VLSI Symposium Technical Digest, 2009*, pp. 192–193.

[20] Kim, W. J.; Choi, S.; Sung, J.; Lee, T.; Park, C.; Ko, H.; Jung, J.; Yoo, I., and Park, Y., Multi-layered vertical gate NAND flash overcoming stacking limit for terabit density storage, *VLSI Symposium Technical Digest, 2009*, pp. 188–189.

[21] Kim, J.; Hong, A. J.; Sung, M. K.; Song, E. B.; Park, J. H.; Han, J.; Choi, S.; Jang, D.; Moon, J.-T.; Wang, K. L. Novel vertical-stacked-array-transistor (VSAT) for ultra-high-density and cost-effective NAND flash memory devices and SSD (solid state drive), *VLSI Technology, 2009 Symposium on*, pp. 186, 187, 16–18 June 2009.

[22] Hsiao, Y.-H.; Lue, H.-T.; Hsu, T.-H.; Hsieh, K.-Y.; Lu, C.-Y., A critical examination of 3D stackable NAND flash memory architectures by simulation study of the scaling capability, *IMW, 2010*, pp. 142–145, 2010.

[23] Hsu, T.-H.; Lue, H.-T.; Hsieh, C.-C.; Lai, E.-K.; Lu, C.-P.; Hong, S.-P.; Wu, M.-T.; Hsu, F. H.; Lien, N. Z.; Hsieh, J.-Y.; Yang, L.-W.; Yang, T.; Chen, K.-C.; Hsieh, K.-Y.; Liu, R.; Lu, C.-Y. Study of sub-30 nm thin film transistor (TFT) charge-trapping (CT) devices for 3D NAND flash application, *IEEE IEDM Technical Digest*, pp. 629–632, 2009.

[24] Whang, S. J.; Lee, K. H.; Shin, D. G.; Kim, B. Y.; Kim, M. S.; Bin, J. H.; Han, J. H.; Kim, S. J.; Lee, B. M.; Jung, Y. K.; Cho, S. Y.; Shin, C. H.; Yoo, H. S.; Choi, S. M.; Hong, K.; Aritome, S.; Park, S. K., Hong, S. J. Novel 3-dimensional dual control-gate with surrounding floating-gate (DC-SF) NAND flash cell for 1tb file storage application, *IEEE IEDM Technical Digest*, pp. 668–671, 2010.

[25] Yoo, H. S. ; Choi, E. S.; Joo, H. S.; Cho, G. S.; Park, S. K.; Aritome, S.; Lee, S. K.; Hong, S. J. New read scheme of variable $V_{pass-read}$ for dual control gate with surrounding floating gate (DC-SF) NAND flash cell, *Memory Workshop (IMW), 2011 3rd IEEE International*, pp. 1–4, 22–25 May 2011.

[26] Aritome, S.; Whang, S. J.; Lee, KiH.; Shin, D. G.; Kim, B. Y.; Kim, M. S.; Bin, J. H.; Han, J. H.; Kim, S. J.; Lee, B. M.; Jung, Y. K.; Cho, S. Y.; Shin, C. H.; Yoo, H. S.; Choi, S. M.; Hong, K.; Park S. K.; Hong, S. J. A novel three-dimensional dual control-gate with

surrounding floating-gate (DC-SF) NAND flash cell original research article, *Solid-State Electronics*, vol. 79, pp. 166–171, Jan. 2013.

[27] Noh, Y.; Ahn, Y.; Yoo, H.; Han, B.; Chung, S.; Shim, K.; Lee, K.; Kwak, S.; Shin, S.; Choi, I.; Nam, S.; Cho, G.; Sheen, D.; Pyi, S.; Choi, J.; Park, S.; Kim, J.; Lee, S.; Aritome, S.; Hong, S.; Park, S. A new metal control gate last process (MCGL process) for high performance DC-SF (dual control gate with surrounding floating gate) 3D NAND flash memory, *VLSI Technology (VLSIT), 2012 Symposium on*, pp. 19–20, 12–14 June 2012.

[28] Aritome, S.; Noh, Y.; Yoo, H.; Choi, E. S.; Joo, H. S.; Ahn, Y.; Han, B.; Chung, S.; Shim, K.; Lee, K.; Kwak, S.; Shin, S.; Choi, I.; Nam, S.; Cho, G.; Sheen, D.; Pyi, S.; Choi, J.; Park, S.; Kim, J.; Lee, S.; Hong, S.; Park, S.; Kikkawa, T. Advanced DC-SF cell technology for 3-D NAND flash, *Electron Devices, IEEE Transactions on*, vol. 60, no. 4, pp. 1327–1333, April 2013.

[29] Choi, E.-S.; Park, S.-K., Device considerations for high density and highly reliable 3D NAND flash cell in near future, *Electron Devices Meeting (IEDM), 2012 IEEE International*, pp. 9.4.1–9.4.4, 10–13 Dec. 2012.

[30] Takato, H.; Sunouchi, K.; Okabe, N.; Nitayama, A.; Hieda, K.; Horiguchi, F.; Masuoka, F. High performance CMOS surrounding gate transistor (SGT) for ultra high density LSIs, *Electron Devices Meeting, 1988. IEDM '88. Technical Digest, International*, pp. 222, 225, 11–14 Dec. 1988.

[31] Pein, H.; Plummer, J. D. "A 3-D sidewall flash EPROM cell and memory array," *Electron Device Letters, IEEE*, vol. 14, no. 8, pp. 415, 417, Aug. 1993.

[32] Pein, H. B.; Plummer, J. D., Performance of the 3-D sidewall flash EPROM cell, *Electron Devices Meeting, 1993. IEDM '93. Technical Digest, International*, pp. 11, 14, 5–8 Dec. 1993.

[33] Pein, H.; Plummer, James D. Performance of the 3-D PENCIL flash EPROM cell and memory array, *Electron Devices, IEEE Transactions on*, vol. 42, no. 11, pp. 1982, 1991, Nov. 1995.

[34] Hsu, T.-H.; Lue, H.-T.; King, Y.-C.; Hsiao, Y.-H.; Lai, S.-C.; Hsieh, K.-Y.; Liu, R.; Lu, C.-Y. Physical model of field enhancement and edge effects of FinFET charge-trapping NAND flash devices, *Electron Devices, IEEE Transactions on*, vol. 56, no. 6, pp. 1235, 1242, June 2009.

[35] Congedo, G.; Arreghini, A.; Liu, L.; Capogreco, E.; Lisoni, J. G.; Huet, K.; Toque-Tresonne, I.; Van Aerde, S.; Toledano-Luque, M.; Tan, C.-L.; Van denbosch, G.; Van Houdt, J. Analysis of performance/variability trade-off in Macaroni-type 3-D NAND memory, *Memory Workshop (IMW), 2014 IEEE 6th International*, pp. 1, 4, 18–21 May 2014.

[36] Toledano-Luque, M.; Degraeve, R.; Roussel, P. J.; Luong, V.; Tang, B.; Lisoni, J. G.; Tan, C.-L.; Arreghini, A.; Van denbosch, G.; Groeseneken, G.; Van Houdt, J. Statistical spectroscopy of switching traps in deeply scaled vertical poly-Si channel for 3D memories, *Electron Devices Meeting (IEDM), 2013 IEEE International*, pp. 21.3.1, 21.3.4, 9–11 Dec. 2013.

[37] Kang, C. S.; Choi, J.; Sim, J.; Lee, C.; Shin, Y.; Park, J.; Sel, J.; Jeon, S.; Park, Y.; Kim, K. Effects of lateral charge spreading on the reliability of TANOS(TaN/AlO/SiN/Oxide/Si) NAND flash memory, *IRPS*, pp. 167–169, 2007.

[38] Cho, W.-s.; Shim, S. I.; Jang, J.; Cho, H.-s.; You, B.-K.; Son, B.-K.; Kim, K.-h.; Shim, J.-J.; Park, C.-m.; Lim, J.-s.; Kim, K.-H.; Chung, D.-w.; Lim, J.-Y.; Moon, H.-C.;

Hwang, S.-m.; Lim, H.-s.; Kim, H.-S.; Choi, J.; Chung, C. Highly reliable vertical NAND technology with biconcave shaped storage layer and leakage controllable offset structure, *VLSI Technology (VLSIT), 2010 Symposium on*, pp. 173, 174, 15–17 June 2010.

[39] Park, K.-T.; Han, J.-m.; Kim, D.; Nam, S.; Choi, K.; Kim, M.-S.; Kwak, P.; Lee, D.; Choi, Y.-H.; Kang, K.-M.; Choi, M.-H.; Kwak, D.-H.; Park, H.-w.; Shim, S.-w.; Yoon, H.-J.; Kim, D.; Park, S.-w.; Lee, K.; Ko, K.; Shim, D.-K.; Ahn, Y.-L.; Park, J.; Ryu, J.; Kim, D.; Yun, K.; Kwon, J.; Shin, S.; Youn, D.; Kim, W.-T.; Kim, T.; Kim, S.-J.; Seo, S.; Kim, H.-G.; Byeon, D.-S.; Yang, H.-J.; Kim, M.; Kim, M.-S.; Yeon, J.; Jang, J.; Kim, H.-S.; Lee, W.; Song, D.; Lee, S.; Kyung, K.-H.; Choi, J.-H., 19.5 Three-dimensional 128 Gb MLC vertical NAND flash-memory with 24-WL stacked layers and 50 MB/s high-speed programming, *Solid-State Circuits Conference Digest of Technical Papers (ISSCC), 2014 IEEE International*, pp. 334, 335, 9–13 Feb. 2014.

[40] Park, K.-T.; Nam, S.; Kim, D.; Kwak, P.; Lee, D.; Choi, Y.-H.; Choi, M.-H.; Kwak, D.-H.; Kim, D.-H.; Kim, M.-S.; Park, H.-W.; Shim, S.-W.; Kang, K.-M.; Park, S.-W.; Lee, K.; Yoon, H.-J.; Ko, K.; Shim, D.-K.; Ahn, Y.-L.; Ryu, J.; Kim, D.; Yun, K.; Kwon, J.; Shin, S.; Byeon, D.-S.; Choi, K.; Han, J.-M.; Kyung, K.-H.; Choi, J.-H.; Kim, K. Three-dimensional 128 Gb MLC vertical nand flash memory with 24-WL stacked layers and 50 MB/s high-speed programming, *Solid-State Circuits, IEEE Journal of*, vol. 50, no. 1, pp. 204, 213, Jan. 2015.

[41] Choi, J.; Seol, K. S., 3D approaches for non-volatile memory, *VLSI Technology (VLSIT), 2011 Symposium on*, pp. 178, 179, 14–16 June 2011.

[42] Im, J.-w., Jeong, W.-P.; Kim, D.-H.; Nam, S.-W.; Shim, D.-K.; Choi, M.-H.; Yoon, H.-J.; Kim, D.-H.; Kim, Y.-S.; Park, H.-W.; Kwak, D.-H.; Park, S.-W.; Yoon, S.-M.; Hahn, W.-G.; Ryu, J.-H.; Shim, S.-W.; Kang, K.-T.; Choi, S.-H.; Ihm, J.-D.; Min, Y.-S.; Kim, I.-M.; Lee, D.-S.; Cho, J.-H.; Kwon, O.-S.; Lee, J.-S.; Kim, M.-S.; Joo, S.-H.; Jang, J.-H.; Hwang, S.-W.; Byeon, D.-S.; Yang, H.-J.; Park, K.-T.; Kyung, K.-H.; Choi, J.-H. A 128 Gb 3b/cell V-NAND flash memory with 1Gb/s I/O rate, *Solid-State Circuits Conference Digest of Technical Papers (ISSCC), 2015 IEEE International*, pp. 23–25 Feb. 2015.

[43] Sako, M.; Watanabe, Y.; Nakajima, T.; Sato, J.; Muraoka, K.; Fujiu, M.; Kono, F.; Nakagawa, M.; Masuda, M.; Kato, K.; Terada, Y.; Shimizu, Y.; Honma, M.; Imamoto, A.; Araya, T.; Konno, H.; Okanaga, T.; Fujimura, T.; Wang, X.; Muramoto, M.; Kamoshida, M.; Kohno, M.; Suzuki, Y.; Hashiguchi, T.; Kobayashi, T.; Yamaoka, M.; Yamashita, R. A low-power 64 Gb MLC NAND-flash memory in 15 nm CMOS technology, *Solid-State Circuits Conference Digest of Technical Papers (ISSCC), 2015 IEEE International*, pp. 23–25, Feb. 2015.

[44] Aritome, S. 3D NAND flash memory—full-scale production from 2015, *Semiconductor Storage 2014*, Nikkei Business Publications (in Japanese), pp. 34–45.

[45] Choi, E.-S.; Yoo, N. S.; Joo, H.-S.; Cho, G.-S.; Park, S.-K.; Lee, S.-K. A novel 3D cell array architecture for terra-bit NAND flash memory, *Memory Workshop (IMW), 2011 3rd IEEE International*, pp. 1, 4, 22–25 May 2011.

[46] Kim, W.; Seo, J. Y.; Kim, Y.; Park, S. H.; Lee, S. H.; Baek, M. H.; Lee, J.-H.; Park, B.-G. Channel-stacked NAND flash memory with layer selection by multi-level operation (LSM), *Electron Devices Meeting (IEDM), 2013 IEEE International*, pp. 3.8.1, 3.8.4, 9–11 Dec. 2013.

[47] Lue, H. T.; Hsu, T.-H.; Hsiao, Y.-H.; Hong, S. P.; Wu, M. T.; Hsu, F. H.; Lien, N. Z.; Wang, S.-Y.; Hsieh, J.-Y.; Yang, L.-W.; Yang, T.; Chen, K.-C.; Hsieh, K.-Y.; Lu, C.-Y. A

highly scalable 8-layer 3D vertical-gate (VG) TFT NAND flash using junction-free buried channel BE-SONOS device, *VLSI Technology (VLSIT), 2010 Symposium on*, pp. 131, 132, 15–17 June 2010.

[48]   Chang, K.-P.; Lue, H.-T.; Chen, C.-P.; Chen, C.-F.; Chen, Y.-R.; Hsiao, Y.-H.; Hsieh, C.-C.; Shih, Y.-H.; Yang, T.; Chen, K.-C.; Hung, C.-H.; Lu, C.-Y. Memory architecture of 3D vertical gate (3DVG) NAND flash using plural island-gate SSL decoding method and study of it's program inhibit characteristics, *Memory Workshop (IMW), 2012 4th IEEE International*, pp. 1, 4, 20–23 May 2012.

[49]   Chen, C.-P.; Lue, H.-T.; Chang, K.-P.; Hsiao, Y.-H.; Hsieh, C.-C.; Chen, S.-H.; Shih, Y.-H.; Hsieh, K.-Y.; Yang, T.; Chen, K.-C.; Lu, C.-Y. A highly pitch scalable 3D vertical gate (VG) NAND flash decoded by a novel self-aligned independently controlled double gate (IDG) string select transistor (SSL), *VLSI Technology (VLSIT), 2012 Symposium on*, pp. 91, 92, 12–14 June 2012.

[50]   Hung, C.-H.; Lue, H.-T.; Hung, S.-N.; Hsieh, C.-C.; Chang, K.-P.; Chen, T.-W.; Huang, S.-L.; Chen, T. S.; Chang, C.-S.; Yeh, W.-W.; Hsiao, Y.-H.; Chen, C.-F.; Huang, S.-C.; Chen, Y.-R.; Lee, G.-R.; Hu, C.-W.; Chen, S.-H.; Chiu, C.-J.; Shih, Y.-H.; Lu, C.-Y., Design innovations to optimize the 3D stackable vertical gate (VG) NAND flash, *Electron Devices Meeting (IEDM), 2012 IEEE International*, pp. 10.1.1, 10.1.4, 10–13 Dec. 2012.

[51]   Chen, S.-H.; Lue, H.-T.; Shih, Y.-H.; Chen, C.-F.; Hsu, T.-H.; Chen, Y.-R.; Hsiao, Y.-H.; Huang, S.-C.; Chang, K.-P.; Hsieh, C.-C.; Lee, G.-R.; Chuang, A.; Hu, C.-W.; Chiu, C.-J.; Lin, L. Y.; Lee, H.-J.; Tsai, F.-N.; Yang, C.-C.; Yang, T.; Lu, C.-Y. A highly scalable 8-layer vertical gate 3D NAND with split-page bit line layout and efficient binary-sum MiLC (minimal incremental layer cost) staircase contacts, *Electron Devices Meeting (IEDM), 2012 IEEE International*, pp. 2.3.1, 2.3.4, 10–13 Dec. 2012.

8

CHALLENGES OF THREE-DIMENSIONAL NAND FLASH MEMORY

8.1 INTRODUCTION

In Chapter 7, several types of three-dimensional (3D) NAND flash memory cells were introduced. In this chapter, challenges of 3D NAND flash memory are discussed.

First, several types of 3D NAND cells are compared in pros and cons of cell structure, process, and memory cell operation in Section 8.2.

The common challenge items of 3D NAND cells are discussed in Sections 8.3–8.9 to clarify key issues of achieving a lower cost, a better performance, and a reasonable reliability.

Many 3D NAND cells are use a SONOS (Silicon(Gate)–Oxide–Nitride–Oxide–Silicon (substrate)) charge storage structure. The data retention characteristic of SONOS cell has problems of quick charge loss and large V_t shift in retention bake because of charge detrapping through thinner tunnel oxide. In Section 8.3, data retention issues are discussed.

The program disturb mechanisms of 3D NAND cells are much different from that of 2D NAND cells, because cell structure and array architecture are totally changed. The analysis results of program disturb in 3D NAND cells are presented in Section 8.4.

WL capacitance is much increased in the stacked word-line (WL) structure in the 3D NAND cell of BiCS, TCAT (V-NAND), and SMArT, because a plane structure of WL has a large parasitic capacitance. However, WL resistance is decreased due to

Nand Flash Memory Technologies, First Edition. Seiichi Aritome.
© 2016 The Institute of Electrical and Electronics Engineers, Inc. Published 2016 by John Wiley & Sons, Inc.

wide WL width. The word-line RC (resistance–capacitance) delay in 3D NAND cell is described in Section 8.5.

The cell current is much decreased in the 3D NAND cell, because channel material is changed from crystal Si (substrate Si) to poly-Si. In Section 8.6, cell current issues in the 3D NAND cell are discussed with channel conduction mechanism, V_G dependence, RTN, back-side trap effect in macaroni channel, and laser anneal process.

In order to reduce bit cost in 3D NAND cell, it is very important to increase the number of stacked cells. In Section 8.7, serious problems of high aspect ratio process and small cell current are discussed in the case of increasing number of stacked cells. And some possible solutions will be presented.

The new structure of the peripheral circuit under cell array is described in Section 8.8. And then, the power consumption issue is presented in Section 8.9.

Finally, the future trend of the 3D NAND cell is discussed in Section 8.10. The lower bit cost can be realized by aggressively increasing number of stacked cells. This will have a big impact on the mass-storage market, such as SSD (solid–state drive) for a consumer and enterprise server for the future.

8.2   COMPARISON OF 3D NAND CELLS

In this chapter, several types of 3D NAND flash memory cells are compared. Figure 8.1 shows the comparison of major 3D NAND flash memory cells, in terms of cell structure, fabrication processes, and operations. For cells, channel structure is

		BICS / P-BiCS	TCAT (V-NAND)	SMArT	VG-NAND	DC-SF
Cell	Channel	Vertical	Vertical	Vertical	Horizontal	Vertical
	Gate Structure	Surrounding (GAA)	Surrounding (GAA)	Surrounding (GAA)	Dual	Surrounding (GAA)
	Charge Storage	SONOS	SONOS (TANOS)	SONOS (TANOS)	SONOS (TANOS)	FG
Process	Stacked Layer	SiO2/ Poly	SiO2/ SiN	SiO2/ SiN	SiO2/ Poly	SiO2/ SiN
	Gate Process	Gate First	Gate Last	Gate Last	Gate Last	Gate Last
	Key Process	Channel Hole RIE	WL Separation	WL Separation	VG Patterning	WL Separation FG Process
Operation	Program	FN	FN	FN	FN	FN
	Erase	FN	FN	FN	FN	FN
	Program Disturb	SG/Dummy-WL Control	SG/Dummy-WL Control	SG/Dummy-WL Control	Z-disturb	Vpass Window
	Erase Speed	Slow	Slow	Slow	Slow	Fair
	P/E cycling	Good	Good	Good	Fair	Fair
	Data Retention	Detrapping / Charge spreading	Detrapping / Charge spreading	Detrapping / Charge spreading	Detrapping / Charge spreading	Detrapping / SILC
	Key	WL RC delay			Layer Selection Scheme	

FIGURE 8.1   Comparison table of major 3D NAND cells for cell structure, process, and operation.

categorized into "vertical" or "horizontal." Most 3D cells have a "vertical" channel that is fabricated by channel plug through multi-stacked gate layers. And "vertical" channels have the "surrounding (GAA; gate all around)" gate structure. The "surrounding" gate structure has a better performance of cell current and cutoff current than "dual" gate structure because of a better controllability of channel potential by gate electrode. And also, the surrounding gate SONOS cell should have a better erase performance due to suppressing the electron back tunneling from control gate by a field relaxation in block oxide.

For processes, in the etching (RIE) viewpoint for a stacked layer, a stacked layer of SiO_2/SiN is easier to set in the RIE condition than that of SiO_2/poly(-Si) because of the similar etching condition of dielectric SiO_2 and SiN. However, in the case of SiO_2/SiN, the gate replacement process from SiN to metal tungsten (W) is needed, as shown in Section 7.3. The tungsten (W) word line has to be separated for each word line ("WL separation"). This is a key process in TCAT (V-NAND), SMArT, and DC-SF cells. For the gate process, The "gate last" process has an advantage that can use the same fabrication order of a gate dielectric film as fabrication order of 2D cell. This means that the gate dielectrics are fabricated in order of tunnel oxide, charge storage SiN, blocking dielectric, and gate tungsten (W). In the case of "gate first," the fabrication order becomes the opposite of blocking dielectric, charge storage SiN, tunnel oxide, and then channel poly-Si. The legacy process of 2D cannot be used in the "gate first" process case.

For operations, the mechanism of program and erase operation is basically the same in all 3D cells, however, erase performance is different between charge trap cells (SONOS(TANOS)) and floating gate (FG) cells. SONOS (TANOS) cells have slower erase than floating gate (FG) cells, even electric field enhancement occurs in tunnel oxide during erase. The program disturb conditions and characteristics of 3D cells are greatly changed from that of 2D cells, as described in detail in Section 8.4. The program/erase cycling performance of SONOS cells is inherently better than that of FG cells; however, data retention of SONOS cells is worse than that of FG cells because of initial data loss, as described in Section 8.3.

The effective memory cell size of several 3D NAND cells is calculated and compared, based on assumptions of Fig. 8.2 and Fig. 8.3 [1]. In SONOS or TANOS 3D cells, thickness of gate dielectric ONO (tono) is assumed to have the fixed value of 20-nm thickness. And the size difference of the channel hole between top and bottom in BiCS/TCAT/SMArT/DC-SF is the fixed value of 10 nm (D), regardless of the number of stacked cells. And the size difference of channel poly of VG-NAND between top and bottom is also a fixed value of 10 nm (D). Minimum width (min W) of hole size of BiCS/TCAT/SMArT/DC-SF at bottom or minimum channel poly-Si space at bottom are the fixed value of 20 nm. In DC-SF cells, the width of floating gate (W_{fg}) of 27 nm is added to obtain a sufficient coupling ratio.

Based on these assumptions, effective cell sizes of each cell type were calculated. Cell size, size of X- and Y-direction, physical cell size, and effective cell size of each cell are shown in Fig. 8.3. In VG-NAND cells, the area of source-side select gate is added 3% of cell size, and the area of drain-side select gate is added 3% for each stacked layer to select the channel layer (see Section 7.5).

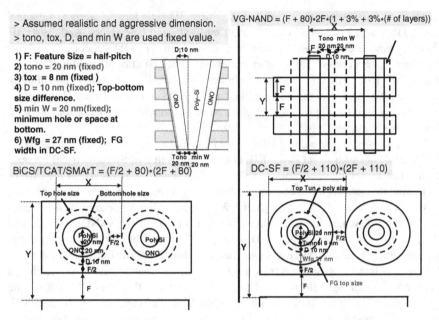

FIGURE 8.2   Assumption of 3D NAND flash memory cell size calculation (1).

Figure 8.4 shows the scaling trend of effective memory cell size in 16 stacked 3D NAND memory cells in comparison with 2D planar FG (P-FG) cells. It can be seen that 3D NAND memory cells cannot be effectively scaled down as feature size is scaled down, in comparison with P-FG MLC (planar FG MLC cells). It means that the increasing number of stacked cells is the only reasonable way to reduce the effective memory cell size in 3D cells.

	Planar-FG	P-BiCS/TCAT/SMArT	VG-NAND	DC-SF
Cell Size	$1.25 \cdot X \cdot Y$ (= 5F2)	$X \cdot Y$	$X \cdot Y \cdot (1 + 3\% + 3\% \cdot (\# \text{ of stacked layer}))$	$X \cdot Y$
X [BL pith]	2F	$F/2 + 2 \cdot t_{ono} + 2 \cdot D +$ min W	$F + 2 \cdot t_{ono} + 2 \cdot D +$ min W	$F/2 + 2 \cdot t_{ox} + 2 \cdot D +$ min W + $2 \cdot Wfg$
Y [WL pitch]	2F	$2F + 2 \cdot t_{ono} + 2 \cdot D +$ min W	2F	$2F + 2 \cdot t_{ox} + 2 \cdot D +$ min W + $2 \cdot Wfg$
Physical Cell Size [nm2] @F = 40 nm	8000	16000	9600	24700
Effective Cell Size @F = 40 nm,16stacked	--------	1000	906	1544 (SLC), 772(MLC)

● 3D memory cell size is strongly limited by tono, D, and minW, which are independent on Feature size, F.

1)   F: Feature Size = half-pitch
2) t_{ono} = 20 nm (fixed)
3) tox = 8 nm (fixed)
4) D = 10 nm (fixed); Top-bottom size difference.
5) min W= 20 nm(fixed); minimum hole or space at bottom.
6) Wfg = 27 nm (fixed); FG width in DC-SF.

FIGURE 8.3   Assumption of 3D NAND flash memory cell size calculation (2).

- BiCS/TCAT/SMArT/DC-SF cannot be scaled down efficiently with feature size (F) scaling.
- Scalability; Planar-FG
  >> VG-NAND
  > BiCS/TCAT/SMArT = DC-SF
- 40-nm Feature size (ArFi limitation) would be reasonable to use 3D cell, not to increase process cost.

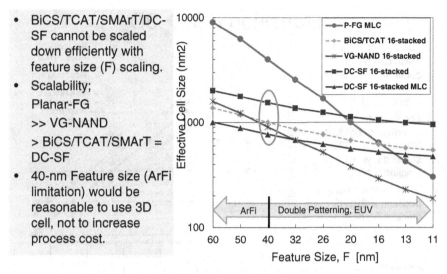

FIGURE 8.4   Scalability of 3D NAND flash memory cell.

Figure 8.5 shows the transition scenario from 2D planar FG NAND cells to 3D NAND cells. The effective cell size of a 1Y-nm 2D cell is nearly equal to that of 16 stacked 3D NAND of BiCS or TCAT or DC-SF cells. This means that a transition from 2D cells to 3D cells is possible from 1Y-nm generation 2D NAND cells if more than 16 stacked 3D NAND cells are used. In fact, 3D NAND flash production was started in 2013 by using 24 stacked cells, and it was extended to 32 stacked cells in 2014, and 48 stacked cells in 2015.

8.3   DATA RETENTION

8.3.1   Quick Initial Charge Loss

The data retention characteristic of a 3D SONOS cell is much different from that of a conventional 2D FG cell. In general knowledge for 2D cells, the 2D SONOS cell shows a larger V_t shift in a retention bake than the 2D FG cell because of a quick charge detrapping through thinner tunnel oxide. Figure 8.6 shows the data retention characteristics after 3K program/erase cycles in both (a) a 3D SONOS SMArT cell and (b) a 2D 2y-nm generation FG cell [2]. The V_t distribution width of a 3D SONOS cell after a 3K cycle is very tight in comparison with that of 2D FG cells. However, after high temperature (HT) data retention bake, V_t distribution width become larger and a large V_t shift-down is observed. These data retention characteristics are similar to well-known data retention characteristics in a 2D SONOS cell (charge trap (CT) cell). In a 3D SONOS cell, the data retention characteristic is one of key challenges that have to be improved or managed.

The quick initial charge loss had been reported in a 2D SONOS cell [3]. Figure 8.7 shows the typical data retention test results for the 2D SONOS cell (a) with cut

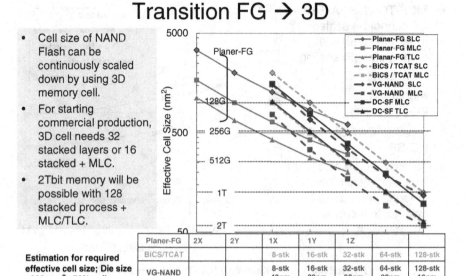

- Cell size of NAND Flash can be continuously scaled down by using 3D memory cell.
- For starting commercial production, 3D cell needs 32 stacked layers or 16 stacked + MLC.
- 2Tbit memory will be possible with 128 stacked process + MLC/TLC.

Estimation for required effective cell size; Die size <200 mm², 70% cell efficiency

Planer-FG	2X	2Y	1X	1Y	1Z
BiCS/TCAT			8-stk	16-stk	32-stk	64-stk	128-stk
VG-NAND			8-stk 40nm	16-stk 32nm	32-stk 26nm	64-stk 20nm	128-stk 16nm
DC-SF			8-stk	16-stk	32-stk	64-stk	128-stk

FIGURE 8.5 Transition scenario from planar FG cell to 3D cell.

ONO and (b) without cut ONO [3]. All cells are first erased to $V_t < 0$ V and then programmed to $V_t > 3$ V, followed by immediate data retention measurements. In cells of $V_t > 4$ V, the quick initial charge loss are observed and quickly saturates to a 200 to 300-mV V_t shift within 1 s. Even in cells of $V_t = 3.4$ V, the quick charge loss of 100 to 200 mV within 1 sec are observed. The cells in both WL-etching processes cut ONO and non-cut ONO show similar behavior. This indicates that the quick initial

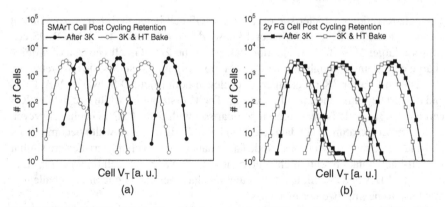

FIGURE 8.6 Cell V_{th} distribution of high temperature (HT) retention bake after P/E cycling of (a)3D SMArT Cell and (b) 2D 2y-nm node FG cell.

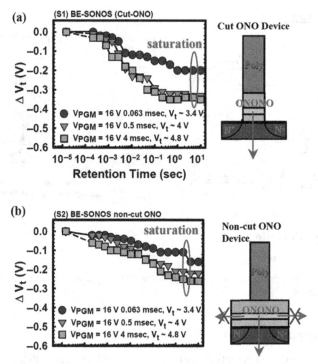

FIGURE 8.7 (a) Quick initial charge loss of 2D SONOS cell with cut ONO, under different program V_t conditions ($V_{t,\text{pgm}} = 3.4, 4, 4.8$ V). Higher programmed state V_t shows a larger charge loss that saturates in < 1 s to a few seconds. (b) Quick charge loss of 2D SONOS with non-cut ONO, under different program V_t conditions. Since the two devices show similar characteristics, the quick charge loss is not caused by charge lateral spreading, but rather through a vertical charge loss mechanism.

charge loss is not related to the charge lateral migration in SiN but to the charge loss through a vertical path.

This quick charge loss phenomenon increases the V_{th} distribution width of program states, and thus it makes damage on the reliability of the programmed data. To minimize the quick charge loss problem, the negative counter pulse scheme was introduced [4,5] in a 3D V-NAND device, as shown in Fig. 8.8. The negative gate voltage is applied to a selected word line just after program pulse (V_{pgm}), while applying V_{read} to unselected word lines to make self-boosting on the channel potential, as shown in Fig. 8.8a. The electron detrapping from charge storage SiN to channel is accelerated by the field between the negative gate voltage and the boosting channel potential. Therefore, during the program sequence, the programmed state's V_{th} distribution can be improved, as shown in Fig. 8.8b. Since this operation is performed in verify read operations, timing penalty for applying this scheme is small.

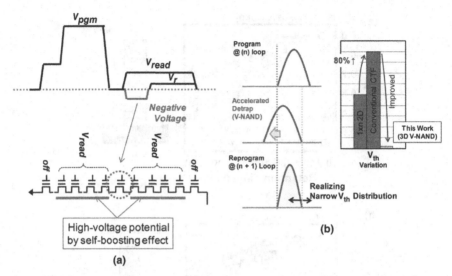

FIGURE 8.8 (a) Negative counter-pulse scheme. (b) Diagram of its mechanism and measured V_{th} variation.

8.3.2   Temperature Dependence

Another important data retention issue in a 3D SONOS cell is that a charge loss mechanism is changed over the temperature range in a regular acceleration test, in contrast with the same mechanism over test temperature in a conventional 2D FG cell, as shown in Fig. 8.9 [2]. In a 2D FG cell, V_t shift is a linear relationship with bake temperature. It means that the data loss mechanism is the same over acceleration bake temperature. However, in a 3D SONOS cell (SMArT cell), V_t shift is a nonlinear relationship with bake temperature. It suggests that the data loss mechanism is different between low temperature and high temperature. The data loss

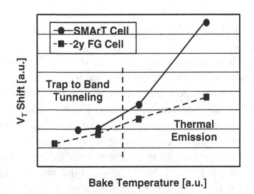

FIGURE 8.9 Temperature dependence of high temperature (HT) retention bake of SMArT cell, compared with 2D 2y-nm FG cell.

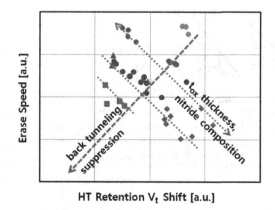

FIGURE 8.10   Trade-off relationship of erase speed and high temperature (HT) data retention V_t shift in a SMArT cell.

mechanisms are considered to be the band-to-band tunneling at low temperature (LT) and the thermal emission at high temperature (HT) [2]. Therefore, the lifetime of data retention cannot be estimated by a simple temperature acceleration test, which has been used as the most general in a 2D FG cell. It is considered that LT lifetime below 90°C has to be evaluated from extrapolation of a long time test in relative LT, at least 3-week V_t shift.

The data retention characteristic of a charge trap cell has trade-off relationship with the erase speed. The typical trade-off relationship between the erase speed and the charge loss at high temperature (HT) is shown in Fig. 8.10 [2]. The way to reduce HT charge loss without the slow erase is to suppress the conduction of the electron from gate to nitride (back tunneling). The suppression of back tunneling can be normally realized by using a large work-function metal gate (e.g., TaN/W) and physically thick Hi-k block dielectric film (e.g., Al_2O_3, or Hf-oxide).

8.4   PROGRAM DISTURB

8.4.1   New Program Disturb Modes

Program disturb phenomena and mechanisms of a 3D NAND cell are much different from that of a 2D NAND cell.

Figure 8.11a shows a schematic view of the cell array architecture of a 3D NAND cell [6, 7] in BiCS, TCAT/V-NAND, and SMArT cell. The NAND string (STR) is located at the intersection point of the drain selection transistor (DSL) and the bit line (BL). Word lines of each string (STR) are connected at a common point in the block, namely Fig. 8.11a shows one physical block. The different point from a 2D NAND cell is that N strings in one block are connected to the same bit line through different select transistors of DSL_1 to DSL_N. In the case of a 2D NAND cell, one string

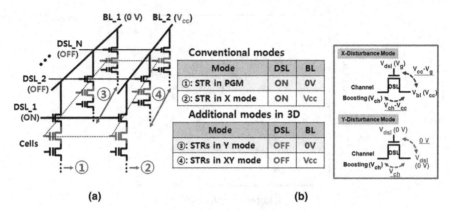

(a)   **(b)**

FIGURE 8.11   (a) Program disturbance modes of 3D NAND flash cell array. N strings are connected to the same bit line in the same block. (b) Comparison of program disturbance modes in a 3D NAND flash cell array. In a 3D NAND cell, two new modes of Y-mode and XY-mode are added.

in the block is connected to one bit line. This 3D NAND array architecture makes a new program disturb mode, as shown in Fig. 8.11b.

When DSL_1 is in the turn-on state (ON; selected), STRs along DSL_1 are in either programming (PGM) or program disturb X mode, depending on the BL bias, where "X mode" has BL in high bias of V_{cc}. Disturb X mode is the same as a conventional program disturb mode in a 2D NAND cell. However, in a 3D NAND cell, the remaining DSL_2 to DSL_N are in the turn-off state (OFF; unselected) so that STRs along DSL_2 to DSL_N are in either program disturb Y or program disturb XY mode, where "Y mode" has BL in 0 V and DSL in turn-off state, and "XY mode" has BL in high bias of V_{cc} and DSL in the turn-off state.

The XY mode is not more severe than the conventional X mode, because the boosting voltage in STR does not cause a leakage current through DSL due to DSL = OFF and BL = V_{cc}. However, the program disturb Y mode has much severer than conventional X mode, because the boosting voltage in STR may cause a leakage current through DSL due to BL = 0 V. In addition, DSL in a 3D NAND cell has the larger subthreshold slope than DSL in a 2D NAND cell. This means that DSL in a 3D NAND cell has large leakage current [2, 6]. Furthermore, in a 2D NAND cell, the leakage current through DSL does not occur in program disturb X mode because the V_t of a DSL transistor becomes high in the program disturb mode due to a strong body effect (strong back gate effect or source bias effect). However, in a 3D NAND cell, the leakage current of DSL cannot be easily suppressed because V_t of surrounding gate transistor of DSL does not become high due to a weak body-effect during program disturb conditions.

In order to suppress a program disturbance in a 3D NAND cell, several approaches to reduce leakage current through DSL were proposed [6]. They are (1) high V_t of DSL, (2) applying negative bias to DSL, and (3) inserting dummy word lines between DSL and edge word line. Both (1) and (2) can decrease a large leakage

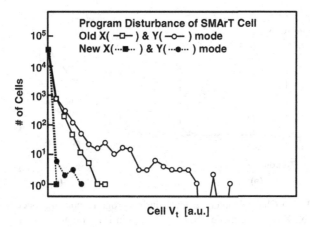

FIGURE 8.12   Suppression of program disturbance fail bit by applying the optimized program inhibit conditions.

current of DSL in a 3D cell. And (3) can control the potential drop from boosting voltage to the voltage applied on DSL, and (3) can also manage the hot carrier generation at edge word-line region by relaxing a high lateral electric field. Therefore, the dummy WL condition (bias, V_t setting, number of dummy WLs, etc.) have to be carefully designed. Figure 8.12 shows the improvement results of program disturb characteristics of X and Y mode by applying conditions of (1), (2), and (3) [2].

8.4.2   Analysis of Program Disturb

Detail mechanism of program disturb in 3D NAND cell had been analyzed [7].

The channel (CH) potential profile of each program disturb modes is calculated by TCAD simulation, as shown in Fig. 8.13a. The CH boosting level of each program disturb modes is determined by the different DSL leakage levels according to bias conditions of BL and DSL. Typically, the CH boosting level of Y mode is the lowest among three modes, and the CH boosting level of XY mode is same or lower than that of X mode. Figure 8.13b shows the measured incremental step pulse programming (ISPP) characteristics of both the program (PGM) mode and three program disturb modes. The CH boosting level of V_{ch} can be derived by the V_G different between the ISPP curves of three program disturb modes and that of the PGM mode. It is confirmed that the V_{ch} of the Y mode is smaller than that of other program disturb modes, which is consistent with the TCAD simulation results in Fig. 8.13a.

In actual array operation, all three program disturb modes occur simultaneously, and thus disturb fail bits appear by statistically complicated circumstance of neighbor cells. Figure 8.14 shows the simulation (model) and measurement (chip) results of the erase (ERS) cell V_t distribution in initial and after programming all pages in a block. The modeling parameters include the ISPP characteristics in Fig. 8.13b, CH boosting variations, RTN, and initial ERS cell V_t distribution. And the modeling parameters

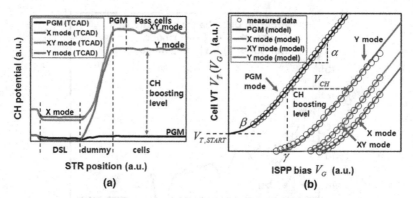

FIGURE 8.13 (a) Channel potential level during program boosting operation by TCAD simulation for X, XY, and Y mode. (b) Cell V_t shift in ISPP program operation of program (PGM) mode and three program disturb modes of X, XY, and Y mode.

were calibrated so that ERS cell V_t distribution after programming is consistent with measured data in cell array. It appears that many bits are over the read voltage (V_r) and then become failure bits.

The effects of three program disturb modes were analyzed by using this model.

In general, program disturb failure bits are caused in ISPP end bias, because cell V_t become highest by highest V_{pgm}, as shown in Fig. 8.13b. Therefore, the probability for a specific cell to be in each program disturb mode at a specific ISPP bias is calculated and shown in Fig. 8.15. In the case of conventional 2D cell, failure bits are caused in only X mode. However, in the case of a 3D cell, failure bits are caused in not only X mode but also two additional modes of Y and XY modes. The probability

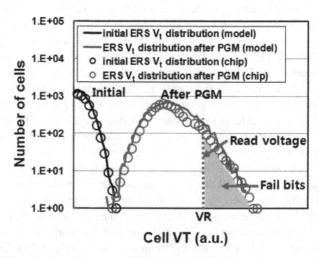

FIGURE 8.14 Change of the erase (ERS) V_t distribution after program operation (PGM).

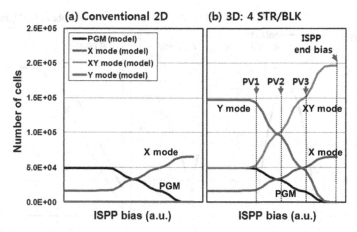

FIGURE 8.15   Probability (number of cells) of each program disturbance mode with increasing ISPP bias.

for a cell in Y mode, which has the lowest CH boosting level, constantly decreases with increasing ISPP bias so that there are a few cells in Y mode at ISPP end bias. However, the portion of cells in XY mode, which has lower CH boosting level than X mode, increases constantly so that there are three times more cells in XY mode than those in X mode at ISPP end bias.

Figure 8.16 shows the effects of Y mode on the ERS cell V_t distribution after programming all pages in a block. The CH boosting level of Y mode is intentionally lowered from the reference level while that of other two modes is fixed on the reference

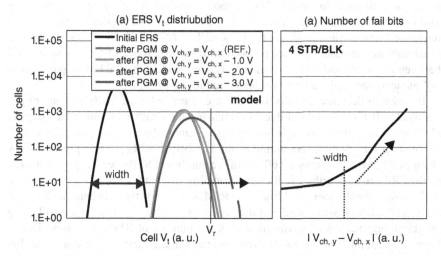

FIGURE 8.16   Effects of Y mode on the erase (ERS) V_t distribution after programming (PGM).

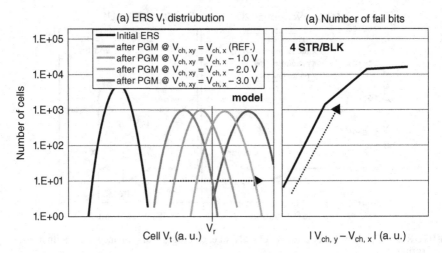

FIGURE 8.17   Effects of XY mode on the erase (ERS) V_t distribution after programming (PGM).

level. It was found that the right-side tail of the ERS cell V_t distribution following programming begins to move up to the positive direction when the difference in CH boosting level between Y mode and other two modes are over 2.0 V. This means that the fail bits (right-side tail bits) originate from Y mode disturb cells in high stress at the end of ISPP, even if these Y mode cells have very low probability of occurrence at the end of ISPP in actual array operation.

Figure 8.17 shows the effects of XY mode on the ERS cell V_t distribution after programing all pages in a block. The CH boosting level of XY mode is intentionally lowered from the reference level while that of other two modes is fixed on the reference level. It was found that the peak of the ERS cell V_t distribution following programming increases in exact proportion to the difference of CH boosting level between XY mode and other two modes. This means that the fail bits originate from XY mode disturb cells in high probability at the end of ISPP.

In order to decrease an effective cell size, the number of STRs per one physical block (BLK) is increased due to decreasing the number of wide gate space in block boundary. Therefore, it is important to estimate the change in the ERS cell V_t distribution following programming as the number of STRs per one physical BLK changes. Figure 8.18 shows the effects of the number of STRs per one physical BLK on the ERS cell V_t distribution after programming. The number of STRs in a physical BLK is intentionally increased from 1 STR/BLK, which is the conventional 2D case, to 16 STR/BLK. It was found that the peak of ERS state V_t distribution following programming increases logarithmically as the number of STRs increases. This is because the number of XY mode stress increases in the exact proportion to number of STRs in a physical BLK. Thus, the resultant fail bits come from the repeated application of ISPP end bias.

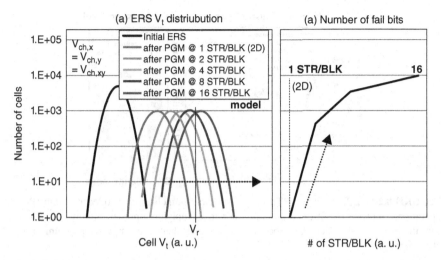

FIGURE 8.18 Effects of cell architecture (number of NAND string in block) on the erase (ERS) V_t distribution after programming (PGM).

Figure 8.19 shows the CH potential profile of cell STR under one of the program disturb modes. There are two major leakage current paths. One is the diffusion current from the bit line (BL) through the DSL. The other is the generation of electron–hole pairs in the area of dummy word lines, where potential drop occurs. The mechanism of electron–hole generation would be trap-assisted generation in a poly silicon channel or in band-to-band tunneling (BTBT) [2, 6]. Therefore, it is important to minimize these currents in order to maintain high CH boosting level of each program disturb mode.

When a STR is under one of the program disturb modes, there is a considerable CH potential drop in the relatively narrow region of dummy WLs, which in turn

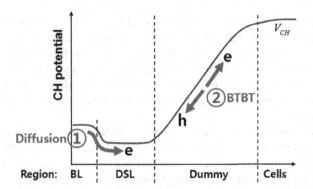

FIGURE 8.19 Two leakage paths that determine the CH boosting level. One is the diffusion current from the bit line (BL) through the DSL. The other is the generation of electron–hole pairs in the area of dummy word lines, where potential drop occurs.

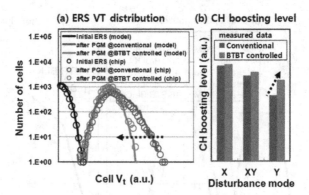

FIGURE 8.20 Effects of the BTBT (band-to-band tunneling) controlled program (PGM) operation. The BTBT-controlled program operation (applying a proper dummy WL scheme) can improve the channel (CH) boosting level and then can improve program disturb characteristics.

generates the electron and hole current. Therefore, it is very important to minimize the electric field in the dummy WL region by applying appropriate bias on dummy WLs together with targeting their V_t. Figure 8.20b shows the measured CH boosting level of conventional case and with applying a proper dummy WL scheme (potential and BTBT controlled). The CH boosting level of X, XY, and Y mode are improved up to 20% due to minimization of electron and hole pair generation in the dummy WL region. Figure 8.20a shows measured ERS cell V_t distribution after program in the array under a proper dummy WL scheme. The failure bits in the right-side tail are well suppressed as a result of a high CH boosting level of the Y mode. Moreover, modeling results are consistent with measured data in array.

8.5 WORD-LINE RC DELAY

The RC (resistance–capacitance) delay of word line has to be minimized for high-speed operation of read and program. In general, the guideline of the RC value is around 1 µs. It means that the time of ramp up and down of word line is about 3 µs (= 3RC), which has less impact on read and program performance. In order to minimize the RC value in 2D cells, the low-resistance materials (CoSi, TiSi, W, etc.) are applied on word line because word-line capacitance cannot be easily reduced due to fixed structure.

In stacked word-line (WL) structure in 3D NAND cells, such as BiCS, TCAT (V-NAND), and SMArT, a WL capacitance is greatly increased from 2D cells. This is because a plane structure of WL in 3D cells has a large parasitic capacitance, while a line structure in a 2D cell has a relatively small parasitic capacitance. Figure 8.21a shows the stacked WL structure of the 3D NAND cell along with resistor and capacitor models [4, 5]. Figure 8.21b shows the WL resistance and capacitance of 3D cells in comparison with those of planar 2D NAND cells [4, 5]. From comparison of 2D and

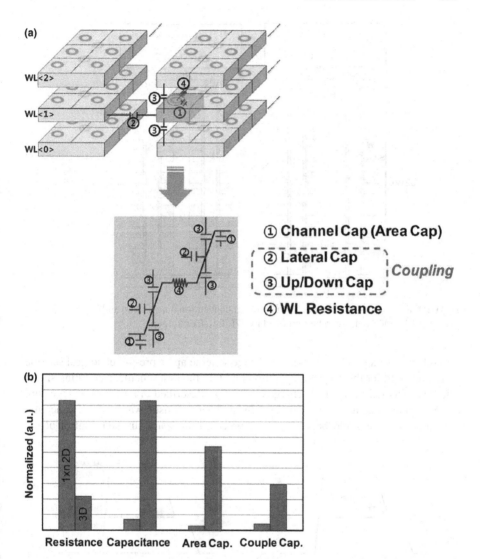

FIGURE 8.21 Word-line (WL) resistance and capacitance. (a) Model of 3D NAND cell (V-NAND array). (b) Comparison with a 2D planar cell.

3D cells in Fig. 8.21b, the RC delay of 3D cell is estimated to be around twice larger than that of 2D cells (resistance, 1/4; capacitance, 8 times). In order to reduce the WL RC delay in future 3D cells, the low-k dielectric or air gap between WLs would be effective, as shown in Fig. 8.22.

In 3D cells, the coupling capacitance between WLs is more than four times larger than that in a planar 2D cell, as shown in Fig. 8.21b. Because of this coupling, a large glitch is caused in the neighboring WLs during program and read operations, and it results in causing an unexpected disturbance problem. In order to resolve the problem,

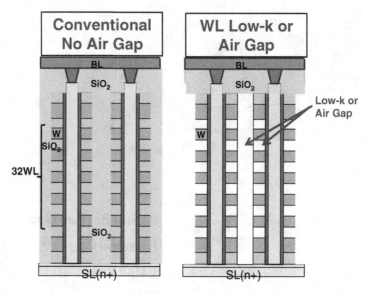

FIGURE 8.22   Schematic 3D cell structures of conventional no air gap and WL low-*k* or air gap. Word-line RC delay can be improved by WL low-*k* or air gap.

two schemes of a glitch-canceling discharge scheme and a pre-offset control scheme were proposed for the program operation [4, 5]. In the first scheme, a coupling signal glitch is canceled by a WL discharge circuitry as described in Fig. 8.23a. Since core circuit signals operate in a very predictable and deterministic way, a glitch-canceling discharge scheme could be achieved by precise timing and amount control of the

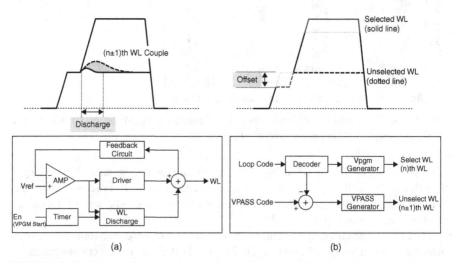

FIGURE 8.23   Bias and block diagrams of (a) a glitch-canceling discharge scheme and (b) a pre-offset control scheme.

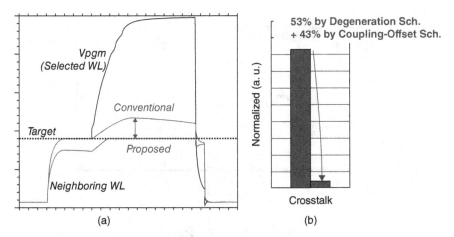

FIGURE 8.24   Simulated waveform of a word-line (WL) signal in a 3D V-NAND array with a glitch-canceling discharge scheme and a pre-offset control scheme.

discharge circuitry. In the pre-offset control scheme, the amount of the coupling glitch is predicted based on the target voltage of the aggressor WL. Then, the target voltage level of the neighbor victim WL could be adjusted according to the prediction, as shown in Fig. 8.23b. Figure 8.24 shows the worst-case simulation result of the proposed two schemes [4,5]. The aggressor WL is the selected WL, while the victim WL is the adjacent unselected WLs. As shown in the figure, the coupling glitch was significantly reduced by using the proposed schemes, which in turn results in eliminating the disturbance caused by WL-to-WL crosstalk.

8.6   CELL CURRENT FLUCTUATION

8.6.1   Conduction Mechanism

The 3D NAND flash cell has a concern to have the larger fluctuation in cell current than the 2D NAND flash cell, because the 3D cell has a different process and structure, such as a poly-Si channel, charge trap cell (thinner tunnel oxide), and tunnel oxide by deposition process. Before developing 3D NAND flash memory, there had been many reports, for example in reference 8, that conduction mechanism and modeling for poly-Si channel was investigated in the planar thin film transistors (TFTs). To understand characteristics of the 3D NAND cell, the very thin poly-silicon (77–185 Å) transistors had been investigated as TFTs [9]. The result showed that the transfer characteristics such as ON current and mobility are enhanced by large grain of poly-Si channel even if it is very thin poly-Si thickness. The 3D cell with a vertical and cylindrical poly-Si channel had been also investigated [10–18]. The variation and fluctuations of conduction property was reported [10–18].

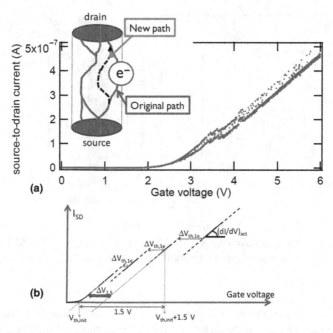

FIGURE 8.25   (a) A selected example of a I_{SD}–V_G characteristic measured with $V_D = 0.1$ V and 1-mV resolution (6000 dots are individual measurement points). This example clearly shows how the curve consists of a discrete number of well-defined curves, each corresponding to a different configuration of percolation paths from source to drain. Trapping of individual electrons (partially) blocks a conduction path as schematically drawn in the inset. A net negative charging is observed with increasing V_G. (b) Schematic reproduction of the I_{SD}–V_G characteristic in (a) with definitions. (i) $\Delta V_{th,1e}$ is the V_{th} shift caused by the trapping of one electron. Three different shifts are indicated. (ii) $(dI/dV)_{act}$ is the transconductance of the linear parts of the curve. (iii) $\Delta V_{1.5}$ is the total $V_{th\text{-}shift}$ between the $V_G = V_{th,init}$ and $V_G = V_{th,init} + 1.5$ V. It is a measure of the charging component in the I_{SD}–V_G curve.

A statistical evaluation of current–voltage characteristics in cylindrical (macaroni) and vertical poly-Si channels transistor had been studied the poly-Si conduction properties and defects [10].

A methodology is unique to extract all components that control the conduction in the poly-Si channel from simple I_{SD}–V_G characteristics. Figure 8.25a shows a 25°C I_{SD}–V_G characteristic up to $V_G = 6$ V measured with a 1-mV V_G resolution. The I_{SD}–V_G curve shows several current drops or shifts to higher V_G in the region of $V_G > V_{th}$. This phenomenon does not appear in transistors on a single-crystalline-Si substrate. The curve is made up of several well-defined curves, each corresponding to a particular current path configuration as illustrated schematically in the inset of Fig. 8.25a. One trap located close to a current path captures a single electron, and the path is partially blocked and then I_{SD} is shifted up. At increasing V_G, the I_{SD}–V_G curves shift to a higher voltage many times, indicating a negative charging by electrons into a poly-Si channel. The V_{th} jumps (I_{SD} shifts up) corresponding to individual electron charging, $\Delta V_{th,1e}$, as defined in Fig. 8.25b, definition (i), can be measured on many devices. The $\Delta V_{th,1e}$ has an exponential distribution and can be fully characterized for a high $\Delta V_{th,1e}$ tail. Because of the charging at increasing V_G, the $I_{SD}-V_G$ curve is stretched out. The actual transconductance in the current paths, $(dI/dV)_{act}$, is defined as the slope of the linear parts of the $I-V$ corresponding to a fixed charge configuration, definition (ii) in Fig. 8.25b. The charging component is defined in Fig. 8.25b, definition (iii), by using $\Delta V_{1.5}$ to determine the total V_{th} shift for applied V_G between $V_{th,init}$ and $V_{th,init} + 1.5$ V. This shift $\Delta V_{1.5}$ can be as large as 0.8 V and depends strongly on the material and the temperature.

Three channel materials of microcrystalline-Si (μc-Si), poly-Si (p-Si), and large grain poly-Si (lgp-Si) were used in single vertical poly-Si channel transistor. The distributions of $\Delta V_{th,1e}$, $\Delta V_{1.5}$, and $(dI/dV)_{act}$ taken at four temperatures (25°C, 60°C, 100°C, 130°C) and for the three channel materials are presented in Fig. 8.26 [10]. The conduction mechanism can be interpreted by using $\Delta V_{1.5}$ for the charging component and $(dI/dV)_{act}$ for current path conduction.

In μc-Si shown in (1a)–(4a), a large charging component ($\Delta V_{1.5}$) at 25°C is observed, as shown in Fig. 8.26 (2a). The charging components ($\Delta V_{1.5}$) at higher temperature are largely decreased. It suggests that the conduction is mainly dependent on shallow energy level traps, which can be easily discharged with limited thermal energy. A temperature dependence of $(dI/dV)_{act}$ in Fig. 8.26 (3a) shows the temperature-activated thermionic emission of electrons over defect-induced barriers, as schematically illustrated in Fig. 8.26 (4a). Reduction or passivation of the traps with hydrogen would be the main challenge for the μc-Si channel.

In p-Si shown in (1b)–(4b), the charging component ($\Delta V_{1.5}$) is not only smaller than $\Delta V_{1.5}$ of μc-Si but also its temperature dependence is reduced, as shown in Fig. 8.26 (2b). No significant difference between 25°C and 60°C is observed. Moreover, between 25°C and 60°C, the $\Delta V_{th,1e}$ distribution drastically reduces and narrows (Fig. 8.26 (1b)). It suggests a redistribution of the percolating current paths. At 25°C the current is confined to a small number of paths that are very sensitive to single electron trapping in shallow states, but above 60°C the current flow becomes more uniform. The temperature dependence of $(dI/dV)_{act}$ of p-Si is identical to μc-Si, and current path is controlled by thermionic emission over defect-induced barriers, as shown in Fig. 8.26 (4b). The impact of temperature on $\Delta V_{1.5}$ in the range 60°C–130°C is very small, as shown in Fig. 8.26 (2b). This indicates that it is more difficult for energetically deeper traps to discharge thermally.

In lgp-Si shown in (1c)–(4c), the large charging component ($\Delta V_{1.5}$) with small temperature dependence (Fig. 8.26 (2c)) indicates a high density of deep traps. The value of $(dI/dV)_{act}$ is much larger than that of μc-Si and p-Si, as shown in Fig. 8.26 (3c). This is very important to obtain a large cell current for a 3D NAND flash. However, the spread of $(dI/dV)_{act}$ is very large in comparison to that of μc-Si and p-Si, as shown in Fig. 8.26 (3c). And there is only a weak temperature dependence. This can be explained by wider current paths in large Si-grains, as illustrated in Fig. 8.26 (4c). The grain boundaries would not act as current barriers, but only as trapping centers with mainly deep energy traps. The large poly-Si grain size causes a large device-to-device variability resulting in a broad spread of both $(dI/dV)_{act}$ and

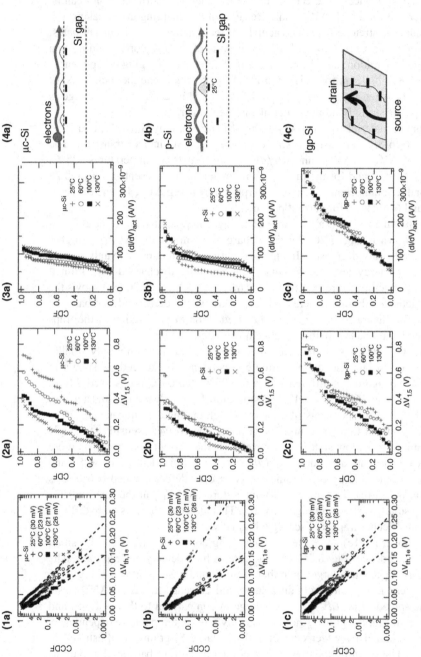

FIGURE 8.26 $\Delta V_{th,1e}$, $\Delta V_{1.5}$, $(dI/dV)_{act}$ of three channel materials of μc-Si (micro-crystalline-Si), p-Si (poly-Si), and lgp-Si large-grain poly-Si). (1) Distributions of $\Delta V_{th,1e}$ for different temperatures and materials. For p-Si (1b) a large difference between 25°C and the other T values indicates a thorough redistribution of current paths. (2) Distributions of the charging component quantified by $\Delta V_{1.5}$. Lowest charging is for p-Si. Largest T-dependence is for μc-Si; indicating shallow states. (3) Distribution of the actual transconductance $(dI/dV)_{act}$. Large spread for lgp-Si indicates few grain boundaries. Very narrow spread in μc-Si indicates the most uniform current flow. (4) Schematic drawing of the conduction model for each of the studied poly-Si options. (4a) and (4b) are band diagram schemes. (4c)

356

$\Delta V_{1.5}$. Improving variability is the main challenge for lpg-Si implementation in a vertical stacked device.

In summary on channel material analysis, the large-grain poly-Si has higher mobility and higher transconductance than microcrystalline-Si and poly-Si; however, variability of a large-grain poly-Si is larger (worse) than that of microcrystalline-Si and poly-Si due to the wider current path in the large poly-Si grain.

More detailed analysis had been performed by using the same methodology and the same channel materials [11]. The values of $\Delta V_{TH,single}$'s $(= \Delta V_{th,1e}$'s) caused by trapping of single electron in poly-Si (see Fig. 8.25b) is significantly larger than those expected from the charge sheet approximation ($\eta_0 \sim q/C_{OX} = 1.2$ mV). The detected values of $\Delta V_{TH,single}$ can reach magnitudes as large as hundreds of times the η_0 value. This implies that the conduction between source and drain is concentrated in a limited number of percolation paths, which can be blocked by trapped electrons [19], as described in the sketch in Fig. 8.25(a). As also observed on deeply scaled planar FETs and FinFETs, the $\Delta V_{th,single}$ distribution follows to the first approximation an exponential distribution, related to the probability of finding a trap at a given distance from the critical point in a percolation path [20]. The average values $\eta = \Delta V_{th,single}$ tend to increase with increasing grain size, as shown in Fig. 8.27a. With an increasing diameter of channel, a reduction of the tail of $\Delta V_{th,single}$ distribution is observed due to the increased number of percolation paths reducing the relative impact of single traps, as shown in Fig. 8.27b.

The I_{READ} $(= I_{SD})$ distribution for lgp-Si stretches out to larger values, but lower I_{READ} tail converges with the lower tail of poly-Si or μc-Si at extremely low percentiles, as shown in Fig. 8.28a [11]. Therefore, even if a large average grain size is used for a channel, the microcrystalline-Si structures remain in a low percentage of less than a few percent, resulting in a similar lower tail of I_{READ} to μc-Si and p-Si. Long N2 annealing slightly shifts the reading current at higher values due to the increase of the average transconductance, as shown in Fig. 8.28b. The lower tail of

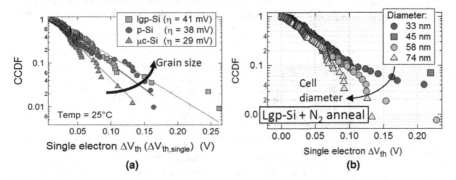

FIGURE 8.27   (a) Complementary cumulative distributions (CCDF = 1-CDF) of threshold voltage shifts $\Delta V_{th,single}$ caused by single electron trapping (see Fig. 8.25) for different poly-Si materials. CCDFs follow an exponential distribution with the average values η given in the inset. (b) A weak reduction of the impact per trap on the $\Delta V_{th,single}$ is observed by increasing the cell diameter.

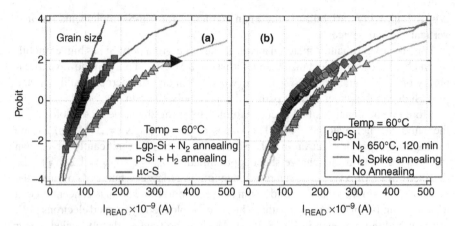

FIGURE 8.28   (a) The highest I_{READ} values are registered for lgp-Si. Interestingly, at low percentiles, the predicted currents converge. (b) Long N2 annealing slightly shifts the reading current at higher values.

I_{READ} in lgp-Si seems to be in sufficient percentage to have an impact on actual read operation in 3D NAND device. It would be a potential issue in a 3D NAND cell.

8.6.2 V_G Dependence

The switching traps were characterized to see their physical properties. A statistical analysis in different channel poly-Si was performed in comparison with monocrystalline planar nFETs. It was confirmed that a significant part of the switching traps were in the poly-Si channel [12].

Figure 8.29 shows two examples of one trap switching with a typical RTN signal in lower V_G of below V_{th}, and one trap switching with a sharply defined switching voltage of $V_{G,switching}$ in higher V_G of above V_{th}, more than 3.5 V in this case [12]. As Fig. 8.30a shows, a large density of switching traps was observed close to V_{th} (band #1)

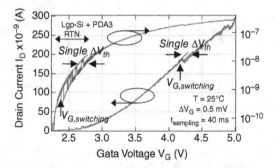

FIGURE 8.29   Random telegraph noise (RTN) and abrupt current drops are clearly observed during the I_D-V_G tracing when high V_G resolution and short sampling time are used. Drain voltage was fixed at 100 mV.

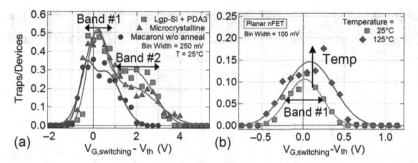

FIGURE 8.30 (a) Independently of the polysilicon option, two distinctive bands appear at the spectra of the number of switching traps normalized by the number of tested devices. (b) For the monocrystalline planar reference, only one band is visible, already indicating that traps switching at higher V_G are polysilicon related.

with a prominent shoulder (band #2) at higher V_G for all the poly-Si processes. Figure 8.30b shows the trap spectrum for monocrystalline planar nFETs obtained in the same procedure. According to the model presented in Fig. 8.31, band #1 is due to the charging of the interface traps and the poly-Si channel traps that shift below the Fermi level E_F when the gate voltage V_G is swept up to V_{th}. For the monocrystalline planar nFETs, the trap spectrum increases with higher temperature (Fig. 8.30b) and is associated with interface traps which are strongly thermally activated. On the other hand, opposite behavior with temperature is observed for poly-silicon channel. The switching traps are reduced at higher temperature. Therefore, it is considered that an important portion of traps are in the poly-Si bulk.

On the other hand, for $V_G > V_{th}$ as shown in Fig 8.31b, the Fermi level E_F in the channel remains at a fixed level, and only the defects that lie within a few kT from E_F can cause I_D fluctuations (band #2). Taking into account that this band was not detected in the reference monocrystalline planar nFETs as shown in Fig. 8.30b, it was concluded that the traps of band #2 are exclusively inside the poly-Si channel.

For the macaroni structure channel, the lowest density of switching traps is observed, as shown in Fig. 8.30a, even though they present the highest interface trap

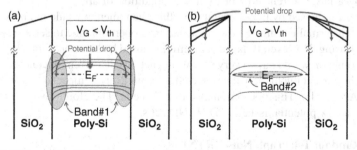

FIGURE 8.31 (a) For $V_G < V_{th}$, the Fermi level E_F sweeps the poly-band gap, progressively charging defects at the interface and in the polysilicon bulk. (b) Afterwards, the E_F in the channel is pinned and only the traps aligned with E_F can cause RTN events.

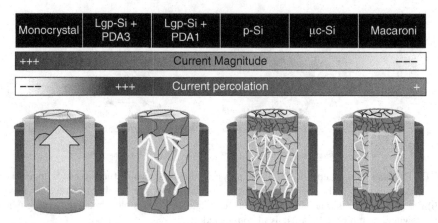

Monocrystal	Lgp-Si + PDA3	Lgp-Si + PDA1	p-Si	µc-Si	Macaroni

+++	Current Magnitude	---

---	+++	Current percolation	+

FIGURE 8.32   Higher current is achieved by enlarging the polysilicon grain size, at the expense of increasing the impact of single traps.

density. This is a direct consequence of the level of percolating conduction path in the channel as sketched in Fig. 8.32. The impact of a single electron-trapping/detrapping event on V_{th} increases with increasing grain size (Fig. 8.27a), intensifying the probability of detecting giant current fluctuations. In the ideal case of a single crystal channel nFET, uniform conduction will be restored and the impact of single electron trapping/detrapping events will be drastically reduced.

In the analysis above, the current drop or shift occurs in ON current region ($V_G > V_{th}$) in poly-Si channel, as shown in Figs. 8.25–8.30. This phenomenon could not be observed in monocrystalline planar nFET and 2D floating-gate NAND flash cells. In 3D NAND flash, the current drop or shift in ON current region ($V_G > V_{th}$) would have a strong impact on cell current fluctuations. This is because unselected cells, which are connected to selected cells in series, operate under the ON current region ($V_G > V_{th}$). If unselected cells have a current drop or shift, a cell current could cause the fluctuations.

However, the analysis above was done in a single vertical poly-Si channel transistor. It is not clear so far whether the current shift (V_t shift, such as $\Delta V_{th,1e}$) is caused by a large current flow or by a bias condition of applied $V_G(> V_{th})$. If the current shift is caused by large current flow, this phenomenon would not be a serious issue, because smaller current flows in actual read operation due to series resistance in NAND string. However, if the current shift is caused by bias condition ($V_G > V_{th}$), this phenomenon would have a very serious impact on cell current fluctuation. This is because the current fluctuation of unselected cells has a direct impact on a read current in 3D NAND cells. This does not happen in 2D cells. Therefore, this phenomenon would be a new potential issue in 3D NAND cells.

8.6.3   Random Telegraph Noise (RTN)

It was also reported that the cell current of 3D NAND cell greatly fluctuated by a trap inside a poly-Si channel [14].

The channel poly-Si is made of silicon grain with different crystalline orientations. Traps at a grain boundary induce cell threshold voltage (V_t) fluctuations that are dependent on their location variations [8,21]. The charge transfer characteristics (ON current) of a planar thin film transistor (TFT) with a poly-Si channel are dominated not by poly-Si thickness but by the grain size of poly-Si [9]. And in the case of the same thickness of a poly-Si channel, the thinner poly-Si channel has better subthreshold characteristics without degradation of ON current and reliabilities. The average grain size integrated is larger than 100 nm in the report [9], however, SEM images show many types of defects, such as micro-subgrain, stacking fault, or multi-twin inside one grain that induces traps sites.

It had been reported that the variation of cell threshold voltage induced by a poly-Si trap was caused by two intrinsic mechanisms of random trap fluctuation (RTF) and random telegraph noise (RTN) [14].

The channel poly-Si had been modeled as a silicon material with high trap density distributed uniformly inside the channel. The trap distribution is evaluated by fine tuning the I_{BL}–V_{WL} curve of a 3D cell with a temperature range of −20°C to 85°C, as shown in Fig. 8.33a [14]. As reported in references 8 and 21, the large trap tail at the band edge was confirmed by the positive current dependence with temperature. The trap density at mid-gap, in the 1–5×10^{18}-cm^{-3} range, was derived by V_t and subthreshold slope temperature dependence. The same temperature dependence of the silicon mobility with an effective mobility calibrated to $130 \, \text{cm}^2 \cdot \text{V}^{-1} \cdot \text{s}^{-1}$ had been used [8]. The cell V_t fluctuations are simulated by defining one trap at one location in space and energy. Thus the energy distribution is separated into a single energy level, as shown in Fig. 8.33b [14]; and for each energy level, at each location, the charge distribution follows a Poisson distribution with a coefficient corresponding to the continuous trap number. The benefit of this approach is not needed for the calibration parameters. The trap distribution is already calibrated from the I_{BL}–V_{WL} curves. As shown in Fig. 8.33a, this model had an excellent agreement with the measurement of the V_t distribution of a 3D NAND cell array with RTF that is described above [14].

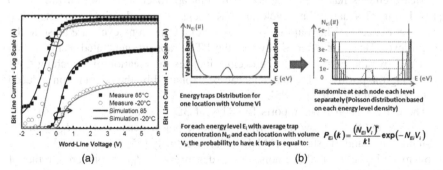

FIGURE 8.33   (a) Measured and simulated cell current at −20°C and 85°C (linear and logarithmic scale). The measured cell is in the median of the array distribution. (b) RTF modeling strategy. The model assumes a poisson distribution from the traps density concentration used for calibrating of (a). No other parameters need to be assumed.

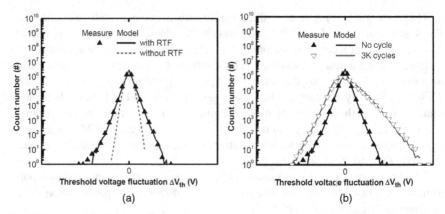

FIGURE 8.34   (a) Measured RTN after no cycles, and simulated RTN with and without RTF. (b) Measured and modeled RTN after no and 3K cycles.

Figure 8.34a shows a measured RTN distribution [14]. For each cell, RTN is evaluated by measuring V_t 200 times and compared to the average cell value [22]. Even if the channel of a 3D NAND cell is undoped poly-Si, RTN distribution presents an exponential tail, similar to 2D NAND cells [22, 23]. The exponential tail has been confirmed due to the presence of traps inside the poly-Si channel. The RTF must be considered when modeling RTN in a 3D NAND structure, as shown in Fig. 8.34a.

As the RTN traps occupancy follows Fermi statistics, their occupancy probability depends on the trap energy level. Thus RTN traps above (below) the Fermi level will be mainly empty (filled) inducing a positive (negative) tail. The probability distribution of the traps at ΔE_T can be evaluated based on the exponential tail coefficient of a single charge. The total RTN distribution is extracted from the sum of the V_t shift for each individual RTN trap. Thus, the total traps distribution is the convolution of the probability distribution over all the possible energy level, considering that for each energy level the trap number inside one cell follows a Poisson distribution [24].

Measurements had been modeled using this approach, as shown in Fig. 8.34b [14]. From RTN data of no cycle and 3K cycles, energy level of generated trap had been extracted. The majority of the switching traps in no cycle are close to the Fermi level. However, after 3K cycles, the RTN traps are generated at more than 0.2 eV above the Fermi level. The cycling induces the creation of switching traps enhancing the positive RTN tail. This can be explained by the degradation induced by the programming step [23, 25].

There are several other reports for RTN of 3D NAND cells [15–18, 26]. Figures 8.35a and 8.35b show the normalized noise power densities (S_{id}/I_{BL}^2) of 2D 32-nm FG cell and 3D stacked NAND cell, respectively [15]. The 3D stack devices show much higher normalized noise power densities (S_{id}/I_{BL}^2) compared to that of 2D 32-nm FG NAND devices because of more traps in poly-Si channel. Differently from the 2D FG NAND cell, 3D stacked devices show the higher S_{id}/I_{BL}^2 of the program state in the SS (subthreshold swing) region than that of erase state, because effective channel length decreases in program state due to higher channel V_{th} than

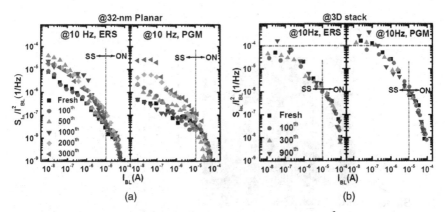

FIGURE 8.35 (a) Normalized noise power spectral density (S_{id}/I_{BL}^{2}) of 32-nm FG NAND flash memory device at 10 Hz with P/E cycling stress. "SS" represents subthreshold region. (b) S_{id}/I_{BL}^{2} of 3D stacked NAND flash memory device at 10 Hz with P/E cycling stress.

V_{th} in gate space region (source/drain region). When cell are erased, the V_{th} in the channel is comparable to that of the gate space region.

8.6.4 Back-Side Trap in Macaroni Channel

The cell current of vertical "macaroni" poly-Si channel is fluctuated by back-side traps, which are located in the interface between the back-insulator and the poly-Si channel, as shown in Fig. 8.36 [27].

The current path was simulated in an erased cell ($V_{th} = -2$ V) and a programmed cell ($V_{th} = 1$ V, 4.5 V), as shown in Fig. 8.37. It was found that the current path of the programmed cell was formed at the back-side of channel poly-Si (Fig. 8.37b and c). On the contrary, the current path of the erased cell was formed at the front-side of channel poly-Si (Fig. 8.37a). From these results, it is recognized that the back-side

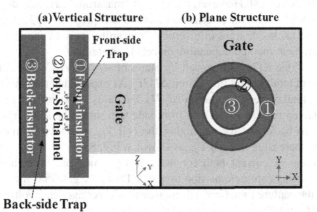

FIGURE 8.36 The schematic pictures for (a) vertical and (b) plane V-NAND structures.

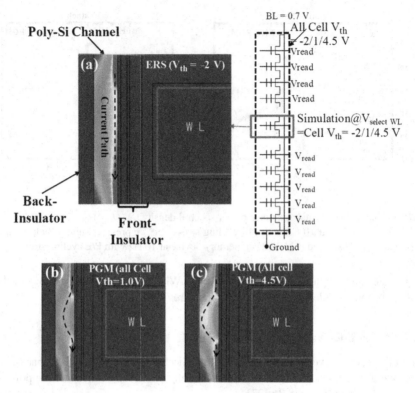

FIGURE 8.37 The simulation results for current path of select WL in V-NAND for (a) $V_{\text{select WL}}$ = cell V_{th} = −2 V, (b) $V_{\text{select WL}}$ = cell V_{th} = 1 V, and (c) $V_{\text{select WL}}$ = cell V_{th} = 4.5 V, respectively.

traps have to be characterized. In general, back-side traps can be analyzed in the back-gate structure [28]. However, the vertical "macaroni" structure does not have a back gate. Then, a new characterization method was proposed [32]. It enables us to investigate the back-side traps of a vertical "macaroni" poly-Si channel by using the RTN measurement method depending on cell V_{th} states.

In Fig. 8.38, the RTN of three cell V_{th} states of −2 V/1 V/4 V shows the different distributions of total current fluctuation $\Delta I_d/I_d$. As expected in the simulation results of Fig. 8.37, total current fluctuation $\Delta I_d/I_d$ decreases as the cell V_{th} increases. This can be explained that the effect of a front-side trap should be smaller as the current path moves to back-side for the higher the cell V_{th}. The total current fluctuation decreases by increasing the cell V_{th}, as shown in Fig. 8.38.

The location of current path for the cell V_{th} states can be clearly confirmed through measuring capture/emission time of RTN as the gate bias increases. For the erase state, the capture time (= τ_1) decreases as WL voltage (gate voltage) increases. Conversely, for the program state, the capture time increases as WL voltage increases, as shown in Fig. 8.39. This is because the capturing probability of traps is opposite

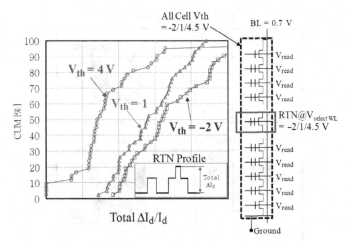

FIGURE 8.38 The cumulative curves of current fluctuation (= total $\Delta I_d/I_d$) at the same current ($I_d = 100$ nA) according to cell states of 40 cells. Inset shows the schematic picture of RTN profile.

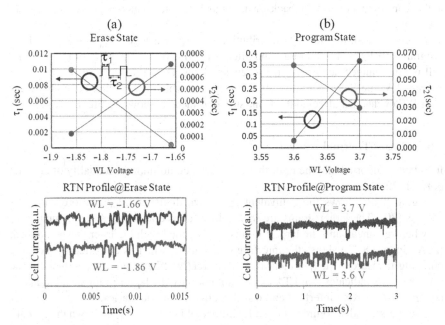

FIGURE 8.39 The RTN profile and change capturing time (τ_1) and emission time (τ_2) as the increasing gate voltage for (a) erase state and (b) program state.

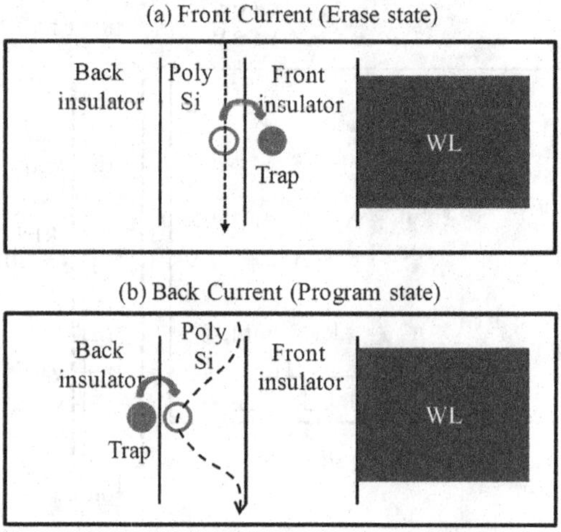

FIGURE 8.40   The schematic pictures for capturing/emission as increasing gate bias. (a) The capturing of electron decreases the current at front insulator (b) The emission of electron increases the current at back insulator.

at the front-insulator and the back-insulator as WL voltage increases, as illustrated in Fig. 8.40.

Figure 8.41 show the RTN distributions for (a) erase cell $V_{th} = -2$ V and (b) program cell $V_{th} = 4.5$ V before and after 10K P/E cycles. The RTN of (b) program cell is not changed according to P/E cycles. However, in the case of (a) erase cell $V_{th} = -2$ V, the RTN increased after P/E cycles, which means that generated traps after P/E cycles just affected the front-side interface, as shown in Fig. 8.41c.

8.6.5   Laser Thermal Anneal

It had been reported that the laser thermal anneal could improve a quality of channel poly-Si [29, 30].

Figure 8.42a shows the equivalent grain size diameter (D_{EQ}) before and after anneals [30]. The larger grains are induced by furnace anneal (FA) as compared to poly before anneal, and the largest grains are obtained with the laser thermal anneal (LTA). Figure 8.42b shows the interface trap density, which is extracted by a charge pumping measurement. It is clear that LTA greatly reduces interface defects in the FA case. This indicates that LTA can obtain not only the larger grain size, but also a better channel–oxide interface and less defective grain boundaries.

The statistical distribution of I_D and subthreshold swing (STS) is shown in Fig. 8.43 for gate-only O (oxide-only) devices [30]. This electrical evaluation is done by sweeping the gate up to 5 V, while keeping 1 V at the drain. A clear improvement in both I_D and STS is observed in LTA devices, leading to up to 10 times higher I_D, 3 times steeper STS, and tighter distributions than the poly-Si case. A clear correlation of the LTA dose with STS and I_D is also confirmed in Fig. 8.43. An LTA value of

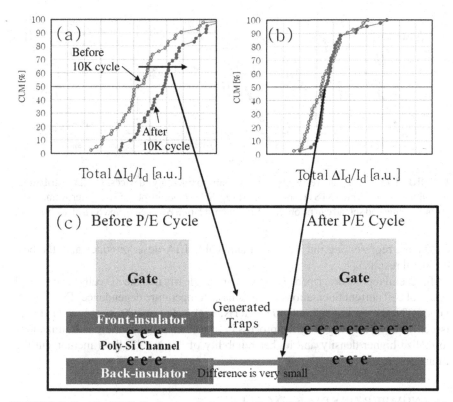

FIGURE 8.41   The RTN distributions before and after 10K P/E cycles at (a) $V_{th} = -2$ V, (b) $V_{th} = 4.5$ V. (c) The location differences of traps before and after 10K P/E cycle. Traps were generated between poly-Si and front-insulator by P/E cycling.

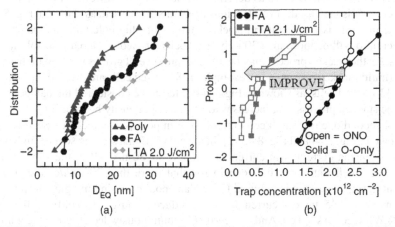

FIGURE 8.42   (a) Equivalent grain size diameter (D_{EQ}) is larger in furnace anneal (FA) and laser thermal anneal (LTA). (b) Trap concentration measured by charge pumping method. LTA clearly reduce interface traps both in O-only (oxide-only) and in ONO.

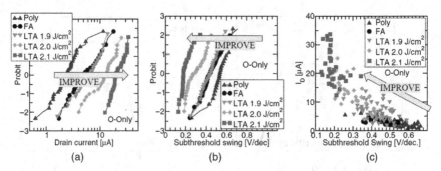

FIGURE 8.43 I_D–V_G curves analysis for O-only (oxide-only) devices. (a) ID distribution, (b) subthreshold swing (STS) distribution, (c) I_D-STS trade-off plot. Clear improvements in absolute value and spread can be observed. Clear correlation between I_D and STS also observed.

2.1 J/cm^2, representing the maximum applicable LTA dose, provides also the best electrical results.

In several reports described above, we can see clearly that the 3D cells have several issues of cell current fluctuations, such as large temperature dependence (Fig. 8.33a), large RTN, the cycling degradation of RTN, back-side trap effect, and so on. These issues have to be managed by the process improvements and operation optimizations to realize higher density and higher reliability of 3D NAND flash memory in the future.

8.7   NUMBER OF STACKED CELLS

In order to reduce the effective cell size, the number of stacked cells has to be increased in 3D NAND cells, as described in Section 8.2. However, if the number of stacked cells is increased, several serious problems are caused, as shown in Fig. 8.44.

The first one is a difficulty in stacked etching of plug hole and gate patterning. Aspect ratio will be more than 30 in over 32 cells stacked. As a realistic solution on this issue, the stacked process may be divided into several groups to avoid high aspect ratio (multi-stacked process). For example, 128 stacked cells are divided into 4 times (×4) 32 stacked cells, as shown in Fig. 8.44. It has to be considered the balance issue that the stacked process cost increases to 4 times, but density become 4 times.

The second one is a small cell current issue in poly-Si channel (see Section 8.6), as shown in Fig. 8.44 and Fig. 8.45 [2]. In the conventional sensing scheme in NAND flash, a sensing current (trip current) is around 60–80 nA/cell in the subthreshold region of a cell transistor. It is considered that more than 200-nA/cell saturation current in the worst case is required to have an enough sensing margin. However, as shown in Fig. 8.45, the cell current is greatly reduced to just ~20% of FG cell even at the 24 WLs (cells) stacked. And cell current is continuously decreased as the number of stacked cells is increased. The low current sensing scheme and/or the material development to enhance the cell current/mobility of the poly-Si channel have to be considered.

How many cells can be stacked ?

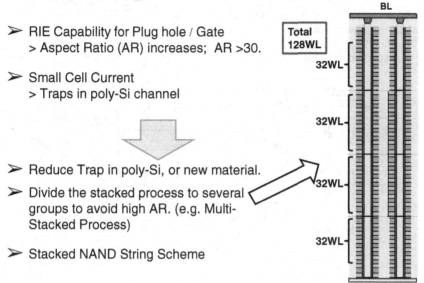

➤ RIE Capability for Plug hole / Gate
  > Aspect Ratio (AR) increases; AR >30.

➤ Small Cell Current
  > Traps in poly-Si channel

➤ Reduce Trap in poly-Si, or new material.

➤ Divide the stacked process to several groups to avoid high AR. (e.g. Multi-Stacked Process)

➤ Stacked NAND String Scheme

FIGURE 8.44   Problems and solutions for increasing the number of stacked layers in 3D NAND flash memory.

In order to solve the problems related to an increasing number of stacked cells, new array architecture would be a solution for future 3D NAND cells. As an example, the stacked NAND string scheme had been proposed, as shown in Fig. 8.46 [31]. The NAND strings of vertical channel 3D cell (BiCS, TCAT, SMArT) are vertically stacked. Bit lines or source lines are fabricated between the NAND strings. This

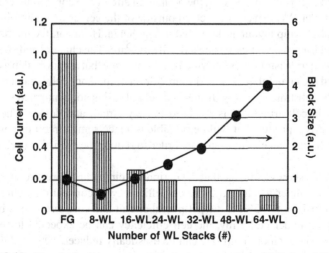

FIGURE 8.45   Trend of cell current and block size as number of WL (word-line) stacks.

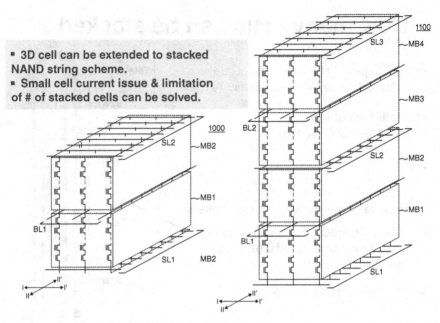

■ 3D cell can be extended to stacked NAND string scheme.
■ Small cell current issue & limitation of # of stacked cells can be solved.

FIGURE 8.46 The stacked NAND string scheme.

architecture can solve the problems of both high aspect ratio etching issue and small cell current issue at the same time.

8.8 PERIPHERAL CIRCUIT UNDER CELL ARRAY

In the first and second generations of 3D NAND flash memory products shown in Fig. 7.35 and Fig. 7.37 in Chapter 7, the peripheral circuit and core circuit (page buffer and row decoder/WL-driver) are located outside of the cell array area, following the conventional 2D chip layout as described in Fig. 8.47a. However, the memory cell of 3D NAND flash has a vertically stacked cell structure. The channel and source/drain of memory cell transistors are not formed on Si substrate, but are formed in a deposited poly-Si. And the channel poly-Si and source/drain of memory cell are not required to connect to Si substrate basically. If the channel poly-Si is not connected to substrate, Si substrate in cell array area is not used for any circuit and device. Therefore, in future 3D NAND products, it will be possible to place some circuit or device on Si substrate in (under) a cell array area, in order to reduce the chip size (i.e., to reduce a bit cost).

Figure 8.47b shows an image that the peripheral circuit and core circuit are formed on Si substrate under a cell array. The cell efficiency of conventional chip layout is normally from 70% to 85%. If the peripheral circuit and core circuit can be formed on Si substrate under cell array, the cell efficiency can be expected to improve to around 95%. The memory chip size can be drastically reduced 10% to 25%, so that the bit cost can be reduced 10% to 25%.

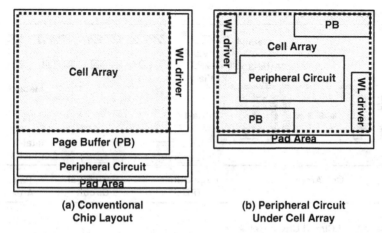

FIGURE 8.47   The memory chip layout image of (a) conventional 3D NAND flash memory and (b) 3D NAND flash memory with the peripheral circuit under cell array.

Figure 8.48 shows the cross-sectional view of (a) a conventional chip layout and (b) a peripheral circuit under a cell array. In (a) conventional chip layout, the page buffer of the core circuit and the peripheral circuit are located outside of the cell array area. However, in (b) the peripheral circuit under a cell array, the core circuit (including page buffer and word-line driver) and the peripheral circuit are located under the cell array area. The metal layers connect the memory cell with a core circuit and a peripheral circuit at the edge of a cell array, as shown in Fig. 8.48b.

In order to realize the peripheral circuit under a cell array, several issues have to be solved. The most important one is that the low-resistance metal layers are required under the cell. For a stable operation of peripheral circuit and core circuit, the low-resistance metal layers are required for the power supply lines (V_{cc}), ground line (V_{ss}), critical signal line, and so on. Normally, the Cu metal layer is used for this purpose in conventional 2D NAND flash memory chips. However, in the case of the peripheral circuit under a cell array, high-temperature processes (>800°C) of 3D memory cell fabrication have a serious damage on the low resistance metal layer (e.g., Cu layer). Therefore, the temperature of 3D cell fabrication has to be greatly decreased, or the high-temperature process immunity of a low-resistance metal layer is required to realize a peripheral circuit under a cell.

8.9   POWER CONSUMPTION

Low power consumption is one of the important requirements for NAND flash storage applications, such as SSD (solid-state drive). In particular, the high-end applications such as datacenters and enterprise SSDs strongly require the low power consumption during high-speed operation.

(a) Conventional Chip Layout

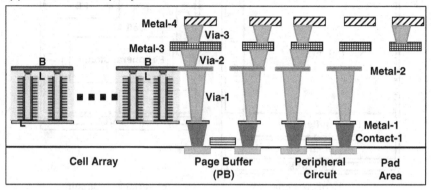

(b) Peripheral Circuit Under Cell Array

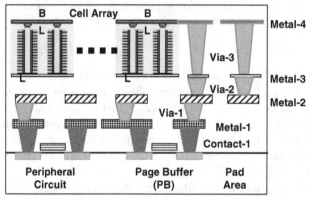

FIGURE 8.48 Cross-sectional image of 3D NAND flash memory in (a) conventional 3D NAND flash memory of cell array, core circuit and peripheral circuit and (b) 3D NAND flash memory with the peripheral circuit under cell array.

Figure 8.49a shows the power breakdown of a typical SSD system in four-way interleaving program operation [4, 5]. The NAND power accounts for about 47% of the entire SSD power, and this portion increases with more way interleaving, as shown in Fig. 8.49b. Figure 8.49c shows that the normalized NAND temperature increases with the increased number of way interleaving in programming. Then, in order not to exceed the temperature limit even with the eight-way interleaving, an SSD is often configured to intentionally reduce its performance and the operation temperature.

The first three-dimensional NAND flash product of the 128-Gb MLC (2-bit/cell) 3D V-NAND flash memory device had used the external high voltage of 12 V, which is available in the SSD board instead of the internal one generated from on-chip pumps to reduce the power without any sacrifice in performance, as shown in Fig. 8.50a [4,5]. When the external high voltage is used, a level detector is implemented so that when

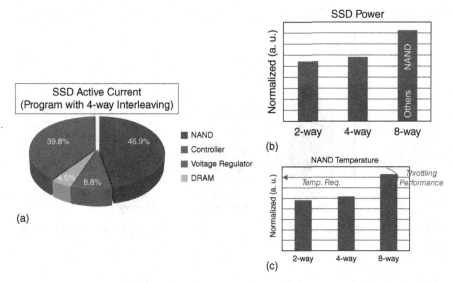

FIGURE 8.49 Diagrams of (a) NAND system power configuration in a four-way program operation. (b) NAND power accounts in SSD. (c) Temperature and way-throttling in an eight-way operation.

the high-voltage level is decreased below a threshold, it discharges internal nodes completely and safely. This operation protects the circuit from malfunctioning even with the unstable high voltage source or in the case of sudden power off. Figure 8.50b shows the simulation result of a level detector.

Figure 8.51 shows a measured active power of the 3D V-NAND at each operation of read, program, and erase. Compared with the consumption without using the external high-voltage scheme, about 50% of the energy consumption was reduced.

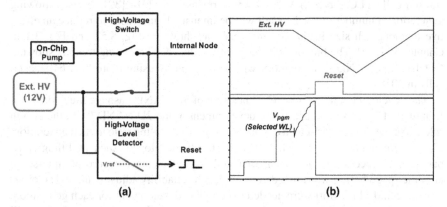

FIGURE 8.50 (a) External high-voltage-supply scheme of 12 V and (b) its simulated waveform.

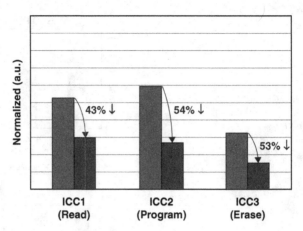

FIGURE 8.51 Measured active power dissipation using an external high-voltage-supply scheme.

As a result, it is possible to use eight-way full interleaving operation without the throttling, thereby increasing the overall SSD performance.

8.10 FUTURE TREND OF 3D NAND FLASH MEMORY

The future trend of 3D NAND flash memory is discussed.

Production of 3D NAND flash memory was started in August 2013. The architecture of the 3D cell was 24 stacked cells MLC (2b/cell) V-NAND as the first generation of 3D NAND [4, 5], as shown in Fig. 8.52. The charge trap (CT) cell with a vertical channel was selected for production because the fabrication process of the vertical channel CT cell is simpler than that of other 3D cells. The second generation of 32 stacked cells TLC (3b/cell) V-NAND was released in 2014 [32]. In the following generation, a number of stacked cell will be intensively increased to reduce an effective memory cell size in the coming years, as shown in Fig. 8.52 and Fig. 3.1 in Chapter 3 [1, 33]. The bit cost will be greatly reduced. The scaling trend is showing that 1-terabit NAND flash memory will be realized by using more than 64 stacked cells in 2018.

The development items for each generation of 3D NAND flash are much different from that of 2D NAND flash. In the development of the 2D NAND flash, the design rule of the memory cell is scaled down by around ×0.8 scaling ratio in each generation. There were many items which had to be newly developed, such as lithography, patterning, process change for smaller cell, layout change in core circuit (sense amplifier/page buffer, word line driver, etc.), cell reliability adjustment, and so on. A long time and a big effort were needed to complete development for each generation. However, in a 3D cell, development items from generation to next generation will be much reduced because the design rule of a memory cell will not be changed so

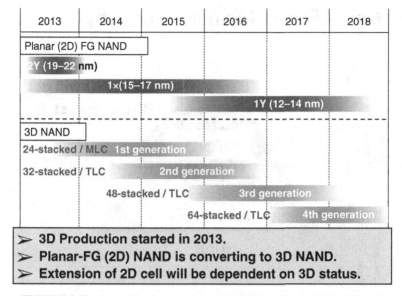

FIGURE 8.52   Transition from a planar 2D NAND cell to a 3D NAND cell.

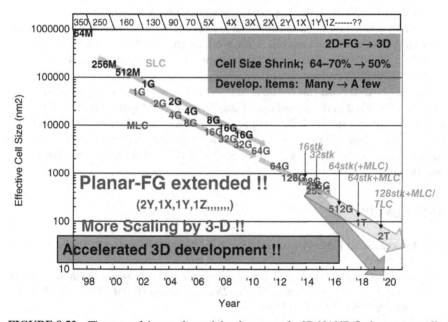

FIGURE 8.53   The case of the accelerated development of a 3D NAND flash memory cell.

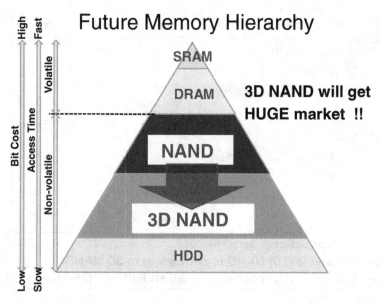

FIGURE 8.54   Future memory hierarchy with a three-dimensional NAND flash cell.

much. Many development items that are related to 2D memory cell scaling may not be developed in 3D flash development. Then, in 3D flash development, it should be focused on a small number of items that are related to process technologies for an increasing number of stacked cells, such as a plug etching, a gate etching, and so on. Therefore, the development speed of generation-by-generation of 3D NAND flash memory would be accelerated, as shown in Fig. 8.53.

The 3D NAND flash can continue to realize the lower bit cost by increasing the number of stacked cells, regardless of accelerating development or not. Then the magnetic memory, such as HDD, will be further replaced by the NAND flash-related product, including SSD (solid-state drive) for consumer and enterprise server, as shown in Fig. 8.54. The market size of SSD will be tremendously expanded in the near future.

REFERENCES

[1] Aritome, S. 3D Flash Memories, International Memory Workshop 2011 (IMW 2011), short course.

[2] Choi, E.-S.; Park, S.-K. Device considerations for high density and highly reliable 3D NAND flash cell in near future, *Electron Devices Meeting (IEDM), 2012 IEEE International*, pp. 9.4.1–9.4.4, 10–13 Dec. 2012.

[3] Chen, C.-P.; Lue, H.-T.; Hsieh, C.-C.; Chang, K.-P.; Hsieh, C.-C.; Lu, C.-Y. Study of fast initial charge loss and it's impact on the programmed states V_t distribution of charge-trapping NAND Flash, *Electron Devices Meeting (IEDM), 2010 IEEE International*, pp. 5.6.1, 5.6.4, 6–8 Dec. 2010.

[4] Park, K.-T.; Nam, S.; Kim, D.; Kwak, P.; LEE, D.; Choi, Y.-H.; Choi, M.-H.; Kwak, D.-H.; Kim, D.-H.; Kim, M.-S.; Park, H.-W.; Shim, S.-W.; Kang, K.-M.; Park, S.-W.; Lee, K.; Yoon, H.-J.; Ko, K.; Shim, D.-K.; Ahn, Y.-L.; Ryu, J.; Kim, D.; Yun, K.; Kwon, J.; Shin, S.; Byeon, D.-S.; Choi, K.; Han, J.-M.; Kyung, K.-H.; Choi, J.-H.; Kim, K. Three-dimensional 128 Gb MLC vertical nand flash memory with 24-WL stacked layers and 50 MB/s high-speed programming, *Solid-State Circuits, IEEE Journal of*, vol. PP, no. 99, pp. 1, 10, 2014.

[5] Park, K.-T.; Han, J.-m.; Kim, D.; Nam, S.; Choi, K.; Kim, M.-S.; Kwak, P.; Lee, D.; Choi, Y.-H.; Kang, K.-M.; Choi, M.-H.; Kwak, D.-H.; Park, H.-w.; Shim, S.-w.; Yoon, H.-J.; Kim, D.; Park, S.-w.; Lee, K.; Ko, K.; Shim, D.-K.; Ahn, Y.-L.; Park, J.; Ryu, J.; Kim, D.; Yun, K.; Kwon, J.; Shin, S.; Youn, D.; Kim, W.-T.; Kim, T.; Kim, S.-J.; Seo, S.; Kim, H.-G.; Byeon, D.-S.; Yang, H.-J.; Kim, M.; Kim, M.-S.; Yeon, J.; Jang, J.; Kim, H.-S.; Lee, W.; Song, D.; Lee, S.; Kyung, K.-H.; Choi, J.-H. 19.5 Three-dimensional 128 Gb MLC vertical NAND Flash-memory with 24-WL stacked layers and 50MB/s high-speed programming, *Solid-State Circuits Conference Digest of Technical Papers (ISSCC), 2014 IEEE International*, pp. 334, 335, 9–13 Feb. 2014.

[6] Shim, K.-S.; Choi, E.-S.; Jung, S.-W.; Kim, S.-H.; Yoo, H.-S.; Jeon, K.-S.; Joo, H.-S.; Oh, J.-S.; Jang, Y.-S.; Park, K.-J.; Choi, S.-M.; Lee, S.-B.; Koh, J.-D.; Lee, K.-H.; Lee, J.-Y.; Oh, S.-H.; Pyi, S.-H.; Cho, G.-S.; Park, S.-K.; Kim, J.-W.; Lee, S.-K.; Hong, S.-J. Inherent Issues and Challenges of Program Disturbance of 3D NAND Flash cell, *Memory Workshop (IMW), 2012 4th IEEE International*, pp. 1, 4, 20–23 May 2012.

[7] Yoo, H. S.; Choi, E.; Oh, J. S. ; Park, K. J.; Jung, S. W.; Kim, S.; Shim, K. S.; Joo, H. S.; Jung, S. W.; Jeon, K. S.; Seo, M. S.; Jang, Y. S. ; Lee, S. B.; Lee, J. Y.; Oh, S. H.; Cho, G. S.; Park, S.; Lee, S.; Hong, S. Modeling and optimization of the chip level program disturbance of 3D NAND flash memory, *Memory Workshop (IMW), 2013 5th IEEE International*, pp. 147, 150, 26–29 May 2013.

[8] Wong, M.; Chow, T.; Wong, C. C.; Zhang, D. A quasi two-dimensional conduction model for polycrystalline silicon thin-film transistor based on discrete grains, *Electron Devices, IEEE Transactions on*, vol. 55, no. 8, pp. 2148, 2156, Aug. 2008.

[9] Kim, B.; Lim, S.-H.; Kim, D. W.; Nakanishi, T.; Yang, S.; Ahn, J.-Y.; Choi, H. M.; Hwang, K.; Ko, Y.; Kang, C.-J. Investigation of ultra thin polycrystalline silicon channel for vertical NAND flash, *Reliability Physics Symposium (IRPS), 2011 IEEE International*, pp. 2E.4.1, 2E.4.4, 10–14 April 2011.

[10] Degraeve, R.; Toledano-Luque, M.; Suhane, A.; Van den Bosch, G.; Arreghini, A.; Tang, B.; Kaczer, B.; Roussel, P.; Kar, G.S.; Van Houdt, J.; Groeseneken, G. Statistical characterization of current paths in narrow poly-Si channels, *Electron Devices Meeting (IEDM), 2011 IEEE International*, pp. 12.4.1, 12.4.4, 5–7 Dec. 2011.

[11] Toledano-Luque, M.; Degraeve, R.; Kaczer, B.; Tang, B.; Roussel, P. J.; Weckx, P.; Franco, J.; Arreghini, A.; Suhane, A.; Kar, G.S.; Van den Bosch, G.; Groeseneken, G.; Van Houdt, J. Quantitative and predictive model of reading current variability in deeply scaled vertical poly-Si channel for 3D memories, *Electron Devices Meeting (IEDM), 2012 IEEE International*, pp. 9.2.1, 9.2.4, 10–13 Dec. 2012.

[12] Toledano-Luque, M.; Degraeve, R.; Roussel, P. J.; Luong, V.; Tang, B.; Lisoni, J. G.; Tan, C.-L.; Arreghini, A.; Van den Bosch, G.; Groeseneken, G.; Van Houdt, J. Statistical spectroscopy of switching traps in deeply scaled vertical poly-Si channel for 3D memories, *Electron Devices Meeting (IEDM), 2013 IEEE International*, pp. 21.3.1, 21.3.4, 9–11 Dec. 2013.

[13] Degraeve, R.; Toledano-Luque, M.; Arreghini, A.; Tang, B.; Capogreco, E.; Lisoni, J.; Roussel, P.; Kaczer, B.; Van den Bosch, G.; Groeseneken, G.; Van Houdt, J. Characterizing grain size and defect energy distribution in vertical SONOS poly-Si channels by means of a resistive network model, *Electron Devices Meeting (IEDM), 2013 IEEE International*, pp. 21.2.1, 21.2.4, 9–11 Dec. 2013.

[14] Nowak, E.; Kim, J.-H.; Kwon, H. Y.; Kim, Y.-G.; Sim, J. S. ; Lim, S.-H.; Kim, D. S.; Lee, K.-H.; Park, Y.-K.; Choi, J.-H.; Chung, C. Intrinsic fluctuations in Vertical NAND flash memories, *VLSI Technology (VLSIT), 2012 Symposium on*, pp. 21, 22, 12–14 June 2012.

[15] Jeong, M.-K.; Joe, S.-M.; Jo, B.-S.; Kang, H.-J.; Bae, J.-H.; Han, K.-R.; Choi, E.; Cho, G.; Park, S.-K.; Park, B.-G.; Lee, J.-H. Characterization of traps in 3-D stacked NAND flash memory devices with tube-type poly-Si channel structure, *Electron Devices Meeting (IEDM), 2012 IEEE International*, pp. 9.3.1, 9.3.4, 10–13 Dec. 2012.

[16] Jeong, M.-K.; Joe, S.-M.; Seo, C.-S.; Han, K.-R.; Choi, E.; Park, S.-K.; Lee, J.-H. Analysis of Random Telegraph Noise and low frequency noise properties in 3-D stacked NAND flash memory with tube-type poly-Si channel structure, *VLSI Technology (VLSIT), 2012 Symposium on*, pp. 55, 56, 12–14 June 2012.

[17] Park, J. K.; Moon, D.-I.; Choi, Y.-K.; Lee, S.-H.; Lee, K.-H.; Pyi, S. H.; Cho, B. J. Origin of transient V_{th} shift after erase and its impact on 2D/3D structure charge trap flash memory cell operations, *Electron Devices Meeting (IEDM), 2012 IEEE International*, pp. 2.4.1, 2.4.4, 10–13 Dec. 2012.

[18] Jeong, M.-K.; Joe, S.-M.; Kang, H.-J.; Han, K.-R.; Cho, G.; Park, S.-K.; Park, B.-G.; Lee, J.-H. A new read method suppressing effect of random telegraph noise in NAND flash memory by using hysteretic characteristic, *VLSI Technology (VLSIT), 2013 Symposium on*, pp. T154, T155, 11–13 June 2013.

[19] Ghetti, A.; Compagnoni, C.M.; Spinelli, A.S.; Visconti, A. Comprehensive analysis of random telegraph noise instability and its scaling in deca–nanometer Flash memories, *IEEE Transactions on Electron Devices*, vol. 56, pp. 1746–1752, 2009.

[20] Franco, J.; Kaczer, B.; Toledano-Luque, M.; Roussel, P. J.; Mitard, J.; Ragnarsson, L.-A.; Witters, L.; Chiarella, T.; Togo, M.; Horiguchi, N.; Groeseneken, G.; Bukhori, M. F.; Grasser, T.; Asenov, A. Impact of single charged gate oxide defects on the performance and scaling of nanoscaled FETs, *Reliability Physics Symposium (IRPS), 2012 IEEE International*, pp. 5A.4.1, 5A.4.6, 15–19 April 2012.

[21] Kimura, M.; Inoue, S.; Shimoda, T.; Eguchi, T. Dependence of polycrystalline silicon thin-film transistor characteristics on the grain-boundary location, *Journal of Applied Physics*, vol. 89, no. 1, pp. 596, 600, Jan 2001.

[22] Kang, D.; Lee, S.; Park, H.-M.; Lee, D.-J.; Kim, J.; Seo, J.; Lee, C.; Song, C.; Lee, C.-S.; Shin, H.; Song, J.; Lee, H.; Choi, J.-H.; Jun, Y.-H. A new approach of NAND flash cell trap analysis using RTN characteristics, *VLSI Technology (VLSIT), 2011 Symposium on*, pp. 206, 207, 14–16 June 2011.

[23] Compagnoni, M. C.; Gusmeroli, R.; Spinelli, A. S.; Lacaita, A. L.; Bonanomi, M.; Visconti, A. Statistical model for random telegraph noise in flash memories, *Electron Devices, IEEE Transactions on*, vol. 55, no. 1, pp. 388, 395, Jan. 2008.

[24] Takeuchi, K.; Nagumo, T.; Yokogawa, S.; Imai, K.; Hayashi, Y. Single-charge-based modeling of transistor characteristics fluctuations based on statistical measurement of RTN amplitude, *VLSI Technology, 2009 Symposium on*, pp. 54, 55, 16–18 June 2009.

[25] Nagumo, T.; Takeuchi, K.; Hase, T.; Hayashi, Y. Statistical characterization of trap position, energy, amplitude and time constants by RTN measurement of multiple individual traps, *Electron Devices Meeting (IEDM), 2010 IEEE International*, pp. 28.3.1, 28.3.4, 6–8 Dec. 2010.

[26] Kang, H.-J.; Jeong, M.-K.; Joe, S.-M.; Seo, J.-H.; Park, S.-K.; Jin, S. H.; Park, B.-G.; Lee, J.-H. Effect of traps on transient bit-line current behavior in word-line stacked nand flash memory with poly-Si body, *VLSI Technology (VLSI-Technology): Digest of Technical Papers, 2014 Symposium on*, pp. 1, 2, 9–12 June 2014.

[27] Kang, D., Lee, C., Hur, S., Song, D., and Choi, J.-H. A new approach for trap analysis of vertical NAND flash cell using RTN characteristics., *Electron Devices Meeting (IEDM), 2014 IEEE International*, pp. 367–370, Dec. 2014.

[28] Kimura, M.; Yoshino, T.; Harada, K. Complete extraction of trap densities in poly-Si thin-film transistors, *Electron Devices, IEEE Transactions on*, vol. 57, no. 12, pp. 3426, 3433, Dec. 2010.

[29] Congedo, G.; Arreghini, A.; Liu, L.; Capogreco, E.; Lisoni, J. G.; Huet, K.; Toque-Tresonne, I.; Van Aerde, S.; Toledano-Luque, M.; Tan, C.-L.; Van den Bosch, G.; Van Houdt, J. Analysis of performance/variability trade-off in Macaroni-type 3-D NAND memory, *Memory Workshop (IMW), 2014 IEEE 6th International*, pp. 1, 4, 18–21 May 2014.

[30] Lisoni, J. G.; Arreghini, A.; Congedo, G.; Toledano-Luque, M.; Toque-Tresonne, I.; Huet, K.; Capogreco, E.; Liu, L.; Tan, C.-L.; Degraeve, R.; Van den Bosch, G.; Van Houdt, J. Laser thermal anneal of polysilicon channel to boost 3D memory performance, *VLSI Technology (VLSI-Technology): Digest of Technical Papers, 2014 Symposium on*, pp. 1, 2, 9–12 June 2014.

[31] Aritome, S. US Patent 8,891,306.

[32] Im, J.-w.; Jeong, W.-P., Kim, D.-H., Nam, S.-W., Shim, D.-K., Choi, M.-H., Yoon, H.-J., Kim, D.-H., Kim, Y.-S., Park, H.-W., Kwak, D.-H., Park, S.-W., Yoon, S.-M., Hahn, W.-G., Ryu, J.-H., Shim, S.-W., Kang, K.-T., Choi, S.-H., Ihm, J.-D., Min, Y.-S., Kim, I.-M., Lee, D.-S., Cho, J.-H., Kwon, O.-S., Lee, J.-S., Kim, M.-S., Joo, S.-H., Jang, J.-H., Hwang, S.-W., Byeon, D.-S., Yang, H.-J., Park, K.-T., Kyung, K.-H., Choi, J.-H. A 128 Gb 3b/cell V-NAND Flash Memory with 1Gb/s I/O rate, *Solid-State Circuits Conference Digest of Technical Papers (ISSCC), 2015 IEEE International*, pp. 23–25 Feb. 2015.

[33] Aritome, S. Scaling challenges beyond 1X-nm DRAM and NAND flash, Joint Rump session in VLSI Symposium 2012.

9

CONCLUSIONS

9.1 DISCUSSIONS AND CONCLUSIONS

The development of NAND flash memory started in 1987 in the R&D center of
Toshiba Corporation [1]. The target market was the replacement of magnetic memory,
such as HDD, and so on. [2]. For this target, the most important requirement to achieve
was "low bit cost." A memory cell size has to be as small as possible to achieve low
bit cost. In general, ideal "physical" two-dimensional memory cell size is $4*F^2$ (F:
feature size), which is defined by $2*F$ pitch for both x- and y-directions. The first
NAND flash memory cells of 1-μm rule [3, 4] was $8*F^2$ cell size by using a wide
x-direction pitch of 4 μm ($4*F$) because LOCOS isolation width was 3 μm wide
due to the limitation of the high-voltage operation. The LOCOS isolation width was
limited by the punch-through of the bit-line junction and threshold voltage of the
parasitic field LOCOS transistor, because a high voltage of ~22 V was applied to
the junction and control gate during programming. In order to reduce the LOCOS
isolation width, a new FTI (field-through-implantation) process had been developed
[5] (Section 3.2). Very narrow LOCOS isolation width of 0.8 μm ($2*F$ in 0.4-μm
feature size) could be realized. Memory cell size could be scaled down to $6*F^2$ by
using a $3*F$ bit-line pitch in the 0.4-μm rule.

For scaling down memory cell size further, the self-aligned shallow trench isolation
cell (the SA-STI cell) had been developed [6] (Sections 3.3 and 3.4). The isolation
width could be scaled down to F using STI, and then the BL pitch could be reduced
to ideal $2*F$ in comparison with that of LOCOS cell of $3*F$ bit-line pitch. Therefore,

Nand Flash Memory Technologies, First Edition. Seiichi Aritome.
© 2016 The Institute of Electrical and Electronics Engineers, Inc. Published 2016 by John Wiley & Sons, Inc.

the cell size of NAND flash memory could be drastically shrunk to 66% (from $6*F^2$ to ideal $4*F^2$; F: feature size). The SA-STI cell process was applied to a NAND flash product with the structure of the SA-STI with an FG wing [7,8] (Section 3.3), because the aspect ratio of the stacked gate could be reduced by the FG wing structure. The SA-STI cell with an FG wing had been used from 0.25-μm generation to 0.12-μm generation. After that, the SA-STI cell without an FG wing had been used from 90-nm generation [6] (Section 3.4). The SA-STI cell has an excellent scalability. The cell size could decrease straightforward as feature-size decreased from 0.25 μm [6] to 1X nm [9] (Figs. 3.1 and 3.2 in Section 3.1). Therefore, the SA-STI cell structure and process have been used more than 10 years and 10 generations due to simple process and structure. Moreover, the SA-STI cell has another advantage, namely, an excellent reliability. The tunnel oxide has no degradation at the STI edge corner because of no sharp STI edge corner by fabricating a floating gate with STI patterning.

The cell size of NAND flash became ideal $4*F^2$ by the SA-STI cell. The feature size (F) is normally determined by the capability of the lithography tool. At present, the most advanced lithography tool is the ArF immersion (ArFi) stepper. Minimum feature size is 38–40 nm. Then the scaling of feature size (F) was limited by 38–40 nm. In order to accelerate to scale down the NAND flash memory cell size further, the double patterning process had started to be used from the 3X-nm generation. The sidewall spacer was used as a patterning mask in the conventional double patterning process. Thanks to double patterning, feature size could be scaled down from 38–40 nm to 19–20 nm. Furthermore, quadruple (×4) patterning had been used beyond 20 nm [9]. Feature size (F) could be scaled down to 10 nm by using ArFi.

As shown in the ITRS roadmap (http://www.itrs.net/), from around the year 2005, the NAND flash memory became a so-called "process driver" device, which has led to the scaling and development of lithography/patterning for minimum device dimension (line/space pitch), by replacing DRAM. This is because the NAND flash memory cell can be easily scaled down as scaling a minimum device dimension, without any electrical, operational, and reliability limitations due to the contribution of key technologies of the SA-STI cell and the uniform program/erase scheme. Therefore, the development of NAND flash memory has had a great impact on leading fine pitch patterning technologies, such as ArF immersion lithography, double patterning, quadruple patterning, and so on.

The multilevel cell is another important technology to reduce "effective" memory cell size without F scaling. In the multilevel cell, V_t distribution width has to be tightly controlled to have enough RWM (read window margin) [10] to prevent from read failure. Several advanced operations for multilevel cells were presented in Chapter 4 to obtain a tight V_t distribution as well as high reliability and performance. The multilevel cells NAND flash product of MLC (2 bits/cell), TLC (3 bits/cell), and QLC (4 bits/cell) are using these advanced operations and architectures, such as ISPP program, bit-by-bit verify, two-step verify scheme, pseudo-pass scheme in page program (Section 4.2), the advanced page program sequence, ABL architecture (Section 4.3), moving read algorithm (Section 4.7), and so on.

The memory cell size of the SA-STI cell could be intensively scaled down by using double and quadruple patterning. However, the SA-STI cell has been facing serious physical limitations, such as floating-gate capacitive coupling interference, electron injection spread, RTN, high field limitation, patterning limitation, and so on. The scaling challenge and limitations were discussed in Chapter 5. Read window margin (RWM) was quantitatively analyzed, and then the solutions to overcome scaling limitations were clarified [10] (Section 5.2). It was concluded that 1Y- to 1Z-nm (13- to 10-nm) SA-STI cells could be realized by using the 60% air-gap process.

The other important requirement for NAND flash is "high reliability." At a term of starting NAND flash development, the program/erase schemes were intensively discussed by an internal development team in order to select a proper scheme. It was not clear how program/erase scheme had an impact on reliability. Then, the reliability of the NAND flash cell was analyzed in several program/erase (P/E) schemes, such as the CHE (channel hot-electron) program scheme [1], the nonuniform P/E scheme [11, 12], and the uniform P/E scheme [13–15] to decide the proper P/E scheme.

The P/E cycling endurance and data retention characteristics were evaluated and analyzed in the two P/E schemes [13, 16] (Sections 6.2 and 6.3). It had been clarified that the uniform P/E scheme, which is used in the NAND flash, had appropriate reliability in comparison with other schemes [13, 16, 17]. And also, the read disturb characteristics had been analyzed in the cells that were subjected to P/E cycling endurance stress [15, 17] (Section 6.4). It had been clarified that the uniform P/E scheme had the better read disturb characteristics because SILC (stress-induced leakage current) could be suppressed by the bipolarity FN (Fowler–Nodheim) program scheme. As a result, the uniform P/E scheme was decided for NAND flash operation.

The uniform P/E scheme had another important advantage. The uniform P/E scheme can realize very low power consumption for programming a large number of memory cells simultaneously (page program). Therefore the programming speed per byte can be quite fast ($\sim$100 Mbyte/s). Due to high reliability and fast programming, the uniform program/erase scheme became de facto standard technology. All of the NAND suppliers (Toshiba/SanDisk, Samsung, Micron/Intel, SK Hynix) have used the uniform P/E scheme for all NAND flash products over 20 years.

In 2007, the new three-dimentional (3D) NAND flash cell of BiCS (bit cost scalable technology) was proposed [18] in order to further scale down the memory cell size of NAND flash. The 3D cells have a vertically stacked structure by the new concept of stacked gate layers. Then effective cell size can be reduced without the scaling feature size of F. After proposal of BiCS, many types of 3D cells were proposed. In Chapter 7, major 3D cells were reviewed and compared. Many of them, including BiCS, TCAT (V-NAND), SMArT, and VG-NAND, are using the SONOS charge trap (CT) cell with SiN charge trap layer. However, the SONOS cell has two serious problems of a slow erase (erase V_t saturation) and a poor data retention. To overcome these SONOS problems, the FG (floating-gate) 3D NAND cell of the DC-SF cell (dual control-gate–surrounding floating-gate cell) was proposed [19, 20] (Sections 7.6 and 7.7). Due to replacing a charge trap cell with a FG cell, the problem related with 3D SONOS could be perfectly solved.

Production of the 3D NAND cell was started in 2013. The 24 stacked cells MLC (2b/cell) V-NAND was the first generation of 3D NAND [21]. The charge trap (CT) cell with a vertical channel was used because of its simple fabrication process. The second generation of 32 stacked cells TLC (3b/cell) V-NAND was released in 2014 [22]. In the following generation, the number of stacked cells will be greatly increased to reduce an effective memory cell size and a bit cost in the coming years. The scaling trend is showing that 1 terabit of NAND flash memory will be realized by using more than 64 stacked cells in 2018.

However, in order to further proceed to the higher-density 3D NAND flash memory, several problems that still remain have to be solved. The challenges of 3D NAND flash memory were discussed in Chapter 8. For 3D NAND cell scaling, it is very important to increase the number of stacked cells without increasing the process cost. The small cell current problem and high-aspect RIE process will be critical. The stacked NAND string scheme [23] or the divided stack process would be one of the solutions for future 3D NAND flash memory.

9.2   PERSPECTIVE

From the production start of NAND flash memory in 1992, the worldwide NAND market has been tremendously expanded due to the boom of a digital camera, USB drive, MP3 player, smartphone, tablet-PC, and SSD (solid-state drive). The overall NAND market is expected to hit $35 billion in 2015. NAND flash memory has created new large volume markets and industries of consumer, computer, mass-storage, and enterprise server. This trend is still so rapidly growing in the world.

The reasons why the NAND flash memory was accepted to the emerging applications were a low bit cost, high reliability, high performance (fast programming), and low power consumption. To achieve these requirements, over 25 years, many indispensable technologies have been developed and implemented to NAND flash products as a de facto standard [24], as described in this book. In order to continue this trend, NAND flash memory has to continuously satisfy market requirements. The most important requirement is the low bit cost. A bit cost of NAND flash memory has to be further reduced. Therefore, the scaling of an effective memory cell size is essential.

The 2D NAND flash has continued mass production by using a 15-nm technology in 2015 (Fig. 8.52). The next generation of 12- to 14-nm technology would come into production on 2016. And the following generation, which is probably close to 10-nm technology, would be possible to implement, as shown in Chapter 5. However, even in 12- to 14-nm technology, it will be tough to realize due to serious scaling limitation. The operation or system solutions will be key technologies to manage a scaling limitation. And the minimum feature size (F) will not exceed over 9.5–10 nm, because it is a limitation of ArFi quadruple patterning. If F is over 9.5–10 nm, process cost greatly increases.

Production of 3D NAND had already started; however, 2D NAND flash will not quickly disappear in the market, because 2D NAND flash is widely accepted in the market and also has a big infrastructure for production. Then 2D NAND flash will continue in production parallel with 3D NAND flash for more than 5 years.

For the 3D NAND flash, the increasing number of stacked cells is very important to realize the smaller effective cell size. If the number of stacked cells is increased to 128 cells, a 2-terabit NAND flash memory product is expected. A critical challenge to increase the number of stacked cells is the extremely high aspect ratio process. The multi-stacked process, which has the divided stacked layers (Fig. 8.44 in Chapter 8), can solve this issue. If the multi-stacked process is realized, the scaling pace of the effective cell size would be much accelerated, as 50% shrinkage of effective cell size for each generation in contrast to 64–70% shrinkage in 2D planar FG cell, as described in Section 8.10 (see Fig. 8.53).

On the other hand, a small cell current is also a serious issue in increasing the number of stacked cells. As one of the solutions, the stacked NAND string scheme, as shown in Fig. 8.46 in Chapter 8, can solve this issue as well as an issue of high aspect ratio process. If the stacked NAND string scheme is successfully developed, low bit cost and high-performance 3D NAND flash memory can be realized.

The low power consumption is other important requirement for NAND flash memory. The power consumption of storage memory has been greatly reduced by using NAND flash memory, compared with magnetic memory of HDD. In the data center, SSD based on NAND flash memory can reduce the power consumption of an enterprise server, replaced HDD, because of low power operation in NAND flash memory as well as low cooling power. The NAND flash is successfully contributing to the ecological environment of the earth in the present and for the future.

REFERENCES

[1] Masuoka, F. Momodomi, M.; Iwata, Y.; Shirota, R. New ultra high density EPROM and flash EEPROM with NAND structure cell, *Electron Devices Meeting, 1987 International,* vol. 33, pp. 552– 555, 1987

[2] Masuoka, F. Flash memory makes a big leap, *Kogyo Chosakai,* vol. 1, pp. 1–172, 1992 (in Japanese).

[3] Itoh, Y.; Momodomi, M.; Shirota, R.; Iwata, Y.; Nakayama, R.; Kirisawa, R.; Tanaka, T.; Toita, K.; Inoue, S.; Masuoka, F. An experimental 4 Mb CMOS EEPROM with a NAND structured cell, *Solid-State Circuits Conference, 1989. Digest of Technical Papers. 36th ISSCC, 1989 IEEE International,* pp. 134–135, 15–17 Feb. 1989.

[4] Momodomi, M.; Iwata, Y.; Tanaka, T.; Itoh, Y.; Shirota, R.; Masuoka, F. A high density NAND EEPROM with block-page programming for microcomputer applications, *Custom Integrated Circuits Conference, 1989, Proceedings of the IEEE 1989,* pp. 10.1/1– 10.1/4, 15–18 May 1989.

[5] Aritome, S.; Hatakeyama, I.; Endoh, T.; Yamaguchi, T.; Shuto, S.; Iizuka, H.; Maruyama, T.; Watanabe, H.; Hemink, G.; Sakui, K.; Tanaka, T.; Momodomi, M., Shirota, R. An advanced NAND-structure cell technology for reliable 3.3V 64 Mb electrically erasable and programmable read only memories (EEPROMs), *Japanese Journal of Applied Physics,* vol. 33, (1994) pp. 524–528, part 1, no. 1B, Jan. 1994.

[6] Aritome, S.; Satoh, S.; Maruyama, T.; Watanabe, H.; Shuto, S.; Hemink, G. J.; Shirota, R.; Watanabe, S.; Masuoka, F. A 0.67 μm^2 self-aligned shallow trench isolation cell

(SA-STI cell) for 3 V-only 256 Mbit NAND EEPROMs, *Electron Devices Meeting, 1994. IEDM '94. Technical Digest, International*, pp. 61–64, 11–14 Dec. 1994.

[7] Shimizu, K.; Narita, K.; Watanabe, H.; Kamiya, E.; Takeuchi, Y.; Yaegashi, T.; Aritome, S.; Watanabe, T. A novel high-density 5F^2 NAND STI cell technology suitable for 256 Mbit and 1 Gbit flash memories, *Electron Devices Meeting, 1997. IEDM '97. Technical Digest, International*, pp. 271–274, 7–10 Dec. 1997.

[8] Takeuchi, Y.; Shimizu, K.; Narita, K.; Kamiya, E.; Yaegashi, T.; Amemiya, K.; Aritome, S. A self-aligned STI process integration for low cost and highly reliable 1 Gbit flash memories, *VLSI Technology, 1998. Digest of Technical Papers. 1998 Symposium on*, pp. 102–103, 9–11 June 1998.

[9] Hwang, J.; Seo, J.; Lee, Y.; Park, S.; Leem, J.; Kim, J.; Hong, T.; Jeong, S.; Lee, K.; Heo, H.; Lee, H.; Jang, P.; Park, K.; Lee, M.; Baik, S.; Kim, J.; Kkang, H.; Jang, M.; Lee, J.; Cho, G.; Lee, J.; Lee, B.; Jang, H.; Park, S.; Kim, J.; Lee, S., Aritome, S.; Hong, S.; Park, S. A middle-1X nm NAND flash memory cell (M1X-NAND) with highly manufacturable integration Technologies, *Electron Devices Meeting (IEDM), 2011 IEEE International*, pp. 199–202, Dec. 2011.

[10] Aritome, S.; Kikkawa, T. Scaling Challenge of Self-Aligned STI cell (SA-STI cell) for NAND Flash Memories, *Solid-State Electronics*, 82, 54–62, 2013.

[11] Shirota, R., Itoh, Y., Nakayama, R., Momodomi, M., Inoue, S., Kirisawa, R., Iwata, Y., Chiba, M., Masuoka, F. New NAND cell for ultra high density 5V-only EEPROMs, *Digest of Technical Papers—Symposium on VLSI Technology*, 1988, pp. 33–34.

[12] Momodomi, M.; Kirisawa, R.; Nakayama, R.; Aritome, S.; Endoh, T.; Itoh, Y.; Iwata, Y.; Oodaira, H.; Tanaka, T.; Chiba, M.; Shirota, R.; Masuoka, F. New device technologies for 5 V-only 4 Mb EEPROM with NAND structure cell, *Electron Devices Meeting, 1988. IEDM '88. Technical Digest, International*, pp. 412–415, 1988.

[13] Aritome, S.; Kirisawa, R.; Endoh, T.; Nakayama, R.; Shirota, R.; Sakui, K.; Ohuchi, K.; Masuoka, F. Extended data retention characteristics after more than 10^4 write and erase cycles in EEPROMs, *International Reliability Physics Symposium, 1990. 28th Annual Proceedings*, pp. 259–264, 1990.

[14] Kirisawa, R.; Aritome, S.; Nakayama, R.; Endoh, T.; Shirota, R.; Masuoka, F. A NAND structured cell with a new programming technology for highly reliable 5 V-only flash EEPROM, *1990 Symposium on VLSI Technology, 1990. Digest of Technical Papers.* pp. 129–130, 1990.

[15] Aritome, S.; Shirota, R.; Kirisawa, R.; Endoh, T.; Nakayama, R.; Sakui, K.; Masuoka, F. A reliable bi-polarity write/erase technology in flash EEPROMs, *International Electron Devices Meeting, 1990. IEDM '90. Technical Digest, 1990*, pp. 111–114, 1990.

[16] Aritome, S.; Shirota, R.; Sakui, K.; Masuoka, F. Data retention characteristics of flash memory cells after write and erase cycling, *IEICE Transaction Electronics*, vol. E77-C, no. 8, pp. 1287–1295, Aug. 1994.

[17] Aritome, S.; Shirota, R.; Hemink, G.; Endoh, T.; Masuoka, F. Reliability issues of flash memory cells, *Proceedings of the IEEE*, vol. 81, no. 5, pp. 776–788, May 1993.

[18] Tanaka, H.; Kido, M.; Yahashi, K.; Oomura, M.; Katsumata, R.; Kito, M.; Fukuzumi, Y.; Sato, M.; Nagata, Y.; Matsuoka, Y.; Iwata, Y.; Aochi, H.; Nitayama, A. Bit cost scalable technology with punch and plug process for ultra high density flash memory, *VLSI Symposium Technical Digest, 2007*, pp. 14–15.

[19] Aritome, S.; Whang, S.J.; Lee, K.H.; Shin, D.G. Kim, B.Y.; Kim, M. S.; Bin, J.H.; Han, J.H.; Kim, S.J.; Lee, B.M.; Jung, Y.K.; Cho, S.Y.; Shin, C.H.; Yoo, H.S.; Choi,

S.M.; Hong, K.; Park, S.K.; Hong, S.J. A novel three-dimensional dual control-gate with surrounding floating-gate (DC-SF) NAND flash cell, *Solid-State Electronics*, vol. 79, pp. 166–171, (2013).

[20] Aritome, S.; Noh, Y.; Yoo, H.; Choi, E.S.; Joo, H.S.; Ahn, Y.; Han, B.; Chung, S.; Shim, K.; Lee, K.; Kwak, S.; Shin, S.; Choi, I.; Nam, S.; Cho, G.; Sheen, D.; Pyi, S.; Choi, J.; Park, S.; Kim, J.; Lee, S.; Hong, S.; Park, S.; Kikkawa, T. Advanced DC-SF Cell Technology for 3-D NAND Flash, *Electron Devices, IEEE Transactions on*, vol. 60, no. 4, pp. 1327–1333, April 2013.

[21] Park, K.-T.; Nam, S.; Kim, D.; Kwak, P.; LEE, D.; Choi, Y.-H.; Choi, M.-H.; Kwak, D.-H.; Kim, D.-H.; Kim, M.-S.; Park, H.-W.; Shim, S.-W.; Kang, K.-M.; Park, S.-W.; Lee, K.; Yoon, H.-J.; Ko, K.; Shim, D.-K.; Ahn, Y.-L.; Ryu, J.; Kim, D.; Yun, K.; Kwon, J.; Shin, S.; Byeon, D.-S.; Choi, K.; Han, J.-M.; Kyung, K.-H.; Choi, J.-H.; Kim, K. Three-dimensional 128 Gb MLC vertical nand flash memory with 24-WL stacked layers and 50 MB/s high-speed programming, *Solid-State Circuits, IEEE Journal of*, vol. 50, no. 1, pp. 204, 213, Jan. 2015.

[22] Im, J.-w.; Jeong, W.-P.; Kim, D.-H.; Nam, S.-W.; Shim, D.-K.; Choi, M.-H.; Yoon, H.-J.; Kim, D.-H.; Kim, Y.-S.; Park, H.-W.; Kwak, D.-H.; Park, S.-W.; Yoon, S.-M.; Hahn, W.-G.; Ryu, J.-H.; Shim, S.-W. Kang, K.-T.; Choi, S.-H.; Ihm, J.-D.; Min, Y.-S.; Kim, I.-M.; Lee, D.-S.; Cho, J.-H.; Kwon, O.-S.; Lee, J.-S.; Kim, M.-S.; Joo, S.-H.; Jang, J.-H.; Hwang, S.-W.; Byeon, D.-S.; Yang, H.-J.; Park, K.-T.; Kyung, K.-H.; Choi, J.-H. A 128 Gb 3b/cell V-NAND flash memory with 1Gb/s I/O rate, *Solid-State Circuits Conference Digest of Technical Papers (ISSCC), 2015 IEEE International*, pp. 23–25 Feb. 2015.

[23] S. Aritome, US Patent 8,891,306.

[24] Aritome, S. NAND Flash Innovations, *Solid-State Circuits Magazine, IEEE*, vol. 5, no. 4, pp. 21, 29, fall 2013.

INDEX

ABL architecture. *See* All-bit-line
    architecture
Advanced DC-SF cell
    charge loss mode with, 326f, 327f
    cross-sectional view of, 324f
    difference among FG/CG/substrate with,
        327f
    effective cell sizes for, 318f
    equivalent capacitor network with,
        324f
    improvement on, 317–18, 319f
    MCGL process for, 319, 320f, 329
    new programming scheme for, 325–28,
        326f, 327f, 328f
    new read scheme for, 319–25, 321f,
        322f, 323f, 324f, 325f, 326t
    profile comparison for, 319f
    program disturbance of neighbor cell,
        328f
    program/erase cycling endurance
        characteristics of, 329f
    program inhibit modes of, 326f
    program scheme of, 328f
    reliability of, 329, 329f

Advanced LOCOS cell, 40–42
    operation of, 42, 42t
    process technology of, 42t
    scaling in, 40f, 41–42, 41f, 42t
Advanced NAND flash device technologies
    dummy word line scheme in, 77–82,
        78f, 79f, 80t, 81f, 82f, 83f
    p-type floating gate, 82–88, 84f, 85f,
        86f, 87f, 88f
Air gap, 149–53
    process flow of, 149, 150f
    program disturb with, 249
    STI, 152, 152f
    threshold voltage shift by FG
        interference with, 151, 151f
    word-line, 152, 152f
    word-line RC delay with, 352f
All-bit-line (ABL) architecture, 7
    floating-gate capacitive coupling
        interference in, 112, 112f
    memory core circuits of, 111–12, 111f
    page program sequence for MLC with,
        111–13, 111f, 112f
    RWM with, 134

Nand Flash Memory Technologies, First Edition. Seiichi Aritome.

IEEE Press Series on Microelectronic Systems

The focus of the series is on all aspects of solid-state circuits and systems including the design, testing, and application of circuits and subsystems, as well as closely related topics in device technology and circuit theory. The series also focuses on scientific, technical and industrial applications, in addition to other activities that contribute to the moving the area of microelectronics forward.

R. Jacob Baker, *Series Editor*

1. *Nonvolatile Semiconductor Memory Technology: A Comprehensive Guide to Understanding and Using NVSM Devices*
   William D. Brown, Joe Brewer

2. *High-Performance System Design: Circuits and Logic*
   Vojin G. Oklobdzija

3. *Low-Voltage/Low-Power Integrated Circuits and Systems: Low-Voltage Mixed-Signal Circuits*
   Sanchez-Sinencio

4. *Advanced Electronic Packaging, Second Edition*
   Richard K. Ulrich and William D. Brown

5. *DRAM Circuit Design: Fundamental and High-Speed Topics*
   Brent Keith, R. Jacob Baker, Brian Johnson, Feng Lin

6. *CMOS: Mixed-Signal Circuit Design, Second Edition*
   R. Jacob Baker

7. *Nonvolatile Memory Technologies with Emphasis on Flash: A Comprehensive Guide to Understanding and Using NVM Devices*
   Joseph E. Brewer, Manzur Gill

8. *Reliability Wearout Mechanisms in Advanced CMOS Technologies*
   Alvin W. Strong, Ernest Y. Wu, Rolf-Peter Vollertsen, Jordi Sune, Giuseppe La Rosa, Timothy D. Sullivan, Stewart Rauch

9. *CMOS: Circuit Design, Layout, and Simulation, Third Edition*
   R. Jacob Baker

10. *Quantum Mechanics for Electrical Engineers*
    Dennis M. Sullivan

11. *Nanometer Frequency Synthesis Beyond Phase-Locked Loop*
    Liming Xiu

12. *Electrical, Electronics, and Digital Hardware Essentials for Scientists and Engineers*
    Ed Lipiansky

13. *Enhanced Phase-Locked Loop Structures for Power and Energy Applications*
    Masoud Karimi-Ghartemani

14. *From Frequency to Time-Averaged-Frequency: A Paradigm Shift in the Design of Electronic Systems*
    Liming Xiu

15. *NAND Flash Memory Technologies*
    Seiichi Aritome

Printed in the United States
By Bookmasters

Printed in the United States
By Bookmasters